PLANT SCIENCE RESEARCH AND PRACTICES

INTRODUCTION TO THE THREE-PLANE CORPUSCULAR-WAVE BIOLOGY OF FOREST TREE SPECIES

PLANT SCIENCE RESEARCH AND PRACTICES

Additional books and e-books in this series can be found
on Nova's website under the Series tab.

PLANT SCIENCE RESEARCH AND PRACTICES

INTRODUCTION TO THE THREE-PLANE CORPUSCULAR-WAVE BIOLOGY OF FOREST TREE SPECIES

NINA F. KUZNETSOVA

nova science publishers
New York

Library of Congress Cataloging-in-Publication Data

ISBN:978-1-53614-411-6

Published by Nova Science Publishers, Inc. † New York

CONTENTS

PREFACE

This book is based on one unifying concept – the concept of a three-plane corpuscular-wave organization of biological systems, using the example of the prokaryotic and eukaryotic cells, and a life cycle of forestry tree plants. The range of questions and problems is not only numerous, but also controversial. Particular attention is paid to the description of plant biology on the basis of the general principles and fundamental laws of physics; the trinity principle of DNA-energy, DNA-particles and DNA information at different levels of their organization; the fundamental unity of living and non-living matter, etc. Presently, these directions of biophysics are still at the stage of formation. Some theoretical basics of the concept have already been described in the authors' previous work, *Corpuscular-Wave Nature and Wave Properties of Plant Cells*, published by Nova Science Publishers in 2013. These concepts include the corpuscular-wave nature of DNA; the notion about a cell as a quantum-mechanical system, an open and closed system, or an equilibrium system; the nature of two internal sources of energy; etc. Many theoretical statements have yet to be formulated by future researchers, and the authors mean to set the stage by setting up the next objectives. Firstly, they hope to show how many biological processes and mechanisms possess wave nature, with special analysis concerning the unique biotechnologies that were created by nature in plants. The possibility of such an analysis is provided by wide ranging experimental material, which has accumulated by studying cell biology, molecular biology, biophysics, development biology, forest and classical genetics, etc. Secondly, and more importantly, the authors wish to promote the expansion of research in this field. Particular attention in this book is given to the ecological issues that include the peculiarities of the formation of genotype-environmental connections at different stages of the life cycle; intra-specific selection; the mechanisms and adaptive devices ensuring the contact of genetic material with the environment and the relative independence from it; the preservation of genotypic diversity; cross- and self-pollination; and other processes, which provide the viability, quality and sustainability of forest ecosystems in a constantly fluctuating environment.

The material presented in the book embraces the different areas of biology and physics (classical and quantum). It consists of two different parts. A three-plane corpuscular-wave model concerning the organization of prokaryotic and eukaryotic cells is presented in the theoretical Part One. The questions connected with the initiation of the reproductive process and the inception of generative organs, the formation of the energetic, corpuscular and informational programs, and the creation of a three-plane matrix of corpuscular-wave organization are analyzed in Part Two (Chapters One through Three). These processes of sexual reproduction are species-specific, and so they are described often in the example of only one species (*Pinus sylvestris* L.). Different aspects of development for a forest tree species, formation of the population gene pool, and the physical models of different functional systems of forest tree species are discussed in Chapters Four and Five.

INTRODUCTION

The idea of dialectical unity of living and non-living matter comes from ancient concepts about the unity and integrity of the world. Therefore, a scientific cognition of the nature of biological systems must be undertaken from the position of the general lows, which are common for the microcosm and macrocosm, living and non-living matter. The basis of our corpuscular-wave models are the most general fundamental laws of physics, which act at all hierarchical levels, do not change with time, reflect the relationships between the properties and attributes of living and nonliving matter. These are: all total conservation laws (energy conservation, momentum conservation, angular momentum conservation, charge conservation, etc.), four types of fundamental interactions (gravitational, electromagnetic, strong and weak), and the law of a corpuscular-wave nature of matter.

Recent biology includes corpuscular biology. The main postulate of cell theory is as follows: the cell is the smallest unit of living matter. By definition, Wolkenstein states that "A living organisms are the open, self-regulating, self-reproducing systems, the most important functioning substances of which are a proteins and a nucleic acids." (1983)It follows that living matter and all its main units of organization are open and non-equilibrium systems. They exchange substance and energy with the environment and do not tend to have equilibrium; the structural order is created during their development, and the entropy is reduced by means of continuous absorption of energy and matter from the outside. During evolution, living matter increases the free energy, which, as is known, is contrary to the laws of thermodynamics.

In the previous work (Kuznetsova, 2013), we discussed the two-plane corpuscular-wave model of plants at all hierarchical levels of their organization. The notion about a cell is given as the quantum-mechanical, equilibrium and open system for prokaryotes; and the quantum-mechanical, equilibrium, and open and closed systems for eukaryotes, which obey the fundamental laws of physics, including the laws of thermodynamics. It is shown that the elementary unit of the wave properties of a cell is a biological harmonic oscillator, for which in prokaryotic cells the cell is together with the oscillating DNA and in eukaryotic cells the nucleus interacts with oscillating DNA are used. The participation of a DNA-wave in such matrix processes as replication, transcription, translation, cell and sexual reproduction were analyzed.

The present work is a continuation and expansion of research in the area of quantum biology. In Part I of this book, a three-plane corpuscular-wave model of prokaryotic and eukaryotic cells is discussed (Chapters 1-2). The questions connected with a participation of DNA-energy, DNA-particle and DNA-information in the life cycle of forest tree species are analyzed in Part II (Chapters 1-4).The purpose of this study is the principle of the trinity genome on the basis of DNA-energy, DNA-particle and DNA-information; to show how these three non-merging and non-dividing essences are differed and interrelated with each other as well as represent a balanced unity with the environment.

Living matter is a self-regulating complex system. According to the general theory of systems, any complex system can be represented in the form of three essences – matter, energy and information, which reflect the nature and general principles of its organization.

All eukaryotes originate from a common ancestor – a prokaryotic species. From the 1960s to the 1970s, scientists had come to the conclusion that prokaryotes and eukaryotes according to their biochemistry, metabolism and internal organization differ more significantly than animals and plants. These differences are so fundamental that in some modern classification systems, there are prokaryotic and eukaryotic supra-kingdoms allocated as two independent ways of evolutionary development. Analysis of their corpuscular-wave organization confirms the legitimacy of this separation. Details of these two cell systems will be discussed below during an analysis of energetic, corpuscular and informational components of the genome of prokaryotic and eukaryotic cells (Chapters 1-2). From the other side, a prokaryotic and eukaryotic type of biological systems unites under many positions: Quantum-mechanical nature, the trinity principle of matter, energy and information at all hierarchical levels, and organization and function of their basic set of systems and universal mechanisms, etc. An energetic system of all species of the prokaryotes and eukaryotes includes four structural elements. These include: 1) Sources of internal energy; 2) sources of external energy; 3) a centripetal energy stream functioning in the ATP-ADP system; and 4) a centrifugal energy stream acting on the basis of processing (Figure 1).

A corpuscular system of prokaryotic and eukaryotic cell works by the same principle (Figure 2). Three universal mechanisms in both types of cells participate in the formation of species-specific DNA-particles: 1) Replication, 2) transcription, and 3) translation. At present, these three matrix processes are the most studied.

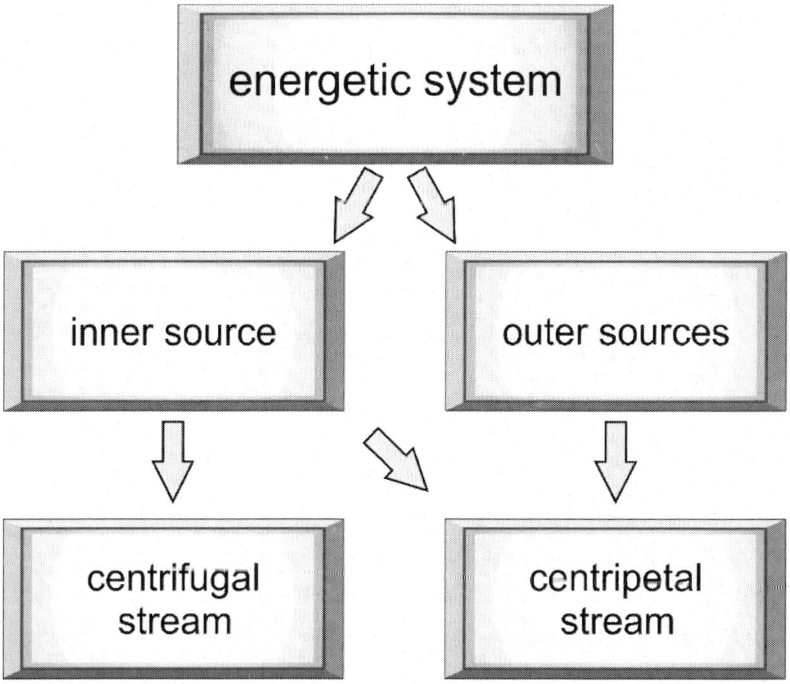

Figure 1. Energetic system of prokaryotic and eukaryotic cell.

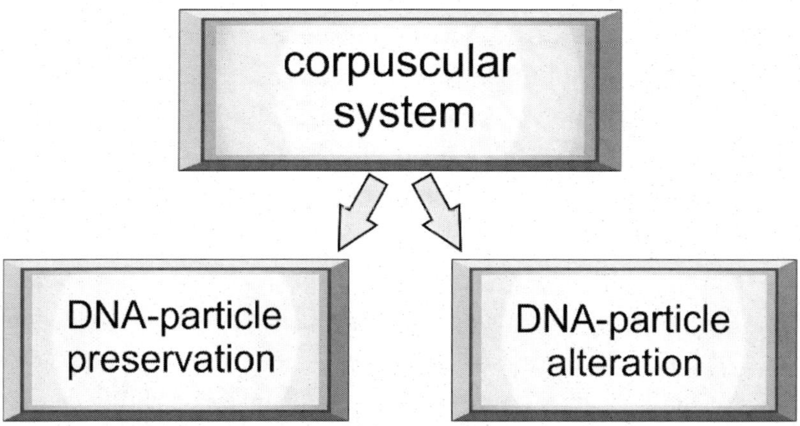

Figure 2. Corpuscular system of prokaryotic and eukaryotic cell.

As the analysis has shown, the physical and biological informational systems have a common principle of organization (Figure 3). DNA-information of prokaryotes and eukaryotes is created with the participation of the same three types of informational systems: 1) An object of class A, as the carrier of the initial information; 2) an object of class B, on the material basis of which the information is created; and 3) an object of class C, for which this information is required (Korogodin, Korogodina, 2000).

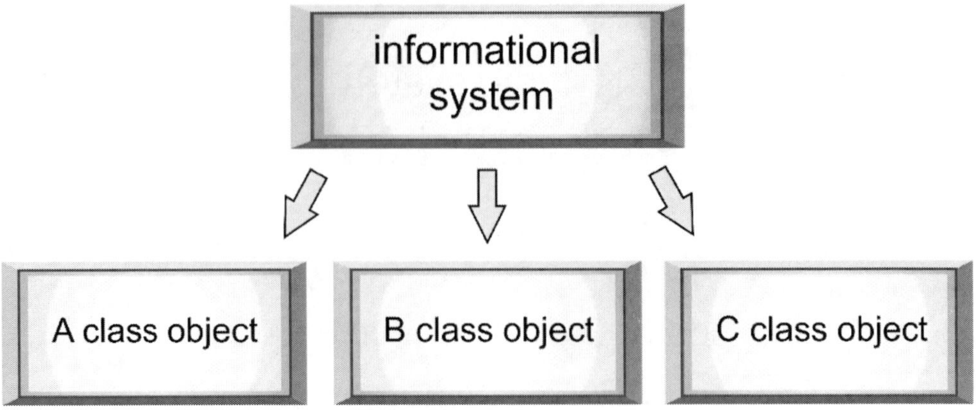

Figure 3. Informational system of prokaryotic and eukaryotic cell.

It should be noted that all of the above-mentioned main elements of the genome cannot be realized separately, but exclusively by the trinity principle – independent and interrelated with each other, and only starting from the cellular level. In Chapters 1-2, the biophysical models of prokaryotic and eukaryotic cells and their three main components of the genome – an energetic, a corpuscular, and an informational – are considered.

In Part II, a three-plane corpuscular-wave model is discussed in the example of the life cycle of forest tree species. Object research is the species edificators *–Pinus sylvestris, Quercus robur, species of Picea, Betula, Populus, Tilia*, etc. – traditionally growing on the territory of the Russian plain, which are characterized by the evolutionary stasigenesis over along period of time. Reaching the genetic and ecological balance with environmental studies, they support it at the optimal level by means of autoregulation systems (Altukhov, 1989; Dynamics of Population, 2004). The plants are stable both in their lifespans and in the transfer of species-specific traits to their progeny. Thanks to this property, a lowland populations of pine, spruce, aspen, etc. become the phenotypic homogeneous forest ecosystems by the molecular-genetic, cytogenetic, morphological characteristics. During many thousand of years, their population gene pools do not change in time and space and correspond to species genotypic and phenotypic norms. It is obvious that there must be the mechanisms to maintain this stability, preservation of genotypic composition constancy at the optimal level.

The strategy of any species is oriented for a maintaining of populations at certain levels of adaptation to the environment. Their population gene pools at the different stages of the life cycle differ by ploidy, morphology, expectancy of life, biological energy, etc. Every gene pool is characterized by a high genetic diversity and superfluous multiplication. The reproduction is regulated by the removal from a renewal genetic diversity of genetic abnormal b energetic weak genotypes of male and female gametes, embryos and plants. Natural selection can exist only in the presence of superfluous multiplication of a genotype number and choice alternative (Stockwell, 1939; Barnes et al., 1962; Zhivotovsky, Duharev 1985; Sato, 1997; Venalainen et al., 1999; Kuznetsova, 2003; Owens & Fernando, 2007; Notov & Zhukova, 2013; etc.).

Scots pine (Pinus sylvestris L.) is one of the more studied species of *Pinopsida*. In Chapters 1-3 from the position of species norm, the three-plane corpuscular-wave model of a program and plant for the development of future trees and the ways of maintaining the population gene pool at the optimal level of adaptation to the environment are analyzed. In Chapter 4,particular attention is given to the questions related to the organization of the main functional systems of forest tree species during ontogenesis (seed germination, age-related changes, systems of spatial coordination, gas exchange systems, supporting and conducting functions, etc.), and the nature and rhythms of reduction and restoration of genetic diversity, the preservation of stability and genotypic and/or a phenotypic species norm depending on the gene pool type, and its state and character of genotype-environment interactions with the environment.

We divided a life cycle of forest tree species into four periods: 1) Initiation; 2) gametogenesis; 3) embryogenesis and 4) ontogenesis. The first three periods are connected with the formation of the individual species-specific corpuscular, energetic, informational programs for development and a common three-plane corpuscular, energetic, informational plan for space organization of future sporophytes. The 4th period is associated with the processes of realization of these three developmental programs and spatial plans during tree ontogenesis. The reproduction initiation period includes the processes of inception of first reproductive organs (microsporangium, megasporangium) and diploid phase of development of generative cells (microsporocytes, megasporocytes) (Chapter 1). The gametogenesis period represents a haploid phase of development of generative cells, which continues from meiosis commencement to the end of fertilization (Chapter 2). The embryogenesis is connected with the processes of formation of the corpuscular matrix (proembryogenesis), energetic matrix (early embryogenesis) and informational matrix (late embryogenesis) (Chapter 3). The ontogenesis is a period of realization of energetic, corpuscular and informational programs into the individual species-specific trees and populations by the principle of "genotype-environmental" interaction (Chapter 4). Formation of a population gene pool is a genetically determined process, which is formed by the principle of a "genotype-environment" interaction and connects with the main regional climatic conditions (hydrothermal regime, vegetative season duration, daytime length, soil type, etc.) (Schwartz, 1980, Happer, 1997; Jorgensen, 2002; Headrick, 2003; etc.). The species and population gene pool has 12 common properties from the positions of "the whole" and "the parts," which are described in Chapter 5 in the example of forest tree species.

PART 1. THREE-PLANE CORPUSCULAR-WAVE MODEL OF PROKARYOTIC CELL AND EUKARYOTIC CELL

Chapter 1

Prokaryotic Cell

In the 19th century, the doctrine about the cell was developed by Schleiden and Schwann. Modern science has proved that a cell is an elementary unit of living matter, beyond which there is no life. Two types of cell systems – prokaryotic and eukaryotic – are found among biological organisms. The lows of conservation and transformation of the energy act in all cell systems. A prokaryotic cell belongs to a simpler pre-nuclear type of organization and structure, a eukaryotic cell to a highly organized one. Prokaryotes include more than 10^8 species. All species are divided into two domains (kingdoms) of organisms – bacteria and archaea, which differ by a structure of cell walls and membranes, and mechanisms of genetic information realization (Modern microbiology…, 2005; Kato, 2011; Ravin & Shestakov, 2013).

From the position of corpuscular biology, the different aspects of the organization and function of prokaryotic cells has been described in many works (Shlegel, 1987; Davis et al., 1990; Alberts et al., 2002; Chentsov, 2004; Modern microbiology…, 2005; Kato, 2011; Kvitko & Zakharov, 2012; Snape et al., 2014; Bonch-Osmolovkaya, 2015; etc.). A prokaryotic cell has a small size. Its cytoplasm is separated from the external environment by a semi-permeable plasma membrane. From the outside, a cell is surrounded with a cell wall, which is fixed in form and may by flexible, tough, and sometimes rigid. Therefore, all microorganisms have a limited ability to secrete and absorb the larger molecules. The nucleus and specialized organelles are absent. Their single circular chromosome is in direct contact with the cytoplasm. The DNA content is much less than in the eukaryotic cell. The processes of replication and transcription occur simultaneously. The main division type is a binary division (or a promitosis).

In our corpuscular-wave model, a prokaryotic cell is a quantum-mechanical, an open and an equilibrium system (Kuznetsova, 2016). Cell reproduction is an indispensable requirement for its existence (Masia, 1961). All prokaryotes have three independent energetic sources (an internal and two external), which give it a significant energy advantage with respect to inanimate matter. A distinctive feature of prokaryotes is that their cell is simultaneously a biological harmonic oscillator, an elementary structural-functional unit of living matter and an independent organism. Its development occurs in three different, indivisible and equally important planes – an energetic, a corpuscular and an informational. During one cell cycle, a DNA molecule (particle and wave) is laid and realized as an energetic, corpuscular and informational program of future plant development.

Characteristics and properties of its genome can be described with the help of three basic states of macromolecules (particle and wave): DNA-energy, DNA-particle and DNA-information. Taking into account that life represents a process connected with the transformation and consumption of energy and the ability of biological systems to perform different works, and the key place among them belongs to the energetic component of a genome. Energy as a measure of interaction has the following forms: An internal, an external, a free and a potential (or related). As was mentioned above, a DNA-energy of all biological systems consists of 4 structural elements: 1) and 2) internal and external energetic sources; and 3) and 4) are centripetal and centrifugal energetic streams (Figure 1).

Energetic Component of Genome (DNA-Energy)

The principle of organization and functioning of energetic systems is the same and relatively simple in all species of prokaryotes (Figure 4). Their energetics is provided by the joint action of one internal and few external types of energetic sources. The central place among them belongs to the internal source – either the biological harmonic oscillator or DNA-wave. A prokaryotic oscillator is the cell inside of which two standing waves with a certain frequency and amplitude move continuously along a ring chromosome (Kuznetsova, 2013). This source has a wave nature, is both a physical and biological body, and its action connects directly with the work of a quantum harmonic oscillator embedded inside the prokaryotic cell. Their external sources of energy may be different. The main mechanisms of their action are complex redox processes during fermentation, photosynthesis and respiration (Alberts et al., 2002; Modern microbiology…, 2005; Kvitko & Zakharov, 2012; Snape et al., 2014; etc.).

A prokaryotic oscillator has a simple inner structure and low energy potential. As was mentioned above, it presents a cell together with oscillating DNA. The polynucleotide chain of its DNA-particle is connected to the plasma membrane by means of special membrane proteins. Each nucleotide is a small dipole, in which the negative charges are concentrated in the region of phosphate groups, and the positive charges are in the region of bases. Dipoles of oscillating DNA are the source of an electromagnetic field (biofield)and have two zones of action: The close-range zone, changing as $1/r^2$, $1/r^3$, etc.; and far-range ones, changing as 1/r. It follows that the area of influence of a DNA-energy is limited by the zone of action of the biofield, and it covers the cell volume and goes beyond it.

The energetic component of a genome is determined in many respects by the operating mode of a biological harmonic oscillator. It is the biofield source, and many intracellular interrelationships of biomolecules are realized with the participation of its forces. The effectiveness of the functioning of external sources also depends on the biofield, which causes a centripetal movement of nutrients from the environment into the cell. In the cell, they are subjected to the decay process, which proceeds with the release of energy. Then, this energy is transferred to the energy of chemical bonds of ATP. The energy accumulated in its molecules is used to perform various works and support the energetic intracellular balance in the ATP-ADP system. DNA-wave radiation and the free energy entering into a cell initiate the processes of a DNA-particle and cell doubling. Apparently, the amount of energy, which enters into the cell from this internal source, is sufficient for maintaining its life at the level of survival. This is evidenced by experiments of long-term cultivation of E. coli under the conditions of substrate starvation (Gusev, 2010). During starvation, the bacteria have preserved their viability during more than two years, and retain the ability to reproduce. However, they were greatly weakened, and their divisions were depressed.

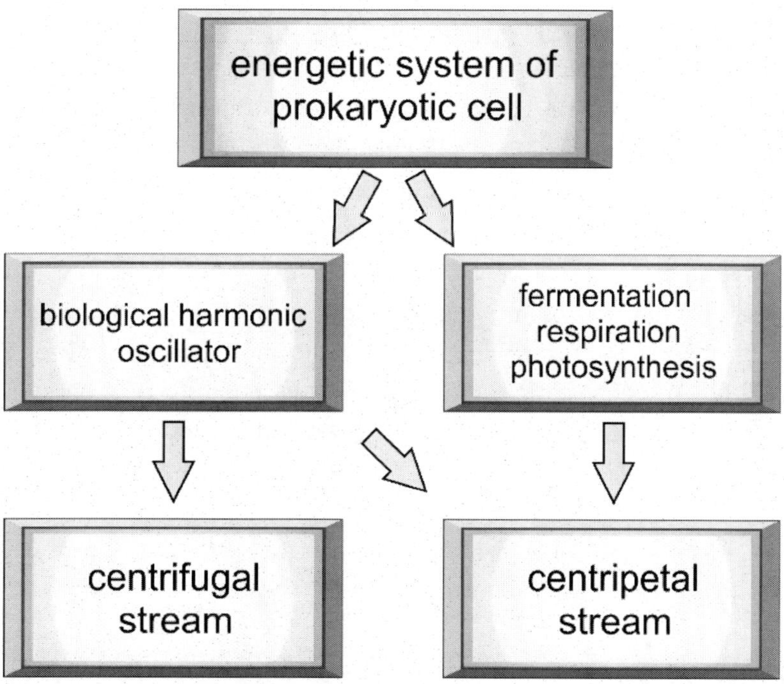

Figure 4. Energetic system of prokaryotic cell.

The biological harmonic oscillator of all prokaryotes works continuously. Therefore, a cell is an open system and it is constantly exchanged with the external environment by matter and energy. This is its distinctive property as a physical system and biological system. The oscillator stops only in two cases: 1) When the cell passes into a metabolic resting state, and 2) when it dies. In the first case, it only suspends the work and this state is reversible, and in the second case, an irreversible destruction occurs. The second important property of a prokaryotic oscillator is that its energy potential never decreases. The energy potential of a eukaryotic oscillator can increase and decrease. For this reason, all species of prokaryotes can only be unicellular organisms.

A prokaryotic cell, as any physical auto oscillation system, has a certain internal energy and biofield. The main part of its life occurs at the lowest energetic level. Therefore, all species of prokaryotes reproduce by vegetative means, using only binary division. Although a life at the lowest level represents an equilibrium and a fairly stable state of the cell system, this position is quite vulnerable. Any energetic losses due to internal or external reasons can lead to cell death. A great deal of energy is lost on friction, because the ring chromosome is directly in contact with the cytoplasm. Therefore, cells for normal functioning require a constant supply of energy from the outside.

Microorganisms use a different way to extract energy from the environment, depending on the substrate, the attitude to oxygen and the type of metabolic process (fermentation, photosynthesis, and respiration). Only some of their species are capable of respiration (facultative aerobes) and photosynthesis (blue-green algae and photosynthetic bacteria). Most parts of microorganisms obtain energy during anaerobic fermentation, which is a less effective way of biological oxidation. Glucose, organic acids, amino acids, alcohols and other substances are subjected to fermentation.

A cell cycle is the time between two radiations of DNA-waves. It is carried out at two energetic levels – the main balanced state (lowest level) and the excited state (the 1st level). In Figure 5, the algorithm of biological energy motion in the alternation of generations is shown. It seems obvious that for the most part of its life, a cell is in equilibrium (0th level). The stability of the zero level is supported by a periodic transition of DNA to an excited state (a “jump” to the 1st level) with a subsequent “fall” to the main state and DNA-wave radiation.

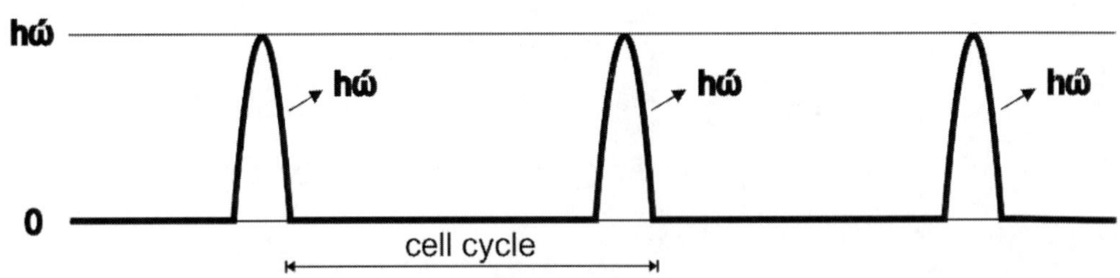

Figure 5. Algorithm of biological motion energy in generations of prokaryotic cells.

An energetic program includes two periods: 1) A formation and 2) a realization. The time interval of its laying down is very short. In fact, it is the moment of DNA-wave transition into an excited state. This transition from the main to the excited state occurs under certain physical conditions, which are created inside the cell during each cell cycle. The excited state is thermodynamically unstable and the DNA-wave very quickly "falls" back to its main zero level. In this case, it emits free energy. Since DNA is a standing wave, two streams of energy enter into the cell during radiation. This energy is spent by the following channels: 1) A replication; 2) a transcription and translation; 3) a cell growth and its division; 4) a support of DNA-wave function; 5) a restitution of ATP from ADP; and 6) a bioluminescence (Figure 6). Since the efficiency cannot be equal to 100%, part of this energy is lost.

The process of energetic program realization continues all cell cycles in the period between two "jumps" of the DNA-wave: During DNA replication, cell division and growth of daughter cells to species-specific sizes. A "jump" in biological energy is the first mechanism that was created by nature to restore intracellular energetic status and rejuvenate cells. During evolution, this mechanism has undergone an essential transformation and gives rise to the specialized stages by transfer of biological energy to higher energetic levels (3rd, 4th, etc. levels).

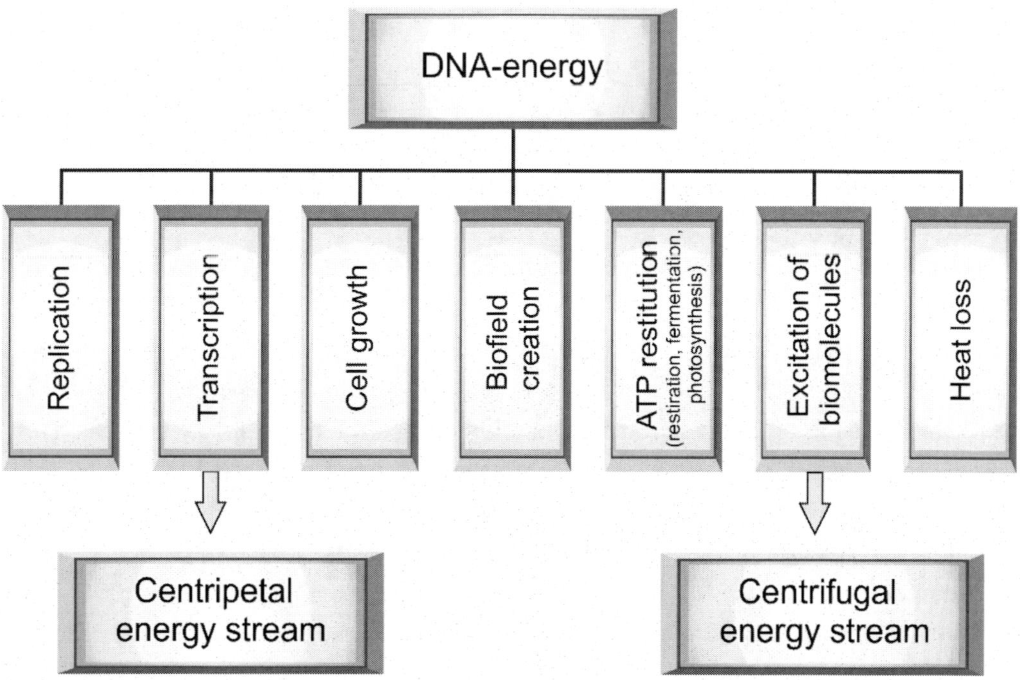

Figure 6. The channels of utilization of energy from a biological harmonic oscillator of a prokaryotic cell.

Now, we will move on to the analysis for a cell cycle of a prokaryotic cell. Let's assume that the initial state of its genome is 1 DNA-particle + 1 DNA-wave. DNA is a neutral molecule (Volkenshtein, 1983). However, its nucleotides have a dipolar structure. Their negative charges are concentrated in the region of phosphate groups, and the positive charges are in the region of bases. A DNA-wave is a standing wave. During radiation, free energy goes into a cell by two different streams. The DNA-particle is tightly attached to the plasma membrane via specific membrane protein which interacts with the macromolecule in the origin zone. The energy stream beats out the electrons from its molecule, after which it becomes positively charged. Excited electrons pass from the membrane protein to the enzyme of DNA-polymerase, and its molecules also receive energy and charge. Thus, the complex with a charge transfer is created during each cell cycle.

Two activated negatively charged DNA-polymerases are attracted to the different threads and contrary waves of the circular macromolecule in the base region and within the origin zone (ORI). Since the enzyme has a negative charge, the electric balance in places of their union is destroyed. As the DNA-polymerase is a negative charge, the negative charged phosphate groups of two DNA threads push off from it, and hydrogenous connections between them are local broke and form a loop or a replication folk. Two molecules of enzymes moving in opposite directions together with their own waves join to each nucleotide and its complementary pair;they carry out the reaction of DNA biosynthesis. During DNA-particle doubling, they gradually spend the energy and inactivate it. After a collision in the ORI zone, two inactive molecules of the enzyme return in the cytoplasm. Then they migrate towards the molecules of positively charged membrane proteins, their charges are neutralized and the complex with a charge transfer ceases to exist.

In prokaryotes, the process of replication occupies 90-95% of the cell cycle duration. At this time, the genome of a cell passes from the 1 DNA-particle + 1 DNA-wave to the 2 DNA-particle + 1 DNA-wave. When the replication forks pass through the origin zone, two identical charged molecules of DNA-polymerase mutually repel each other. This leads to the division of the membrane protein in half and the formation of two positively charged molecules. They, as similar charged particles, mutually repel each other and diverge to the distance of the Coulomb repulsion. Double DNA is split into two identical DNA-particles. According to experimental data, the nucleoids of bacteria always go away to the same distance after division – 1/4 of the cell length (Chentsov, 2004). The continuity of the plasma membrane and cell wall in this moment is disrupted, and the mother cell begins to grow rapidly. After the divergence of nucleoids, the formation of septum and division of the body of cells begins, and the daughter cells are completely separated from each other. All these processes are cyclical. They require mach energy and the same cyclic character of its distribution during every cell cycle.

The distribution of energy from internal and external sources is an ordered process connected with certain intracellular processes and biomolecules. The main energetic processes that determine the differences between living and non-living matter occur at the cellular level. Prokaryotic and all types of eukaryotic cells have the same intracellular mechanisms of energy transformation and a scheme of energy transfer pathways. There are two energetic channels - centrifugal and centripetal streams, which participate in the processes of storing, converting, delivering and using energy within a cell in accordance with definite laws (Figure 4). The centripetal stream works in the ATP-ADP system, and the centrifugal stream operates on the basis of processing.

The macroergic phosphate bonds of ATP molecules contain a lot of energy. ATP acts as an energy accumulator and "energetic shuttle," taking part in the different biosynthetic reactions and various kinds of works. A system of intracellular transport of chemical energy is a pair of ATP-ADP molecules. The mechanism of their functioning are reversible: The reaction of ADP phosphorylation connects with the accumulation of energy, and the reaction of ATP dephosphorilation with its releasing. Recharging of the ATP-ADP system takes place continuously. This process in a cell can be repeated with a frequency of 50 sec. (Svenson, Webster, 1980). The resynthesis of ATP from ADP and inorganic phosphate is a complex endergonic reaction and it constantly requires free energy. We think that the energy from the internal source (biological harmonic oscillator) and one or few external sources (fermentation, respiration and photosynthesis) are involved in its passing.

The ability of microorganisms to survive in conditions of substrate hunger confirms the participation of the biological harmonic oscillator in the creation of a centrifugal energetic stream (Gusev, 2001). DNA-wave radiation occurs once in the cell cycle, and part of free energy from the internal source is spent on the ATP synthesis from ADP. The resulting pool of ATP molecules is able to maintain intracellular metabolism only at a minimal level. This energy for normal vital function of a cell under constantly changing conditions is not enough. Microorganisms require the energy from another source.

The prokaryotic cells are an open system and they connect with the environment. Different external sources participate in the creating and functioning of their centripetal stream (Figure 4). Prokaryotes, depending on the type of metabolic pathways and methods of energy recovery, are polymorphic organisms. They are able to use the energy contained in the chemical bonds of organic (organotrophs) and inorganic (lithotrophs) substances, both under anaerobic and aerobic conditions. The reaction of species with respect to oxygen is very broad; from this inability, oxygen is absorbed into the partial and absolute dependence on it. Some of them include the autotrophs (blue-green algae and photosynthetic bacteria), and others are the heterotrophs, with part of them capable of fixing molecular nitrogen. Nevertheless, the end product of all these reactions is ATP. The emerging ATP molecules continuously replenish the intracellular pool so that the

ATP concentration into the cell and, accordingly, energetic capacity of centripetal stream is maintained at a constant level.

In prokaryotes, the transcription is carried out on the DNA matrix simultaneously with the replication process. Molecules of big primary transcripts of pre-RNA in the course of transcription are formed: mRNA, tRNA and rRNA. rRNA is the main structural unit of the ribosomes. The centrifugal energetic stream functions on the basis of posttranscriptional processing, during which a huge molecule of primary RNA transcripts (mainly of heavy pre-rRNA) are split into separate fragments of mature rRNA. In the previous work, we assumed that the physical nature of processing is due to the exponential disintegration of RNA molecules (Kuznetsova, 2013). The process of their spontaneous disintegration proceeds with the liberation of free energy. In ribosomes, this energy is captured and used as an energy supply of protein biosynthesis.

Now it seems obvious that the energetic stream created during RNA processing is internal by origin, prolonged by time, centrifugal by the direction of energy motion (from the nucleoid to the cytoplasm) and acts during all cell life. Since the number of synthesized pre-RNA molecules and the probability of their exponential disintegration are approximately the same (after the first half of the disintegration period there remains a half of primary molecules, after the second, a quarter, etc.), the centrifugal energetic stream created during processing is characterized by constancy in the cellular lines of one species.

To summarize the above, it should be noted that the energetic system of a prokaryotic cell is sufficiently reliable (joint action of internal and one or few external energy sources, centrifugal and centripetal energetic streams). Continuous work of a biological harmonic oscillator as a source of internal energy and biofield is synonymous with its life. The quality of an oscillator is significantly reduced from direct contact with the cytoplasm and the environment. Protection from the environment is provided by a rigid cell wall and a semi-permeable plasma membrane. Connection with the nutrient sources occurs by means of biofield. Division initiates the DNA-wave radiation, which leads to cell doubling every 20 minutes. This gives an opportunity for the rapid proliferation of species, and creates a material basis for natural selection and speciation.

Corpuscular Component of Genome (DNA-Particle)

The corpuscular system of prokaryotic and eukaryotic cells consists of two basic elements, which connect with the preservation of a DNA-particle and its changing (Figure 7). Four universal mechanisms take part in the conservation of a species-specific DNA-particle. Three of them are the intracellular complex matrix processes: 1) A replication; 2) a transcription and 3) a translation. An analysis of these reactions in prokaryotes is described in a number of works (Davis et al., 1990; Alberts et al., 2002;

Chentsov, 2004; Ohniwa et al., 2010; Kvitko & Zakharov, 2012; al., 2014; Pierce, 2014, Kuznetsova, 2016, etc.). A separate mechanism of a corpuscular system is the phenomenon of chirality.

The chirality of biological molecules is another important property, which distinguishes the corpuscular field of living matter from the inanimate ones. It is known that only one optical isomer of many organic molecules is involved in the basic intracellular metabolism. For example, the majority of amino acids are the L-forms, and all sugars have the D-form. The homochiral structure is characterized proteins, nucleic acids, carbohydrates, etc. We think that one reason for optical isomery can be the plasma membrane of the cell, which acts with respect to light as an "intracellular polaroid" (Figure 8). A plasma membrane is an obligatory component of all cells. Lipids by its intra-membrane organization represent two parallel rows elongated in length molecules. Light, passing through them, becomes circularly polarized. Circularly polarized light is absorbed differently by the left and right stereo-isomers from the racemic mixture of dissymmetrical biomolecules. Unidirectional selection led to the optical isomery of an intracellular molecular compound, which during evolution was fixed in living matter.

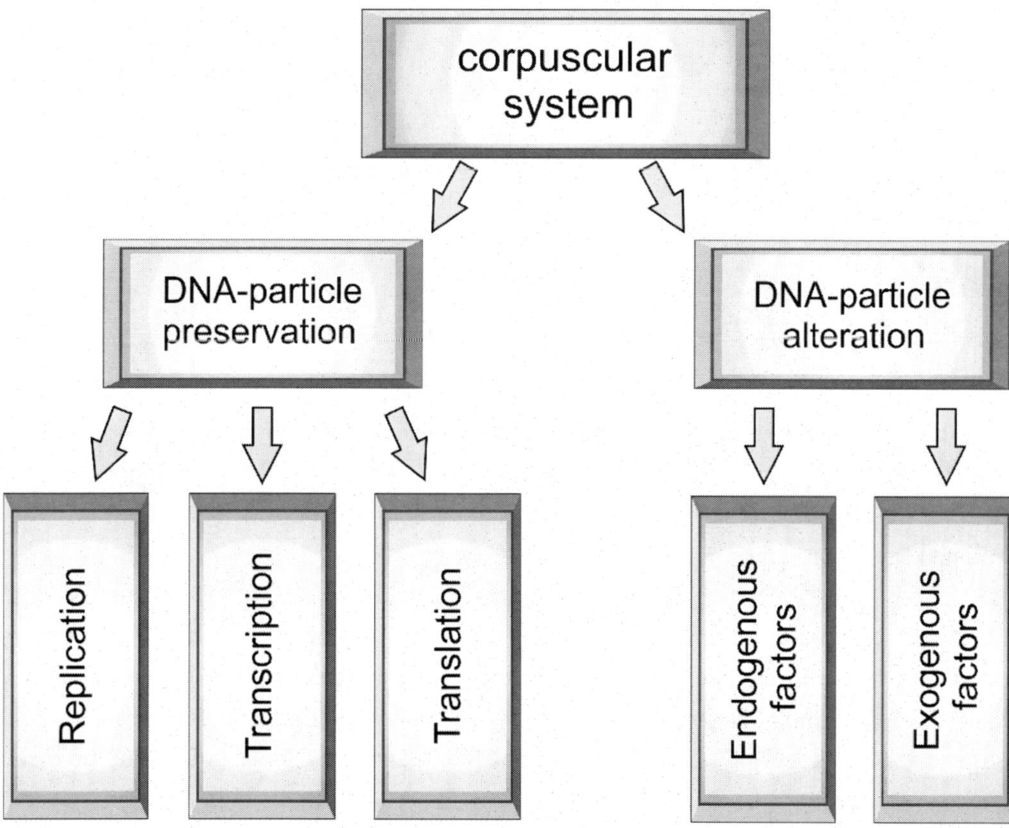

Figure 7. Corpuscular system of prokaryotic cell.

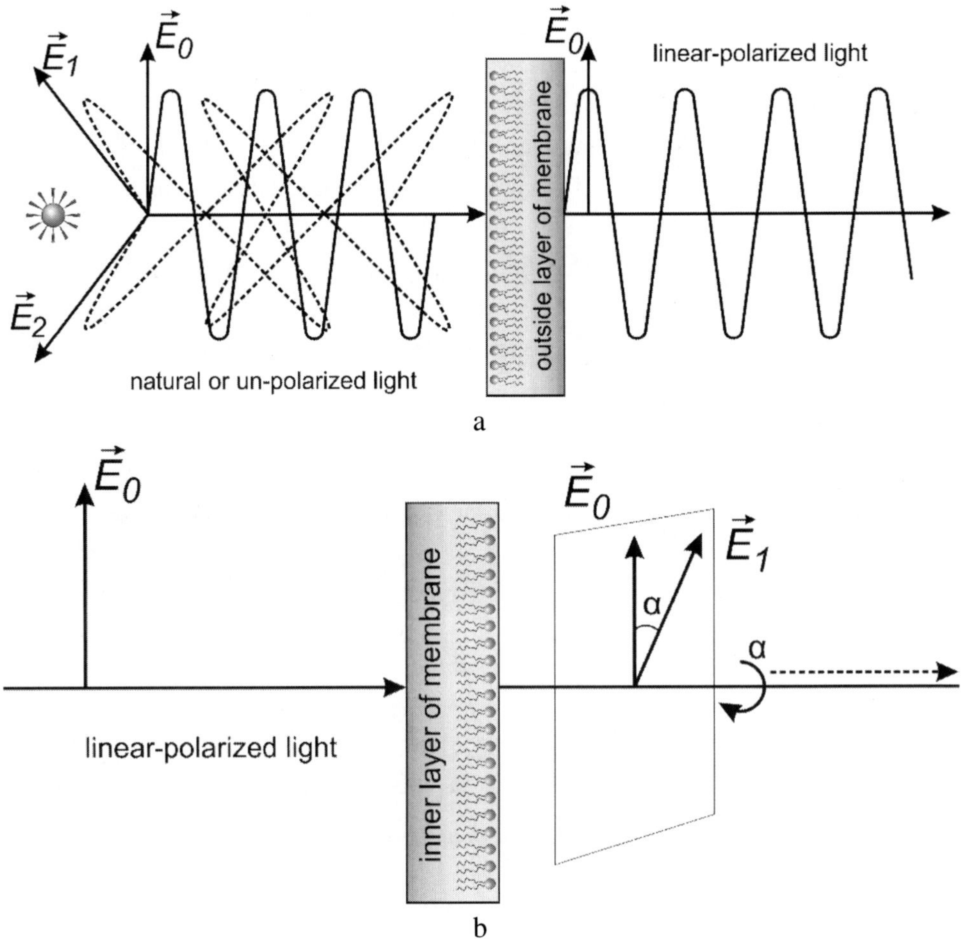

Figure 8. The trasformation of un-polarized sunlight into linearly-polarized light (9a) and the linearly-polarized light into circularly polarized light.

In prokaryotic and eukaryotic organisms, the change of a DNA-particle occurs under the influence of both external (climatic conditions, ecology of habitat) and internal factors (spontaneous mutations, thermal motion of molecules). After transition to a multi-cellular level of eukaryotes, the special mechanisms (crossing-over, fertilization) were developed to increase their intraspecific variability.

Since all prokaryotes are unicellular organisms, the action of the corpuscular system is limited to the frames of a cell. A species-specific intracellular corpuscular field is created during each cell cycle. The corpuscular field is the nucleotide sequence of a DNA-particle, the molecular composition of a cell, the mechanisms for its formation and change, and the intracellular distribution of the biomolecules. 80-90% of a genome is the sequences of structural genes encoding proteins, DNA and RNA. This is a conservative set of genes responsible for their biosynthesis and the species/genus detection of the individuals.

A prokaryotic cell is characterized by a specific and different chemical composition, analogues of which are not found in the inanimate matter. Its main components are nucleic acids, carbohydrates, lipids and proteins that participate in the formation of all enzyme systems and constitutional proteins. Three universal matrix mechanisms – a replication, a transcription and a translation – take part in their formation.

Replication

The cell division of bacteria in the stage of logarithmic growth occurs on average every 20-30 min. During the cell cycle, their genome has two different states: 1 DNA-particle + 1 DNA-wave; and after replication, 2 DNA-particle + 1 DNA-wave. Information about the molecular composition of future cells is laid in the chemical formula of a DNA-particle as a nucleotide sequence of a ring chromosome. The formation of an intracellular corpuscular field begins with the DNA doubling. The main place in the replication process belongs to the DNA molecule (particle and wave): a DNA-particle as the matrix from which the species-specific corpuscular information is read, a DNA-wave, which supports high accuracy of biosynthesis.

As is known, a DNA-particle by its physico-chemical properties is a stable and neutral molecule. But its nucleotides have an asymmetric structure: The positive charges are concentrated in the base region, and the negative charges are in the phosphate group. A DNA-wave is the standing wave, or two waves moving along a circular chromosome in opposite directions.

Two molecules of DNA-polymerase received an energy and negative charge, and are attracted to different threads of DNA (particles and waves) in the region of bases and begin to passively diverge with its waves in the opposite directions from the origin zone. The negative charge of an enzyme breaks the electrical equilibrium in the place of a molecular union. As the association is negative, the negatively charged phosphate groups of two threads of DNA push off from it. Increased inner asymmetry leads to local destruction of hydrogen bonds between two of its threads. They straighten, move away from each other, and form the loop on the macromolecule. This is the point of origin and the zone of DNA-replication folk, where a process of DNA biosynthesis takes place.

Then, the active DNA-polymerase moves together with the DNA-wave and automatically untwists two of its treads. At that time, the nucleotides from the opposite wave pass by turns. The enzyme catalyzes the additional reaction to each nucleotide of its complementary pair and carries out a DNA biosynthesis in the 5' $\rightarrow$ 3' direction. The wave nature of the process provides the same reaction rate $\approx$ 30 μm per minute or 1,000 nucleotides per sec. (Rees and Sternberg, 2002) and very high exactness readings – one error per 10^9 base pairs (Drake et al., 1998).

Transcription

Transcription is the next step in the creation of an intracellular corpuscular field. This is the complex system of matrix reactions, which includes two main processes: The transfer of corpuscular information from the DNA-particle, the RNA-particles of three RNA types (mRNA, rRNA and tRNA), and the relocation of energy from a biological harmonic oscillator in the chemical bonds of a great number of RNA biomolecules. In prokaryotes, the RNA synthesis is carried out simultaneously with the replication and by the principle of complementarity.

The corpuscular-wave principle of action concerning replication and transcription mechanisms is quite similar and at the same time they have a number of principal differences. Firstly, the whole DNA-particle is a matrix for the replication process, and the transcription occurs only in certain parts. Secondly, a DNA-polymerase and RNA-polymerase receive the energy and are activated during a DNA transition from the 1st to the lowest energetic level. Both these enzymes are of a rather large size and are initially characterized by an internal asymmetry of charges. However, a DNA-polymerase after activation becomes a negatively charged molecule, and the electric field of RNA-polymerase is created only by increasing its polarity. This is necessary to increase the reactivity of monomer units (nucleotides) and attach them to the new RNA molecule.

Then the activated RNA-polymerase due to the internal asymmetry of charges is attracted to one of the threads of macromolecules. Moreover, the enzyme forms a stable bond only in certain regions of DNA-particles – the promoters. These are the special places for the initiation of transcription processes, characterized by a specific nucleotide composition and size.

The corpuscular-wave mechanism of DNA and RNA biosynthesis is quite similar. The electric equilibrium within the enzyme-macromolecule association is disrupted, leading to a local divergence of two DNA threads and the appearance of a loop on the DNA-particle. The transcriptional bubble has a small size, because the breakage of hydrogen bonds is caused only by the asymmetry of enzyme molecules. The RNA-polymerase joining the DNA (particle and wave) begins to move along the copied site with the rate of 40 nucleotides per second. They untwist the threads of DNA-particles and build the new RNA-particles on the counter DNA-wave by the principle of complementarity. During biosynthesis, the enzyme molecules gradually expend energy, inactivate, and, after a collision with the termination area, passes into the cell.

Translation

Translation is the process of protein synthesis on the matrix of mRNA. It is the last stage by the formation of intracellular corpuscular field of the prokaryotic cell. In fact, a translation is the decoding of corpuscular information laid in the DNA-particle into the nucleotide sequence of proteins. The biosynthesis of proteins proceeds in three stages: 1) Codon recognition; 2) formation of peptide bonds and 3) termination. It should be noted that the mechanism of initiation and termination of these reactions are highly specific processes. At present, the structural and functional peculiarities of the organization of these stages at the molecular level are well described in many monographs (Alberts et al., 2002; Chentsov, 2004; Modern microbiology..., 2005; Snape et al., 2014; etc.). Now we will move to the analysis of the problem regarding energy of transmission, or a sustainable energy of thousands of different endergonic biochemical reactions taking place simultaneously in the prokaryotic cell.

According to present concepts, the source of energy for protein biosynthesis is an ATP molecule, the chemical bonds of which contain a significant amount of energy. In this regard, we drew attention to a few other facts. We list some of them. It is not clear what the source of energy for the biosynthesis of constitutional proteins in conditions of substrate hunger. The translation process takes place mainly in the ribosome. As ATP molecules determine the intracellular location of ribosome is not known. The volume of ribosome's can reach 40% of the prokaryotic cell mass (Rees and Sternberg, 2002). For what purposes is their excess amount created in the cells. All this indicates that for the biosynthesis of protein should be another source of energy – a highly specialized one independent of the external environment.

As was mentioned above, such a special energetic source is the processing. The free energy, which is gradually liberated during exponential disintegration of RNA molecules, represents a centrifugal energetic stream stretched in time. The results obtained by different authors indicate, that in prokaryotes this stream originates only from two primary transcription products – pre-rRNA and pre-tRNA. Biosynthesis of RNA occurs on the matrix of DNA-particle due to the energy of internal source - a biological harmonic oscillator. The energy of the centrifugal stream is used for energy supply of the translation process, and partly for maintain of energetic potential of a cells during the cell cycle. Ultimately, all types of RNA after protein synthesis are gradually destroyed at constituent components that are involved in the formation of next-generation RNA molecules.

The major part of cellular RNA is the ribosomal RNA. rRNA are part of the ribosomes, where protein biosynthesis takes place. Prokaryotic ribosome's have the same structure and perform the same function therefore they are the intracellular organoid (Chentsov, 2004). Prokaryotic ribosomes are small (20 x 17 x 17 nm), have a sedimentation coefficient of 70S, are freely located into the cytoplasm and constitute a significant part of the cell mass. They consist of two subunits: 30S and 50S. A small subunit contains one rRNA molecule, and a large one contains two molecules. Now, the dynamics of its molecules are represented as the next process. Initially, a giant pre-rRNA molecule on the matrix of a DNA-particle is synthesized. Then it split into smaller fragments of 5S, 16S, 23S, which are part of a common primary transcript. Molecules of pre-rRNA are most affected by processing, during which more than 1/3 of its molecule is destroyed. The pro-tRNA molecule is approximately 20% longer than the mature tRNA.

In prokaryotes, a translation begins, when the process of transcription of the mRNA has not yet ended. However, a main part of protein biosynthesis occurs in the ribosomes. Their volume can reach 40% of the cell mass (Rees and Sternberg, 2002). The translation process is not random. The protein biosynthesis begins if the certain requirements are satisfied, namely, mRNA, tRNA, enzymes, amino acids and energy are simultaneously localized within the ribosome. It seems obvious that all types of RNA are involved in biosynthesis of proteins. mRNA serves as a matrix from which the corpuscular information about amino acid sequence of proteins is read. tRNA performs two functions – a transport system for amino acids and a reading system. rRNA is the main source of energy for protein biosynthesis. The joint work of mRNA, rRNA and tRNA is organized so, that the amount of proteins in the cell doubles before each division.

Variability

The variability is a universal property of living matter. It is inherent to all microorganisms. There are very small unicellular organisms that have a huge surface relative to their volume, through which not only gas exchange, water and nutrients entrance occurs, but the direct effect of the environment on their metabolism takes place. For reduce the intraspecific variability, they must expend additional energy to maintain of homeostasis of internal medium.

Prokaryotic cell is an open system that constantly exchanges energy and matter with the environment. Its DNA-particle is directly in contact with the cytoplasm, therefore the thermal motion of molecules also increase the level of intraspecific variability. In the population, there are no two individuals with the same phenotype and genotype. Fluctuations of different endogenous and exogenous factors (temperature, concentration of mineral salts, pH of environment, etc.) is the reasons of genotypic and modification or phenotypic variability, the change of qualitative and quantitative traits within the reaction norm of species. Modification variability is not stable, it does not affect the genotype, and does not significantly influence on the processes of reproduction and development. Genotypic variability has often the spontaneous character. Mutations and recombinations of genes are the basis of genotypic variability of microorganisms, which in turn changes the molecular (corpuscular) composition of cell. It induces a mutational process, leading to the appearance of new hereditary traits and different mutations, most of which are removed during natural selection.

INFORMATIONAL COMPONENT OF GENOME (DNA-INFORMATION)

Many works have been devoted to the phenomenon of life and information (Shannon, 1949; Ashby, 1956; Korogodin & Korogodina, 2000; Küppers, 2000; Yockey, 2005; Thilms, 2012; etc.). We think that all morphological diversity represented in the world of prokaryotes is due to the function of DNA-information. Microorganisms would have the same spherical shape, if there was no the informational component of genome. The scheme of their informational system organization is shown in the Figure 9. The common principle of its operation is in many respects similar to the action of radio communication means. It includes the same three basic elements: A class object, B class object and C class object. However, the biological systems are characterized a number of design features that distinguish them from the physical ones. Firstly, the DNA-information has the terrestrial and cosmic origin. Secondly, the role and importance of the informational component of genome is determined by two its main destinations: 1) the creation of the morphological diversity of biological world, and 2) the ability of species to adapt to different environmental conditions (Kuznetsova, 2016).

The ability to carry out the purposeful actions is a distinctive property of living matter (Korogodin, 1983; Korogodin & Korogodina, 2000). Purposefully arranged species cannot arise accidentally, due to a successful combination of circumstances. The environment can change information, but it is not its sources. It follows that the biological information beyond living organisms cannot arise (Eigen, 1982). A further search of theories and hypotheses to explain the species-specific morphological heterogeneity and feasibility led us to the conception of the cosmic nature of DNA-information (Kuznetsova, 2016).

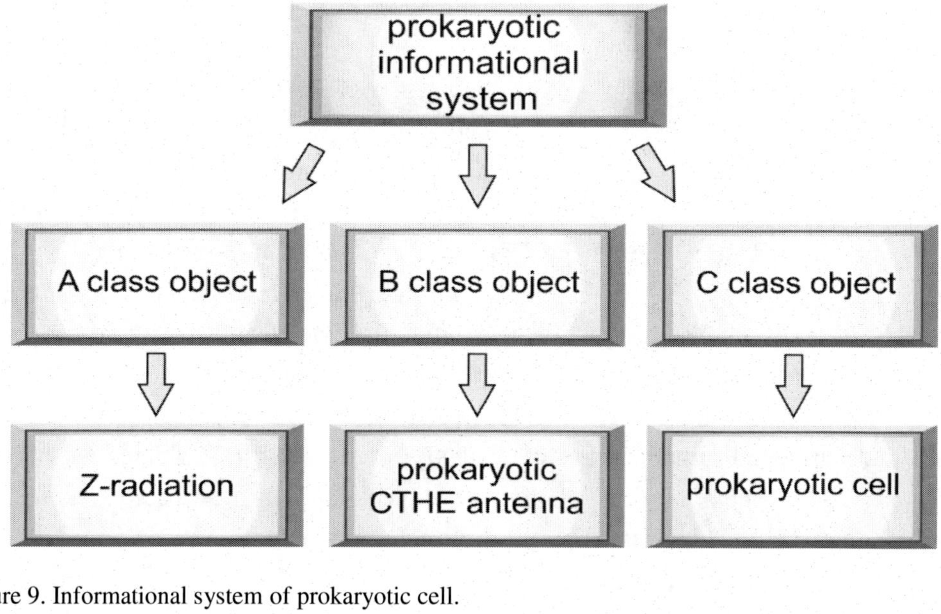

Figure 9. Informational system of prokaryotic cell.

It is known that a continuous exchange of energy, matter and information takes place between a living matter and an environment. Information as energy and matter is uncreated, indestructible and eternal in time and space. It is proved that the Earth's surface is constantly permeated by cosmic streams of particles and electromagnetic radiations, in which light waves play only a very little part. Chizhevsky considered that our planet is connected with the cosmos by millions of "invisible treads." It had been supported that the special bioactive radiation together with cosmic streams is present, which he called the Z-radiation (Chizhevsky, 1972). In the view of a number of sciences, the progressive evolution on the Earth is caused by universal radiation (Yagodinsky, 2015).

Let us assume that Z-radiation really exists and this bioactive radiation is an A class object with regard to the informational system of prokaryotic and eukaryotic organisms. In this case, a Z-radiation or a set of species signals, if the favorable environmental conditions and certain level of the DNA-energy and DNA-particle organization are present, should lead to the emergence of new species and their evolution.

Thereby, we have an A class object – the Z-radiation, which by its origin is the cosmic information. As we remember, information is not matter or energy (Wiener, 1983). For example, for receiving of the cosmic information, the species signal primarily must be fixed on some intracellular physical carrier inside a prokaryotic cell. It therefore means, if an A class object is the cosmic information, then in the cell there must be devices that are able to catch only their own species signal and convert it into an intracellular operator or a B class object. In order to understand how the informational system of prokaryotic organisms works, let's look once again at their cells from the physics standpoint and try to understand the specific features of the inner structure in terms of this function.

Prokaryotes are small unicellular organisms without a nucleus, and they reproduce only via the binary type of division. The type of cell reproduction is only binary division (or promitosis). Most species have a globular or oval form, less often a tortuous, filamentous, etc. Their DNA-particle has a continuous ring structure and is connected with the plasma membrane by means of special membrane proteins. A cell wall is located outside the plasma membrane, giving it a certain shape and rigidity. Microorganisms retain their specific identity (size, shape, lifestyle) over tens of millions of years.

In corpuscular biology, a species-specific DNA-particle (size, chemical structure) is the main source of genetic information and intraspecific diversity of microorganisms. Prokaryotic genomes are very dynamic structures, even within a species. Differences by the gene number in *E. coli* can reach 30% (Ravin & Shestakov, 2013). They, as open systems and very small organisms, are constantly in contact with a fluctuating habitat. During 3.8 billion years of their evolution as the result of intraspecific variability, mutations, intragenomic rearrangements, different internal and external impacts and noises, a large part of corpuscular information would have long been lost or greatly altered. Mechanisms for its recovery are still not known or are the subject of discussion. Consequently, there must be special mechanisms that protect the initial species information from degradation.

Now, we turn again to the prokaryotic cell, as a B class object, which must catch the Z-radiation and fix it on the special intracellular physical carrier. It is known that an antenna is used to receive the electromagnetic waves in all radio communication. We think that a ring chromosome can be such an antenna in a prokaryotic cell. In that regard, we have noted that the prokaryotic DNA-particle has a shape and an internal structure reminiscent of the Contrawound Toroidal Helical Antenna (CTHE), only much more perfect. Let's compare their "technical characteristics."

Spiral type antennas of the CTHE type are employed on the stationary and mobile radio communication devices, provided the uninterrupted reception of signals of certain frequency. They are characterized by a weak influence of the environment on the input impedance, high Q-factor and range of communication, narrow bandwidth for the selected frequency of a considerable flexibility without breaking electrical contact, and a high sensitivity to the magnetic component of the electromagnetic field. In this type of receiving antennas, two spiral curved conductors are used as the frame. Such antennas are called the retarding antennas, because the velocity of wave propagation along the helix axis is less than light speed. Conductors along the length are interconnected by a system of transverse clamps and are not linked in the intersection points. The ring is the optimal shape for all types of loop antennas. Compact size is important for their advantage. When moving, the high communication quality is ensured by the rigid construction of their carrier.

A prokaryotic chromosome or the spiral micro-antenna of the CTHE type represents a continuous ring chain of a tightly twisted DNA-particle. The perimeter of its ring in *E. coli* is about 1.6 mm (Chentsov, 2004). The DNA consists of two polynucleotide chains that are twisted one around the other in the form of a double helix (or two spirally twisted conductors). They are connected together by a large number of hydrogen bonds (or a large number of transverse clamps). The threads of DNA-particles are combined with the plasma membrane by means of its ends, which also do not unite with each other. It seems obvious that the Contrawound Toroidal Helical Antenna and the prokaryotic chromosomes are similar in their construction. Therefore, we will describe the chromosome of the prokaryotic cell in the analyzing of an information component of the genome as a "prokaryotic CTHE antenna" (Kuznetsova, 2016).

Now we will consider at what moment of the cell cycle this occurs and how the reception of the species cosmic signal by the prokaryotic CTHE antenna takes place. For this, we carried out an analysis of the process of cell division as a part of the informational system and the main mechanism of action for the B class object. Note that maximum reception of an antenna can occur only when the plane of its frame is perpendicular to the direction of the Z-radiation.

Bacterial cells are divided every 20-30 minutes, by means of only one type of cell division – a binary division or promitosis. It is preceded by the DNA replication, after which the state of the genome becomes 2 DNA particles + 1 DNA-wave. The process of cell division begins with the separation of positively charged plasma membrane protein in half at the point of origin (the place of DNA association with the plasma membrane) at the moment of two negatively charged inactivated molecules of DNA-polymerase passing through (Kuznetsova, 2013). Two positively charged molecules of protein mutually are repelled from each other and take away daughter chromosomes. The distance between nucleoids is always the same: A quarter of the cell length. The state of the genome becomes 1 DNA-particle + 1 DNA-wave. Cytokinesis and formation of two new cells complete the process of every binary division.

In the above-described corpuscular-wave mechanism of division, we are particularly interested in the moment of separation of double ringed chromosomes, which present a prokaryotic CTHE antenna (Figure 11). The process of its splitting begins from the origin point. The area of the doubled DNA-particle gradually decreases up to the top of the rings. After the destruction of this last binding piece, the nucleoids are completely separated from each other. As each daughter chromosome is a frame antenna, then during the separation of two DNA-particles, the angle of inclination of their plane relative to the Earth's surface should gradually change (Figure 10). In this moment, apparently, the automatic configuration of two prokaryotic CTHE antennas to the frequency of a certain species signal occurs. After division, the daughter chromosomes return to their initial position, and the reception of cosmic information ceases. And if that's true, then the prokaryotic antenna works in the mode of the reception only in the very narrow period of splitting of the doubled chromosome. Since the wave function on the DNA during division does not stop, the electromagnetic waves of a new species cosmic signal cause forced oscillations of free electrons in the prokaryotic CTHE antenna with a frequency equal to its frequency. The frequency, amplitude and phase of DNA-wave oscillations are modulated by this incoming signal.

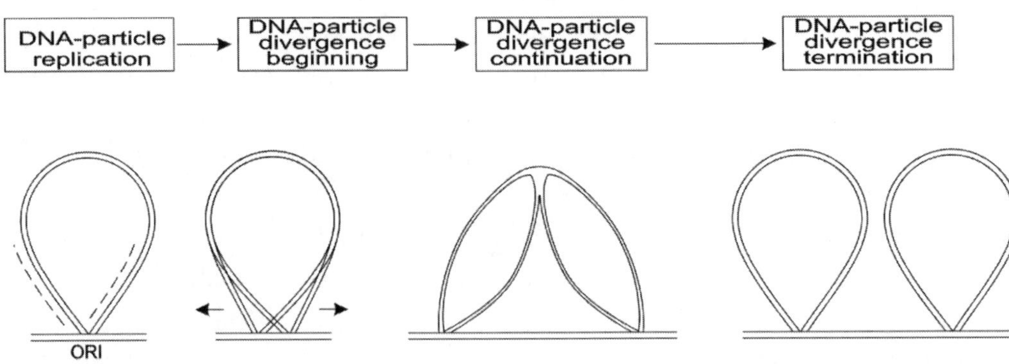

Figure 10. The scheme of configuring of prokaryotic CTHE antenna for receiving of species cosmic signal during cell division.

It should be emphasized that the DNA-wave in the moment of reception of a new cosmic signal already carries the old information signal, or the cellular DNA-information (Kuznetsova, 2016). Cellular DNA-information is a modulated wave, whose frequency and amplitude is close to the characteristics of the species cosmic signal, but not identical to it. Initially, it arises as a result of the superposition of two waves: An unchanging cosmic signal and a cellular DNA-wave of the previous generations. Then, this resulting wave is exposed to various (internal and external) factors causing its distortion during the next cell cycle. Noises and disturbances can be random or directed. Directed changes occur for each cell division and are associated with the natural and climatic specificity of the environment. In this case, these changes of cellular DNA-information are amplified in certain segments of a DNA-wave and pass from generation to generation. Therefore, they can be considered as an adaptive mechanism, as the summary "memory" of the same genotype-environmental interactions, the result of adapting different generations to their ecological niche, helping the species to adapt to its conditions.

Undoubtedly, the invariable species signal from the cosmos with respect to prokaryotic and eukaryotic biological systems acts as a restoring factor for their genome. The property of frailty is inherent in the DNA-information, as is any piece of information. It can be lost or distorted due to mutations, intra-genomic variability, different noises and interferences. The analysis of the complete genomes of prokaryotes showed that their genetic material is a very dynamic formation (Ravin & Shestakovб 2013; Bonch-Osmolovskaya & Atomi, 2015; etc.). Intraspecific variability occurs in different directions and leads to mutations, loss or acquisition of genes (segments of the genome), to changes in genome size.

The frailty of DNA-information is especially important for microorganisms. Firstly, the prokaryotic cell is an open system, and the stream of matter and energy is constantly going through it. Secondly, a cell cycle proceeds mainly at the lowest energetic level, therefore, the intracellular metabolism depends heavily on environmental conditions. Thirdly, the ring chromosome-wave is in direct contact with the cytoplasm. Biochemical processes and thermal motion of molecules also affect the information component of the genome. Fourth, the DNA-wave changes as a result of the different environmental fluctuations (temperature, pressure, and chemical composition). Despite the constant damage of information under the influence of noises, all microorganisms retain their species identity. The basic morphological, physiological and biological characteristics of each species remain always unchanged. This is a confirmation that the restoration of their DNA-information occurs in every cell generation.

We think that another ordering agent of DNA-information, which limits intraspecific heterogeneity, is a prokaryotic CTHE antenna. Technically, its characteristics (nucleotide composition, shape, size and diameter of frame antenna) should not exceed the variability of a species norm reaction. If the intraspecific variability is higher, such a cell will not be able to take its own cosmic signal. DNA-information without restoration is doomed in terms of degeneration. This leads to an increase of different morphological and genetic abnormalities and cell death.

Applying what has been said to radio engineering, we must take into account the following. In modern radio receivers for tuning to receive only one station, an internal generator of electromagnetic oscillations is used, which is called a heterodyne. This device allows you to tune the receiver to a narrow range of received waves, including the desired frequency range. A receiver with such a generator is called the superheterodyne radio receiver.

A B class object and a superheterodyne radio receiver are similar by the construction and principle of work. A prokaryotic cell is a self-oscillating system where a modulated DNA-wave (i.e., a cellular DNA-information) performs the undamped oscillations with a frequency close to the frequency of a species cosmic signal. Therefore, the cellular DNA-information with respect to the bioactive Z-radiation represents the in-build generator of electromagnetic oscillations inside a cell or an intracellular heterodyne. This heterodyne is used to configure the prokaryotic CTHE antenna to receive only a certain cosmic signal. In the time reception of a new cosmic signal, the waves of two closest in frequency species-specific signals (cosmic and cellular) are superimposed and an intermediate frequency appears on which the complete vector addition of waves and the enhancement of the species signal occur.

Since the main morphogenetic traits of each species are repeated in their descendants, we have a typical case of information transmission. In prokaryotes, a biological harmonic oscillator, and an organism, a cell is found. The material carrier of cellular DNA-information is a modulated wave of the oscillating ring chromosome. Therefore, outside cellular systems no DNA-information can exist. Accordingly, an object of the C class, where the informational process is realized in the phenotype of the species-specific individuals is a cell too.

Obviously, the objects of the C class in the radio receiver and biological systems are different. In prokaryotic organisms, an object of the C class for which the species-specific information is required is two daughter cells. After reception, a new cosmic signal, the maternal carrier and the source of information of their future development is two modulated DNA-waves with a strengthened individual species informational signal each. Then, the decoding process or the stage of transmission of information from a source (new cellular DNA-information) in the phenotype of two independent microorganisms begins.

The nature and mechanisms of decoding information have long been the discussion. The main role belongs to devices and operators, which transform the information of a DNA-wave in the intracellular morphogenetic processes. Therefore, morphogenesis has a corpuscular-wave nature. By the laws of physics, the modulated DNA-wave generates a non-uniform electromagnetic field or a morphogenetic field according to Weiss's terminology (Weiss, 1928). The zone of its action is inside and beyond the cell. This field serves as an information matrix that interacts with the cytoplasm, creating an intracellular ordered heterogeneity. Then, the complex of biophysical and biochemical form-building processes in accordance with the morphogenetic information are initiated. A morphogenetic field under its influence orderly changes, and the individuals of certain shapes (coccoid, rod-shaped, etc.) and size are developed. The daughter cells after division have the same DNA-energy, DNA-particle and DNA-information. Then, the DNA-information undergoes changes, because their development is carried out independently from each other and through channels with different noises. Therefore, cells take a close but not identical phenotype.

SUMMARY

To summarize the above, it should be noted that the cosmic nature of DNA-information makes it possible to understand how and why new species of animals, plants and microorganisms appear during evolution. Essentially, a speciation presents many new variants of ordered changes of the DNA-energy, DNA-particles, and DNA-information at a given time. A genetic apparatus, which is able to capture an information signal of more evolutionary advanced species, can be created as a result of intraspecific variability. The probability of this event is very small, but it takes place. Then, the individuals with distinct differences with individuals of the original species appear and multiply. In this case, a new independent species originates in certain ecological conditions.

The horizontal evolution of prokaryotes is due to a change of two components of the genome – a DNA-particle and DNA-information, and does not touch DNA-energy. The main energetic level is the lowest level, with a periodic transition to the 1st level. Apparently, the stability of the energy component of a genome and the high variability of corpuscular ones may be the reason that speciation in prokaryotes occurs as a horizontal evolution, which is controversial until now. Mutations and changing of DNA-particle lead to a horizontal transfer of genes, a creation in the great number of the prokaryotic CTHE antennas with different technical characteristics. This explains their high diversity, conservatism of evolution and tolerance to various environmental conditions (Knoll & Bauld, 1989). We think that the most important mechanisms of bacterial adaptation (rapid mutations, new cycles of metabolism, decreased metabolic rate, sporulation, lateral transfer of "useful" genes) and DNA-information restoration during every cell cycle ensure high stability of their ecosystems.

Chapter 2

EUKARYOTIC CELL

Living matter for further evolutionary development was necessary to increase the Q-factor and energetic capacity of a biological harmonic oscillator, as a physical system. Such perfected type of oscillator was the nucleus together with the oscillating DNA. For this, the DNA (particle and wave) has moved from the cellular cavity inside a nucleus, or inside a closed cavity with reflective walls.

Between the appearance of the prokaryotic and eukaryotic cell there was a very long time interval. By this time, prokaryotes have already gone a long way of evolution and have undergone many changes. They have mastered different ecological niches, a huge number of species have originated, differing by the morphology, metabolism, ways of obtaining energy, types of nutrient substrates, etc. In the Precambrian era, an important event that occurred was the appearance of photosynthesis, during which oxygen is released as a by-product. Therefore, an atmosphere rich in oxygen existed on the Earth by the time the eukaryotic cell appeared (Schopf, 1981). Therefore, in the course of prokaryotic evolution, the likelihood of origination of a eukaryotic cell gradually increased. When this transition occurred (≈ 1.0-1.4 billion years ago), it gave a powerful impetus for the living matter development: The appearance of multicellular animals and plants, the directed evolutionary development of species without complete "repetitions," their specialization and complication by the lines of DNA-energy, DNA-particle and DNA-information, and the perfection of connections with the environment.

However, the most cardinal changes occurred from the line of DNA-information. This is due to the fact that the ring chromosome or prokaryotic CTHE antenna was replaced. Instead, a set of V-shaped chromosomes associated with the nucleus enveloped in three points (centromeric and two telomeric sites) originated. This entailed an appearance of a principal new mechanism of cell division – mitosis. The probability of simultaneous realization of these two events is extremely small. Apparently, these processes went through short-lived intermediate forms, which left no traces.

Eukaryotic cells are much larger than the prokaryotic cells. They differ significantly by the morphology and functional destination and are characterized by sufficiently strict intracellular compartmentalization, which is achieved by means of membranes. The number, length and chemical composition of chromosomes are changed. The set of chromosomes are located inside the nucleus. All eukaryotic cells have two internal energy sources: a biological harmonic oscillator and metaphase chromosomes. The specialized organelles (endoplasmic reticulum, mitochondria, plastids, lysosomes, Golgi apparatus, etc.) are in the cytoplasm. The plant cell is bounded outside by the plasma membrane and cell envelope. The processes of replication and transcription are separated.

The genome of a eukaryotic cell has a corpuscular-wave nature. It is arranged and functioned by the trinity principle of DNA-energy, DNA-particle and DNA-information. In eukaryotes, the actual status of any cell can be characterized by four basic statuses of a DNA molecule (particle and wave): DNA-energy, DNA-particle, DNA-information and spin direction. Now, we will analyze the genetic information, the maternal carrier of which is the biological harmonic oscillator or the DNA-energy.

ENERGETIC COMPONENT OF GENOME (DNA-ENERGY)

The common principle of organization of an energetic system in a prokaryotic and eukaryotic cell is the same (Figure 11).These are the sources of internal and external energy and two intracellular energetic streams – a centripetal and a centrifugal. However, each structural element of a eukaryotic cell underwent significant changes during evolutionary development. Figure 11 shows the main structural elements of the energetic system of a eukaryotic cell and their connections during the cell cycle.

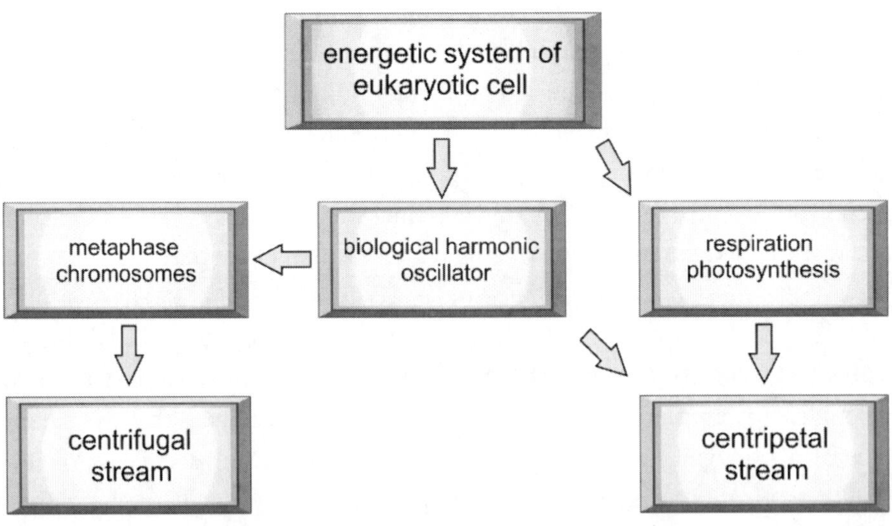

Figure 11. Energetic system of eukaryotic cell.

Sources of Internal Energy

The eukaryotic cell has two sources of internal energy – a biological harmonic oscillator and metaphase chromosomes. As was mentioned above, the main internal source is the biological harmonic oscillator – the nucleus together with the oscillating DNA. Its energetic capacity and Q-factor has been significantly enhanced due to an increase of chromosome number and the displacement of DNA-wave into the nucleus. As a result, the eukaryotic species began to master level after level in the course of evolutional development of plants and animals.

The law on mandatory cell reproduction extends to all prokaryotic and eukaryotic organisms (Maszia, 1963). The inductor of cell doubling is the free energy, which is released at the DNA-wave transition to the lower energetic level. The principle of biological harmonic oscillator function after displacement into the volume of a nucleus has changed. If the prokaryotic oscillator does not stop during cell divisions and the cell is an open system, then the eukaryotic cell and its cell cycle have a more complex organization. The oscillator of a eukaryotic type acts during an interphase and it is broken down during mitosis. Accordingly, the cell is an open system and a closed system during this time. Figure 12 shows the main channels of free energy utilization from a 1st main source of internal energy. These include: DNA replication, ATP restriction, cell growth, centrioles formation, chromosome coiling during G_2 period and energy losses. Note that the level of these losses is so great that the energetic balance of interphase cell becomes the negative (Zholkevich et al., 1973). This comforts that the contribution of the biological harmonic oscillator to the energetic system of a eukaryotic cell is very significant.

The main distinguishing feature of work of a eukaryotic oscillator is as follow: it is filled by the biological energy to the maximal species-specific level, and then a DNA-energy spontaneously descends to the zero level in accordance with 2nd law of thermodynamics. Therefore, the life cycle of a vast majority of the eukaryotic organisms is divided into two periods – sexual reproduction and ontogenesis (Figure 13). “Charging” of oscillator and transition of DNA-wave to higher levels is a strictly genetically determined process, which takes place on the basis of an energetic line of the species reproductive cycle, under certain physical conditions with energy consumption. Its “discharge” and spontaneous motion to lower energy levels occurs during ontogenesis. Development of a multicellular plant from the initial sporophyte up to its natural death represents a directed downward change of biological energy from the maximum species-specific level to level zero, which occurs within the framework of the plant body as a closed system.

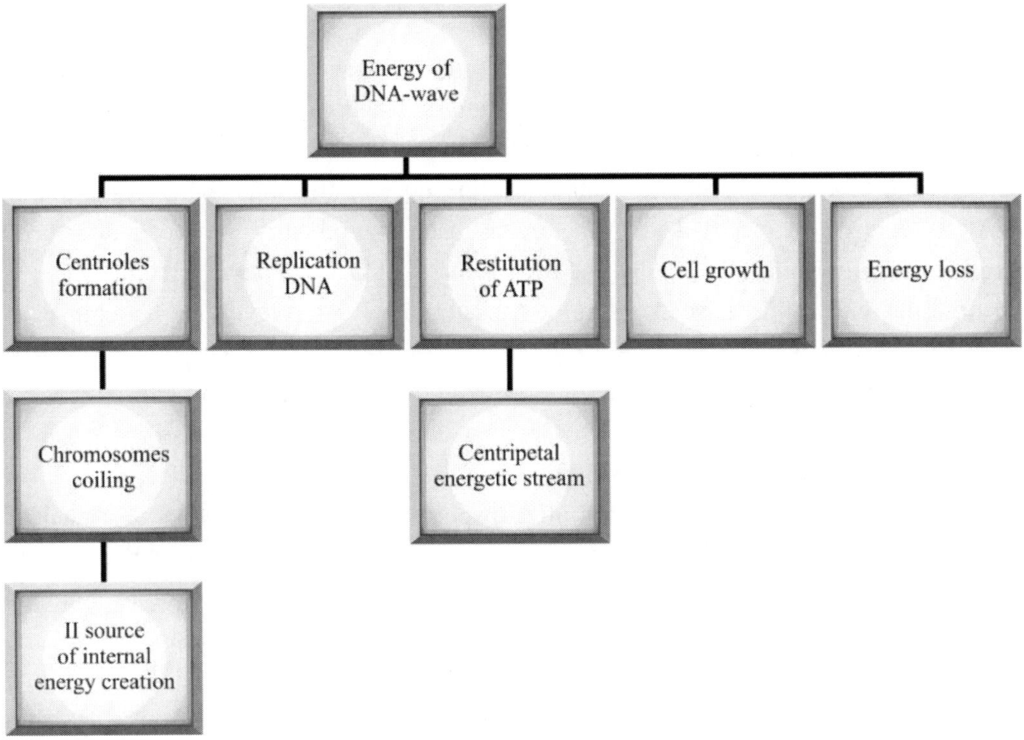

Figure 12. The channels of utilization of energy from biological harmonic oscillator of eukaryotic cell.

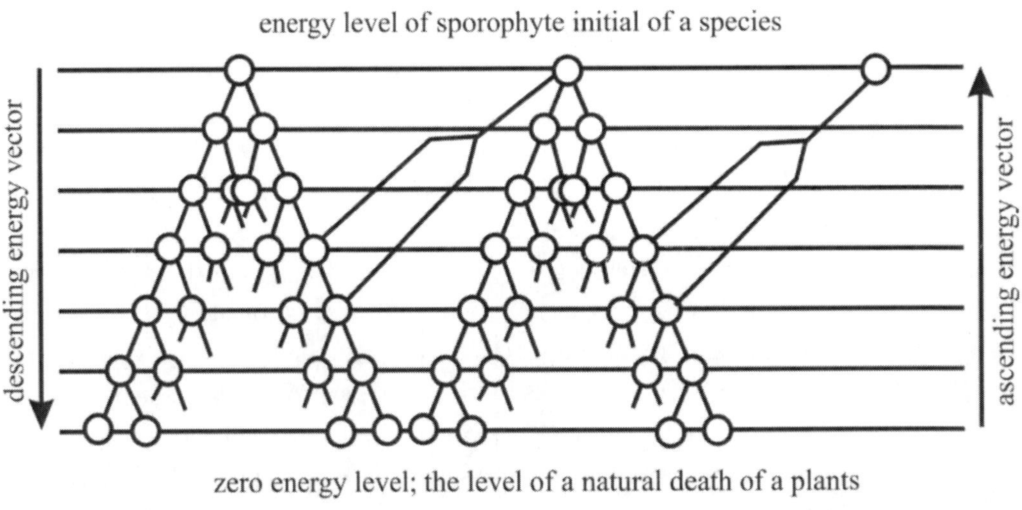

Figure 13. The motion of biological energy in the life cycle of eukaryotic multicellular species during periods of sexual reproduction and ontogenesis.

The set of metaphase chromosomes is a 2nd source of internal energy. This source is detected only in the proliferating cells of animals and plants. By physical nature, this is the elastic energy of very compressed metaphase chromosomes. The process of chromosome spiralization and accumulation of elastic energy begins in the G_2 period of the interphase (internal coiling), continues in the prophase of mitosis (external coiling), and reaches a maximal energy in the metaphase (metaphase chromosomes). The liberation of free energy takes place during chromosome uncoiling during telophase. This source gives rise to a centrifugal and centripetal energetic stream. The energy of these streams is utilized by the several channels, raising the energetic state of daughter cells before their ontogenesis (Figure 14). The metaphase chromosomes, as the independent internal source separates the processes of replication and transcription in a space and in an amount of time. As a consequence, the energetics of DNA biosynthesis is provided by the energy of a biological harmonic oscillator (S period), and the RNA and protein biosynthesis due to the energy of metaphase chromosome (telophase, interphase).

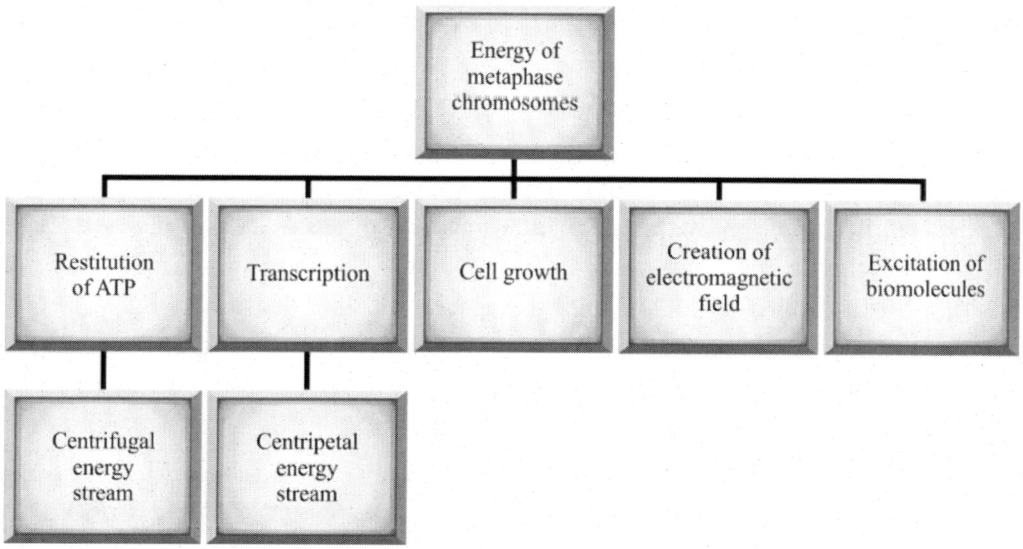

Figure 14. The channels of utilization of energy from metaphase chromosomes of a eukaryotic cell.

Sources of External Energy

In plants, both external sources (respiration and photosynthesis) have more efficient mechanisms for extracting and utilizing energy from the environment. In all metabolic pathways of prokaryotic organisms, they use only one – respiration. Both these redox reactions are carried out in the specialized organelles: photosynthesis in chloroplasts, respiration in mitochondria. The final product is ATP molecules, which serve as a source of energy for many endergonic biochemical reactions and intracellular processes. The biophysical models of photosynthesis and respiration are described in more detail in Chapter 5.

Centrifugal Energetic Stream

Centrifugal energetic stream is internal by origin. It originates from the 2nd internal source (metaphase chromosomes) in the telophase and acts during an interphase. The vector of energy motion is directed from the nucleus to the cytoplasm. Centrifugal energetic stream functions on the basis of transcription and processing.

In previous works (Kuznetsova, 2013; Kuznetsova, 2016), a hypothesis was proposed that the physical nature of a processing and centrifugal energetic stream of prokaryotic and eukaryotic cells is the RNA exponential disintegration. During spontaneous division of RNA molecules, the probability of their disintegration never changes, and therefore by the end of the first half of the disintegration period half of the pre-rRNA remains, the second – a quarter, etc. The exponential disintegration of pre-RNA occurs with the releasing energy, and this process is stretched in a time.

During transcription, many large molecules of primary RNA transcripts (pre-mRNA, pre-tRNA and pre-rRNA) are formed. The main part of products primary transcripts is pre-rRNA. In the course of processing, the mature rRNA molecules are formed and about half of initial RNA precursors are destroyed. Other types of RNA undergo processing too. The centrifugal energy stream in cellular metabolism has a specific purpose. Most parts of its energy is spent on energetic supply of translation. This process occurs in ribosomes, where mRNA, rRNA and tRNA are an informational, energetic and transport base for protein biosynthesis. During each cell cycle, the amount of protein is doubled.

Centripetal Energetic Stream

Centripetal energetic stream works in the ATP-ADP system also during the interphase. Cell life requires much energy. The ATP pool is constantly recharged in the ATP-ADP system. ATP is a molecule of a wide spectrum action. It serves as a main energetic source for many enzymatic reactions that underlie different metabolic reactions and intracellular works.

This stream is the internal and external by origin. Both internal sources (biological harmonic oscillator and metaphase chromosomes) and external sources (photosynthesis and respiration) take part in its creation and maintenance. The centrifugal stream originates from 2nd internal sources. The part of free energy from uncoiling chromosomes in the closed system of a telophase cell is expended on the restoration of ATP from ADP. Then, this stream is continuously replenished with energy from both external sources. An interphase cell is an open system. Energy enters into a cell during photosynthesis (sunlight energy) and respiration (chemical energy of carbohydrates, lipids, etc.).

ENERGETIC SYSTEMS OF CELL CYCLE

The cell cycle of eukaryotic cells consists of an interphase (G_1 pre-synthetic, S synthetic and G_2 post-synthetic periods) and mitosis. The interphase duration is ≈ 90% from the cell cycle (Epifanova, 2003). Mitosis is subdivided at the prophase, metaphase, anaphase and telophase. During the cell cycle, the cell genome passes through 4 corpuscular-wave basis states: 1) 2 DNA-particle –1 DNA-wave; 2) 2 DNA-particle – 0 DNA-wave; 3) 1 DNA-particle–0 DNA-wave; and 4) 1 DNA-particle – 1 DNA-wave + 2 DNA-particle – 1 DNA-wave. Each of these states corresponds to a certain cellular stage: 1) Latent development; 2) mitotic development; 3) cell birth; and 4) cell ontogenesis. All of these stages are characterized by a strict sequence of intracellular events, the main components of which are as follows.

Latent Development

DNA-wave radiation in the end of the G_1 period and the entry of two energy quanta into the nucleus initiates the development of cells of a new generation. The genetic status of a maternal cell in this moment is written as the following: 1 DNA-particle– 1 DNA-wave. The cell is an open system. Its energetic potential is very high and consists of a biological harmonic oscillator, photosynthesis and respiration. Duc to the energy of an oscillator, a charge transfer complex is created: Two positively charged centrioles and a great number of activated negatively charged molecules of DNA-polymerase. This is the background for the beginning of 2nd energy source creation. The activated molecules of enzymes are attracted to the counter waves of a standing DNA-wave and the biosynthesis of a DNA-particle is carried out by the complementarity principle. During the S period, the state of the genome becomes: 2 DNA-particle–1 DNA-wave. After replication, a free electron from the molecules of a DNA-polymerase pass into the centromere regions of chromosomes.

2nd internal source creation occurs from the G_2 period and up to telophase mitosis. In the S- and G_2 periods, two positively charged centrioles are held by the biofield forces inside the nucleus, where its intensity is the maximal (nearby zone). Doubling the DNA-particle during the passage of its nucleotides between centrioles undergoes the internal coiling or the conformational changes of macromolecules by a one-side nucleotide turn (Kuznetsova, 2013). Chromosomes in the time of complex isomer formation gradually shortens, and their internal energy increases.

Photosynthesis, respiration, biosynthesis of proteins, carbohydrates, etc. continue up to mitosis. However, the intensity of centrifugal and centripetal streams by this time is greatly reduced.

Mitotic Development

A DNA-wave breaks down and the biofield disappears, as soon as the length of the chromosomes becomes smaller than the size of the nucleus. After biofield disappearance, two positively charged centrioles mutuality repulse and fly away in opposite directions, giving rise to two cell centers. Chromosomes split into chromatids up to the centromere. The nucleus envelope without the biofield forces is destroyed too. This completely changes the physical situation in a cell and it passes on to mitotic division. Its genome state becomes: 2 DNA-particles–0 DNA-waves. The cell is a closed system. Photosynthesis, respiration and all biosynthetic processes are absent, or are supported at a very low level. The energetic system (eukaryotic oscillator, centrifugal and centripetal streams) is destroyed. Intracellular processes go in the direction of entropy increase.

However, the formation of a 2nd internal energy source continues as a physical process. Between the centrioles, as between two positive charged particles, an electric field appears. Chromosomes under its influence continue the process of spiralization (external coiling of chromatids). Progressive coiling leads to the maximal accumulation of elastic energy inside each chromatide separately. In metaphase, the chromosomes are lined up along equatorial plate.

Cell Birth

Anaphase begins with the separation of two sets of daughter chromosomes as a result of the attraction of negatively charged centromeres and positively charged centrioles. The charges are neutralized after their collision. The charge transfer complex, which appeared during DNA-wave radiation, cease to exist and the metaphase chromosomes begin to uncoil and extend. Around each set of chromosomes, an electromagnetic field is created, which attracts a maternal cytoplasm, sorts its molecular composition in accordance with electrochemical structure and size of biomolecules, spatially separates the daughter cells. The nucleus envelope is restored as result of centripetal movement of phospholipids.

In telophase, the genome of two new cells is 1 DNA-particle–0 DNA-wave. They are inside the envelope of the mother cell, which is a closed system. The uncoiling of metaphase chromosomes occurs simultaneously with the liberation of free energy, which are used by the different channels (Figure 14). This energy activates the RNA-polymerase molecules and raises the energy potential of future cells. Restitution of ATP molecules and creation of centripetal energetic stream takes place. Kinetic energy of biomolecules increases. The biological harmonic oscillator is “turned on” after a chromosome collision with the envelope of the nucleus.

Cell Ontogenesis

The independent life of daughter cells begins with a restoration of wave function on DNA, biofield and spatial separation of cells from each other. The state of their genome is 1 DNA-particle –1 DNA-wave (G_1 period). The cell again becomes an open system. Its energy potential is high. The typical ultrastructure of cytoplasm with the chloroplasts, mitochondria, endoplasmic reticulum, Golgi apparatus and other organelles is restored. The active exchange takes place with the environment by matter and energy. A cell grows rapidly.

In the moment of DNA-wave appearance, the molecules of activated RNA-polymerase are attracted to one of the standing DNA-waves and the biosynthesis of primary RNA transcripts (pre-mRNA, pre-tRNA and pre-rRNA) begins, which gives rise to a centrifugal energetic stream. A centripetal energetic stream is also replenished with ATP molecules, which are formed during photosynthesis and respiration. At the end of the G_1 period, the next radiation of the DNA-wave takes place, and the process of cell formation of a next generation begins in the same sequence. A biological harmonic oscillator of the mother cell before each division spontaneously descends to the lower level. Therefore, the energy potential of subsequent cellular generation is lower than the previous generation by one energetic level.

Energy movement in the cell cycle has a cyclic character. Its change represents a two-vertex curve with right-side asymmetry (Figure 15). The first energetic maximum is created by the energy eukaryotic oscillator in the G_1 period end, the second – by the energy of metaphase chromosomes in the telophase of mitosis. The two most energy-consuming reactions of cell cycle – a replication and a transcription – are provided by energy from these two internal sources. The minimal energy level is observed in the prophase, metaphase and anaphase of mitosis.

The passing from a proliferating cell to a differentiated one is caused by a change in the energetic state of DNA-energy. The cell after reaching the Hayflick limit is forced to move to a more economical way of development (Shay, Wright, 2000). Preservation of cell viability is possible after its transition to a state of a differentiated cell. For this, a replication, a mitotic division and 2nd source of internal energy are removed from a cell cycle. It is believed that the DNA synthesis and mitosis are the most energetically dependent processes of a cell cycle (Webster & Vant'Hof, 1969). The eukaryotic oscillator of a differentiated cell switches to an incessant regime of operation. A cell becomes the open system. The external energy sources (photosynthesis, respiration) operate constantly. However, the oscillator of a differentiated cell continues to emit the quanta of free energy with the same periodicity and tends to reach equilibrium in accordance with the 2nd law of thermodynamics up to its death. Its energy in this time is mainly spent on the energy supply of the transcription process, and on the replenished component the centripetal and centrifugal energetic stream.

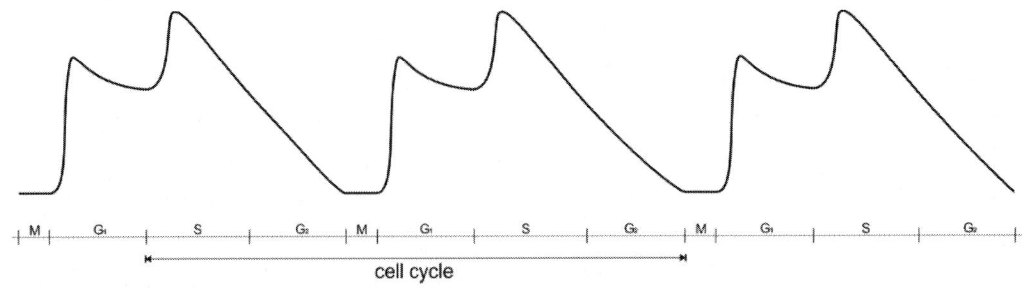

Figure 15. Cyclic change of the energetic level of eukaryotic cell in several adjacent cell cycles.

ENERGETIC SYSTEMS OF MULTICELLULAR ORGANISMS

Now let us analyze the energetic aspects of the transition from unicellular to multicellular organisms of plants. Based on the corpuscular-wave nature of process, the energy capacity of the eukaryotic oscillator of multicellular species must be 3 or more energetic levels, and their life cycle should be divided into two periods: a sexual reproduction and ontogenesis.

In prokaryotes, the main physical factor of DNA transition into an excited state is an increased intracellular pressure, which is achieved during cell growth in fixed cell volume. In some lower plants, the transition of a biological harmonic oscillator to the higher energetic levels is carried out by means of modifying prokaryotic mechanisms. The life cycle of such algae as a *Chlorococcales, Volvocacalles* by this principle is organized.

Activation Stages

The reproductive cycle of *Chlorella* consists of only one stage by the DNA transition to the higher energetic levels, or the activation stage (ᚨ). Cells of this algae has a spherical shape, outside they are covered with fairly tough envelope. It has a nucleus and only one chloroplast. *Chlorella* reproduces by means of division, during which the autospores are formed. We noticed that transition autospores to the higher energetic levels resembles a mechanism which uses a prokaryotic cell for this purpose. Only *Chlorella* increases intracellular pressure in the process of cell divisions (2-4), which occur into a fixed volume of a maternal cell. Therefore, the sizes of daughter cells become smaller, and the common molecule number of proteins, lipids, etc. in them is doubled. Because the concentration of molecules is a direct function of pressure, a DNA-energy will move to higher levels after each division. Rejuvenated autospores after exiting from the envelope of mother cells grow rapidly and reach a species-specific shape and size. During their vegetative development, the DNA-wave spontaneously descends from the level at a level in accordance with 2nd law of thermodynamics. After reaching the zero level (equilibrium state), *Chlorella* again turns to sexual reproduction, and the cycle of its development is repeated.

A prokaryotic mechanism is the most primitive bio-technique for the rejuvenation of eukaryotic cells. Its essential disadvantage is that the unicellular organism (*Chlorella, Chlamydomonas*), or all or part of cells (*Gonium, Pandorina, Eudorina*) for the DNA transition to the higher levels should develop in a fixed volume of the initial cell, which leads to the death of a maternal organism. A prokaryotic mechanism became a dead-end branch in the evolution of multicellular animals and plants. It was succeeded by the activation stage, which acts on the basis of special divisions by cell rejuvenation.

In the previous work, we suggested that the age-related changes of cells and their morphogenetic activity are related with the wave nature of DNA and energetic state of biological harmonic oscillators (Kuznetsova, 2013).Its transition to the maximal species-specific energetic level occurs on an energetic line of reproductive cycles and requires certain energy costs from a mother plant. Therefore, the number of generative cells involving the process of sexual reproduction is significantly less than the total number of its somatic cells.

The deciphering of the reproductive cycle matrices showed that the energetic line of lower plants has 1-3, and the higher plants 3-6 activation stages, which participate in the process of biological harmonic oscillator lifting to higher energetic levels. It therefore means that the process of plant speciation went in the direction of biological energy increase. Energetically strong species have a more significant advantage over the immature matter, over less evolutionarily advanced species. This allowed them to come on to land and mastered various ecological niches.

Activating Mitosis

The activation stage is a genetically determined mechanism for the rejuvenation of generative cells, the process of which occurs together with special cell division – activating mitosis. This type of mitosis proceeds strictly in accordance with the law of the energy of conservation and leads to differentiation of daughter cells by the DNA-energy level.

Probably, the change of a genome state by the DNA-energy occurs in the metaphase of activating mitosis (Kuznetsova, 2013). The polar axis of the mother cell and spindle of division coincides with the gravity direction. The protoplast is also differentiated according to the density gradient. Metaphase chromosomes form an equatorial plate, which are located across the gravity. Chromatids of doubled chromosomes are maximally condensed and are linked only in the centromere region. Centromeres after division into two parts immediately go towards the cellular centers, carrying with them the chromosomes. In this moment, the chromosomes take the V-shape, and the sister chromatids collide with each other and exchange energy. Since one part of centromeres moves into the region with a less dense cytoplasm, and the second part into the denser cytoplasm, the chromosomes of the apical pole receive energy upon impact, and the chromosomes of the basal pole lose it.

Let us assume that the initial position of the genome of the mother cell corresponds to the 4th energetic level and 3rd level after DNA radiation (Figure 16a). During activating mitosis, the chromosomes of the apical pole receive 3 quanta of energy upon impact and the DNA-energy of one daughter cell rises to the 6th level ($E_1 = 3h\omega + 3h\omega = 6h\omega$). The chromosomes of the basal pole, on the contrary, lose 3 quanta of energy and the energetic potential of second daughter cell decrease sharply ($E_2 = 3h\omega - 3h\omega = 0$). A rejuvenated cell passes into the reproductive process, and the second cell will die.

Note that certain limitations exist for the initiation of activating mitosis and the reproductive process. It can begin only from the 3rd energetic level. If the mother cell has the 2nd level, it goes down to the 1st level after DNA radiation (Figure 16b). In accordance with the law of conservation of energy, a DNA-energy of one daughter cell during division will move to the lowest level, because there are no energetic states below zero. Another daughter cell in this time will return to the initial 2nd level, or the level of the mother cell. This type of cell division is usually described as asymmetric mitosis. Asymmetric mitosis is widespread in plants. The cells of an apical meristem of a root and shoot are divided in this way.

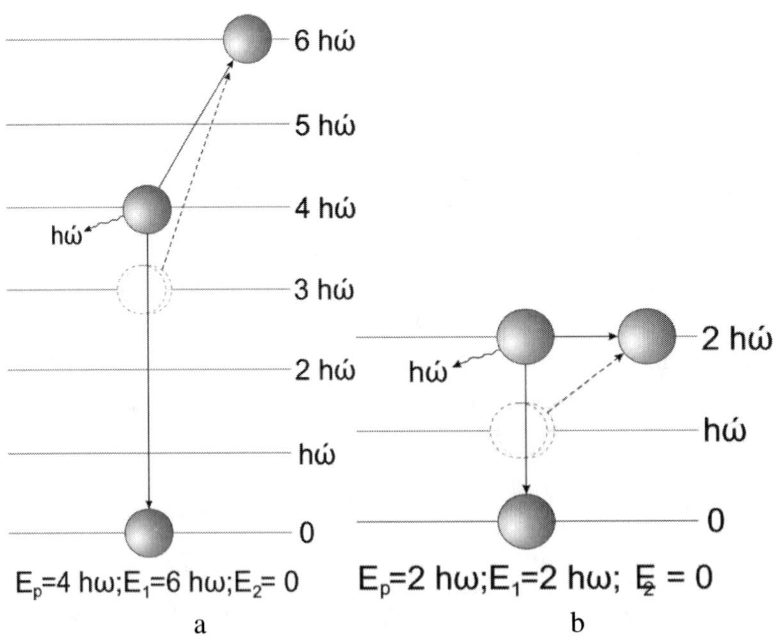

Figure 16. The scheme of activating mitosis (a) and asymmetric mitosis (b).

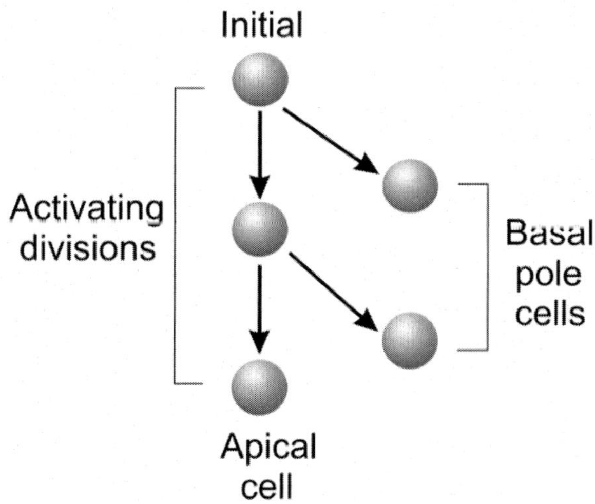

Figure 17. Activation stage with the length in 2 divisions of activating mitosis.

In the higher plants, several activating mitosis cells within one activating stage takes place often one after another. As a rule, they are combined with other types of divisions (mitosis, meiosis, free-nuclear mitosis, etc.). Together, they represent a certain stage of the reproductive cycle. The activation stage (↱) is a strictly genetically determined process. It can begin only in certain time and in the certain generative structures (cells, organs). Cells (vegetative, generative) having a definite energetic level for each activation stage participate in it. The polar axis and the planes of their divisions must coincide with the direction of gravity. In space, each activation stage is also organized in a certain manner (Figure 17).

The first cell (vegetative, generative), from the division of which the activation stage is beginning, is the initial stage. A stage is ended by the appearance of the energetically strong apical generative cell. The divisions of activating mitosis between the initial and the apical direction of which coincides with the gravity, are the cells of the basal pole. The divisions of generative cells, which activating mitosis depart to the basal pole, are the secondary. Activation stages are characterized by: 1) A strict sequence of cell divisions; 2) a fixed number of primary and secondary divisions; and 3) a polarity of generative cells and their division planes. The main purpose of stage is the rejuvenation of generative cells through the uneven distribution of biological energy between the daughter cells.

Stages of Reproductive Initial Emergence

This type of reproductive stage is part of the energetic line of the reproductive cycle of all higher plants and some lower plants with a complex structure of the thallus (brown algae, mosses, ferns, etc.). As a rule, it is located before the activation stage. On a diagram, it is denoted as ⬭.

It is noted that the development of generative cells often occurs within a complex of undifferentiated vegetative or generative cells (Allen, 1946; Owens & Molder, 1979; Kozubov et al., 1982; Kuznetsova, 2002). It originates from one or more meristematic cells, which are characterized by the increased mitotic activity (Allen, 1946). Their divisions occur only by the usual mitosis, can be accompanied by a cytokinesis (primordium of microsporophyll, primordium of megasporophyll, archesporium), and in some instances only the nuclei are divided and the cell envelopes between them are formed later (female gametophyte, free-nuclear proembryo). The local cellular complex on a cut has the form of an oval or a circle. Therefore, the directions of longitudinal axes in the surface cell layer are different. All of them are undifferentiated, identical by an ultrastructure and a form. The total number of cells is fixed by a species-specific limit. In higher plants, the male and female generative lines of reproductive cycles began with the stage of reproductive initial emergence. Apparently, the formation of a local cellular complex is an indispensable requirement for the initiation of the reproductive process, and this causes the transition of a meristematic cell from a vegetative to a reproductive development. The initial, or the first cell of activation stage (microsporangium initial, ovule initial, archegonium initial, embryonic initial, etc.) becomes one or more cells of a complex, in which the longitudinal axis of symmetry coincides with the gravity direction.

Since the DNA-energy is decreased by one energetic level after each mitotic division, then the ⬭ stage formation is associated with biological energy losses. Consequently, the two opposite processes proceeds on the energetic line of the reproductive cycle. The level of the biological energy rises at the activation stages, and it decreases at the stages of reproductive initial emergence.

The energetic line of the reproductive cycle of gymnosperms and angiosperms ends with the formation of the energetic program or DNA-energy of the future plant. During early embryogenesis, the energetic matrix on the base of an undifferentiated multicellular embryo is formed: Two meristematic zones (apical shoots meristem, apical root meristem), a zone of differentiation and a root cap. The corpuscular-wave process of the energetic matrix formation is described in detail in Part II, Chapter 3.

The plant world has undergone a long path of evolutionary development. Speciation of eukaryotic plants was carried by changing the energetic, corpuscular and informational components of the genome. Initially, the capacity of the energetic line has been rising due to an increase in the number of the reproductive stage of both types – ↾ and ⬭. This is clearly seen by the evolutionary changes of reproductive cycles in different systematic groups of lower and higher plants (Table 1, 2). Then, the optimization of an energetic line structure began. The number of activation stages did not change, but the ⬭ stages, at which energy losses occur, began to disappear. The number of ⬭ stages and a cell amount in them decreases during gymnosperm evolution. For example, a stage of free-nuclear proembryo in *Ginkgo biloba* falls out. Then, a stage of megagametophytogenesis transition from gymnosperms to angiosperms disappears.

Biological Protection System, Its Mechanisms and Evolution

Live matter soon after the origination of sexual reproduction began to develop by the principle of bipolarity. Plants do not only have different mechanisms and ways of DNA transition to higher energetic levels, but also various effective means for their protection from the high biological energy of generative cells. These two systems are fundamentally different. In the course of plant and animal evolution, the speciation proceeds in the direction of biological energy increase. Biological protection systems, which regulate the relationships between the vegetative and generative sphere, on the contrary, do not change until the old mechanisms of energy protection exhausts their resources. After the level of DNA-energy begins to exceed their threshold, the old protection systems are disintegrated, the process of speciation on its basis ceases, and a new type of reproductive cycle and a more progressive way of biological protection arises. The replacement of the protection system radically changes the matrix of the reproductive cycle, and, as rule, a life cycle of species (generation number, their ploidy).

The action of the mechanisms of energy protection is caused by the wave nature of the energetic component of the genome. As is known, first generative cells originate from the meristematic cells of the mother plant. After receiving a portion of biological energy, they become energetically strong cells, and take advantage over the somatic energetically weak cells and tissues. The nearby cells equally react to the appearance of a source of a powerful biofield. They turn to it, “forget” their own program of development, go into an increased metabolic activity regime and are transformed into secretory type cells. New synthesized biomolecules begin to move by the vector of a stronger biofield and are used for the nutrition of quickly growing and multiplying generative cells. Such an intense function regime leads them to death. Then the cells of the next layer enter the zone of the powerful biofield, and then the next cellular layers. This process can continue until the plant death. But this is not what has happened, because the multicellular plants have effective mechanisms for regulation of relations between generative and vegetative spheres.

There are two universal mechanisms (a removal of reproductive structures from a maternal plant and a transition of generative cells in a resting state) and 4 ways (haplont method, external fertilization, bimorth type, double protection) of biological protection (Figure 18).

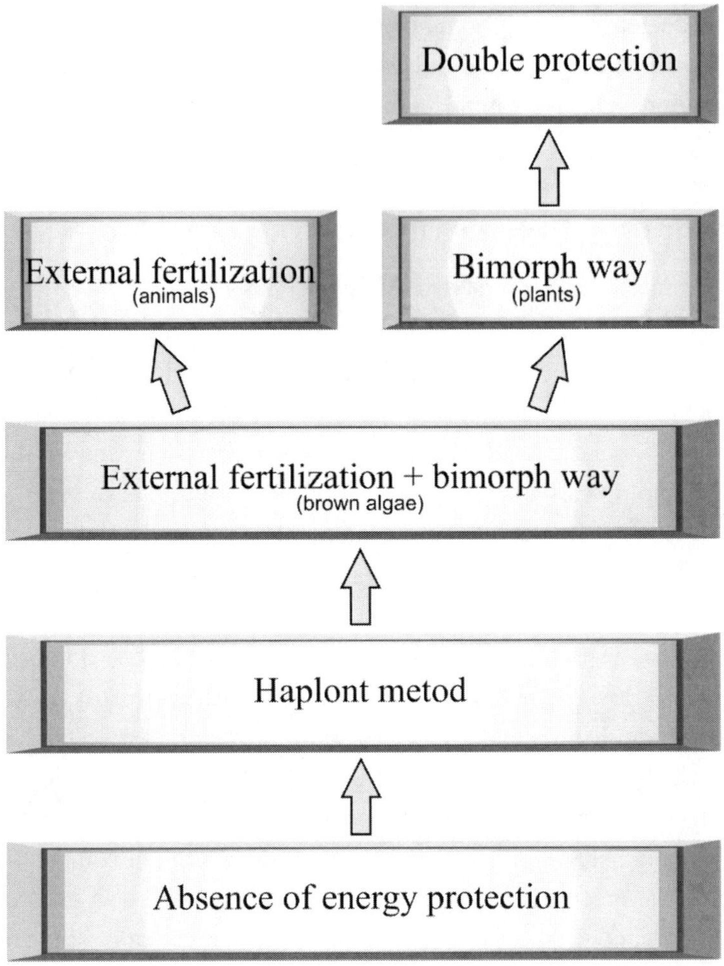

Figure 18. Evolution of biological protection system in the plants.

As was mentioned above, the biological protection system in such classes of green algae as a *Chlorococcales, Volvocacalles*are absent. Therefore, each sexual process ended by the appearance of a rejuvenated sexual progeny and the death of the initial plant of algae. For the description of the mechanisms and types of biological protection system, we will be using the factual material of embryology, systematic methods and evolution of plants (Life of plants, 1978; Singh, 1978; Smith, 1980; Kozubov, 1982; Embryology of angiosperm, 1984; Grant, 1984; Raven et al., 2005; Gilbert, 2010; Evert & Eichhorn, 2013; etc.).

Haplont Method

The haplont method was the first way of energetic protection, which ensured the simultaneous normal development and differentiation of the vegetative haploid body of the algae and process of their sexual reproduction. This method is inherent in many haploid species of lower plants: Most species of green algae (Conjugatophyceae, Ulotrichophyceae, Siphonophyceae, etc.), red algae (Rhodophyta), brown algae (Cutleriales) and mosses (Bryophyta). Their reproductive cycle consists of 1-3 activation stages. The vegetative body presents the haploid stage of development or a gametophyte. Only the zygote is a diploid cell, which is the organ of settlement. The zygote exits from a resting state, undergoes meiosis and develops a new haploid generation outside the mother plant.

The main mechanism, which regulates the relationship between the vegetative and generative sphere at the haplont method of the biological protection, is a durable outer wall of the zygote. Its extracellular polymeric substances are characterized by a high stability and insolubility in water. Most likely it is an insulator, which decreases a tension of electromagnetic field between energetically strong zygote and energetically weak cells of algae. Its thickness and chemical composition guarantee the energy protection and viability of vegetative and generative generations during their joint development. This durable wall passes the zygote into a resting state.

Signs of an apparent failure of the haplont method are observed in Rhodophyta. Its reproductive cycle contains 3 activation stages. The sexual process of red algae is the oogamy, which proceeds on the mother plant from initiation and up to carpospore maturation. The zygote is the last cell of the reproductive cycle, which germinates without a dormancy period. Probably, the thickness and hardness of its wall is not sufficient for the hold down development of a zygote by the frameworks of one cell. It undergoes meiosis and gives rise to an energetically strong haploid of undifferentiated generation, which parasitically develops on the algae body until the formation of carpospores. Then, they leave the carposporangium.

The greatest crisis of the haplont method is observed in Bryophyta. In mosses, alternation of generations is observed. Their vegetative generation is the haploid, and the reproductive cycle also has 3 activation stages. Archegonium appears in mosses. It is is a specialized multicellular organ, inside which ovicells are formed and conditions are created for DNA-wave transition to higher energetic levels. Male gametes are developed inside the antheridium. The sexual process is carried out on the haploid mother plant, which ends by the formation of an energetically strong zygote. The zygote passes to the development on the body of mosses without a resting period, and gives rise to an energetically strong diploid differentiated generation, or a sporogon. Its nutrition is provided by the mother plant. During their joint development, the incompatibility processes on the border of two tissues (haploid and diploid) take place. The forming abscission layer gradually separates the sporogon from the mother plant, and protects it from death. However, the sporophyte generation has enough time for the completion of the reproductive process on the gametophyte body: The creation of a capsule, origination and maturation of the meiospores.

The haplont method was replaced by the external fertilization and bimorph way of biological protection, which already existed in brown algae (Phaeophyta). Both these ways act with the alternation of haploid-diploid generations, when the sporophyte has a more potent development. The bimorph way has received further evolutionary development in plants, and an external fertilization in animals.

To summarize the above, it should be noted that red algae and mosses according to the ways of the biological protection system are included in the deadlock branch. However, by the reproductive cycle structure, they are a more highly organized systematic group than the brown algae. Their energetic line includes 3 activation stages, and archegonium (specialized female generative organ) appears.

Bimorth Way

Ferns (Polypodiophyta) are characterized by the alternation of generations with a domination of diploid sporophyte. The initiation and beginning of reproductive development is carried out on the sporophyte up to meiosis. The energetic line has 3-5 activation stages.

The bimorth way is the main manner of energy protection in Equisetophyta, Lycopodiophyta, and Polypodiophyta. In essence, it is a mechanical removal of energetically strong generative cells (spores) from a sporophyte body until the end of reproductive development. Characteristic traits of a bimorph way of biological protection are the break of the reproductive cycle, sporophyte dominance and alternation of generations, which lead to their independent existence.

Spores by their morphology may be the same in a form and size, or differentiated into micro- and megaspores. After maturation, they transition into a resting state and leave the mother plant. Spores after getting into favorable conditions germinate give rise to the energetically strong haploid gametophytes (prothallium). The prothallium exists or in the form of bisexual biomorth, or as the two (male and female) plants living independently in biomorth. The duration of their life cycle is short. Then the reproductive process on the prothallium body resumes: An antheridium, archegonium, spermatozoids and ovicells are formed, which participate in the fertilization. A zygote gives rise to a new fern plant, and the prothallium dies. The zygote development begins with the mitotic division, and not with meiosis, as in mosses. Therefore, the diploid plants in ontogenesis are formed.

Tables 1 and 2 show the reproductive cycles of mosses (Bryophyta) and isosporous ferns (*Filicales*).It is obvious that their matrices have the same construction by the number and sequence of stages. But reproductive processes on their basis proceed in opposite directions. Accordingly, the last generative cells of the moss reproductive cycle are a haploid meiospore and a diploid zygote in a *Filicales.* We conclude that the matrix of the Polypodiophyta reproductive cycle is a continuation of the evolutionary development of the Bryophyta matrix. Nevertheless, these systematic groups of plants differ radically by the direction of reproductive processes and energy protection.

Double Protection

The role of a biological protection system during plant evolution is constantly raised and becomes an increasingly more rigidly genetically regulated process. The mechanism of double protection acts by the principle of "nesting dolls." Its main advantage lies in the fact that between the energetically strong embryo and mother plant there is endosperm tissue (a haploid in gymnospermae and a triploid in angiospermae), whose energetic potential is higher than in the meristematic tissue of a sporophyte, but lower than in the embryo. The energetic gradient between them is regulated by means of an embryonic chamber.

The seed is the main organ of breeding for all seed plants. Its reproductive development is fully carried out on the body of the sporophyte. The alternation of generations is preserved, but the gametophyte is strongly reduced and has lost its independence. The energetic line consists of 6 activation stages. The top of the reproductive development is the diploid genome of a sporophyte initial.

The process of seed formation includes gametogenesis on the basis of the matrix of a species' reproductive cycle and embryogenesis. Mitosis, asymmetric mitosis, activating mitosis, meiosis, activating meiosis, endomitosis, cleavage and free-nuclear mitosis participate in the seed development. During gametogenesis, the energetic, corpuscular and informational programs of future plants are created. During embryogenesis, the three-plane plan of a spatial organization of the plant body is formed. Three matrices are produced and superimposed on each other by layers: A corpuscular – in the proembryogenesis (gymnosperms) and fertilization (angiosperms); an energetic –in the early embryogenesis; and an informational – in the late embryogenesis (Part II, Chapter 3). A part of the embryo cells during this time loses the property of totipotency.

The fixed energetic gradient between embryo and endosperm is the advantage of a double protection mechanism. This is due to the fact that it is formed on the basis of the energetic line of a reproductive cycle. DNA-energy of the haploid endosperm of gymnosperms is increased only in two activation stages (↾) by the female generative line: Megasporangiogenesis and female meiosis. In angiosperms, the initial cells of the endosperm and embryo arise simultaneously on the final stage of the reproductive cycle during double fertilization (after 6 activation stages): A diploid embryo as a result of the syngamy of more energetically strong sperm with an ovicell nucleus; the triploid endosperm – a second sperm with two polar nuclei of central cell in descending order of energetic gradient. For this reason, the energy gradient between the embryo and endosperm should be relatively stable.

Thus, in all species of gymnosperms and angiosperms between the energetic levels of the sporophyte, embryo and endosperm there is a genetically determined ratio, or the energetic parity. An endosperm contains the cells in such quantity, which can provide embryogenesis with nutrients, limit the embryo growth with a fixed volume and supply it with nutrients during embryogenesis and seed germination.

Conclusion

After the nucleus's appearance, the energetic system of a eukaryotic cell, plants and animals reaches a higher level of development. The cell becomes an open and a closed system. Its genetic apparatus is spatially separated from the cytoplasm. The 2nd source of internal energy (metaphase chromosomes) has appeared. The processes of replication and transcription in the cell cycle have become isolated. Specialized organelles (chloroplasts, mitochondria) participate in the work of a centrifugal energetic stream in the ATP-ADP system. Mitosis as the main type of cell division, and an asymmetric mitosis and activating mitosis, which lead to differentiation of daughter cells by the biological energy level, have arisen. A eukaryotic oscillator began to master higher energetic levels, and living matter comes out on a multicellular level of organization. The life of plants and animals is divided into two periods: Ontogenesis and sexual reproduction. The movement of biological energy goes by a descending vector in ontogenesis and by an ascending vector in sexual reproduction. DNA-wave transition to higher energy levels is a multi-stage strictly deterministic process. It occurs on an energetic line of the reproductive cycle in accordance with the law of conservation of energy.

We think that the speciation takes place in the direction of continuing biological energy growth. The number of activation stages is increased from 1 to 6. Living matter passes from horizontal evolution to progressive (vertical) evolution. Its division on the two branches – plants and animals – occurred in accordance with the law of linear momentum conservation.

The importance of a biological protection system in the course of plant evolutionary development has constantly increased and reached its maximum in the angiosperms. The biological protection system has undergone 4 stages of transformation: 1. Absence of energy protection, when the emergence of each daughter generation ends with the death of the mother plant; 2. the haplont method, when a hard envelope of diploid zygote is the mechanism for energy protection of the haploid mother plant; 3. the bimorph way, when two mechanisms in the processes of energy protection participate (a hard envelope of energetically strong spores and spore removal from a sporophyte body). The alternation of two generations (equal or different biomorphs) takes place, which carry out an independent existence; and 4. double protection, when the formation of the energetic potential of an embryo and endosperm occurs on the basis of the reproductive cycle and there is a certain energetic gradient between these two tissues. In gymnosperms, the DNA-energy of endosperm increases on the 2 activation stages (part of the female generative line); in angiosperms – on the 6 activation stages (all reproductive cycle).

CORPUSCULAR COMPONENT OF GENOME (DNA-PARTICLE)

The field of action of the corpuscular system of eukaryotic organisms is limited by the frames of a single cell, and the frames of a unicellular and multicellular organism like plants and animals. In the first case, the species-specific DNA-particle, with the participation of three matrix processes (replication, transcription and translation), creates a corpuscular intracellular field during the cell cycle, and transmits this corpuscular information to the next cellular generations. In multicellular organisms, the corpuscular component of a genome is formed during sexual reproduction on the basis of a corpuscular line of the reproductive cycle, and preserves this information during ontogenesis.

The scheme of the corpuscular system organization of a eukaryotic cell is shown in Figure 19.

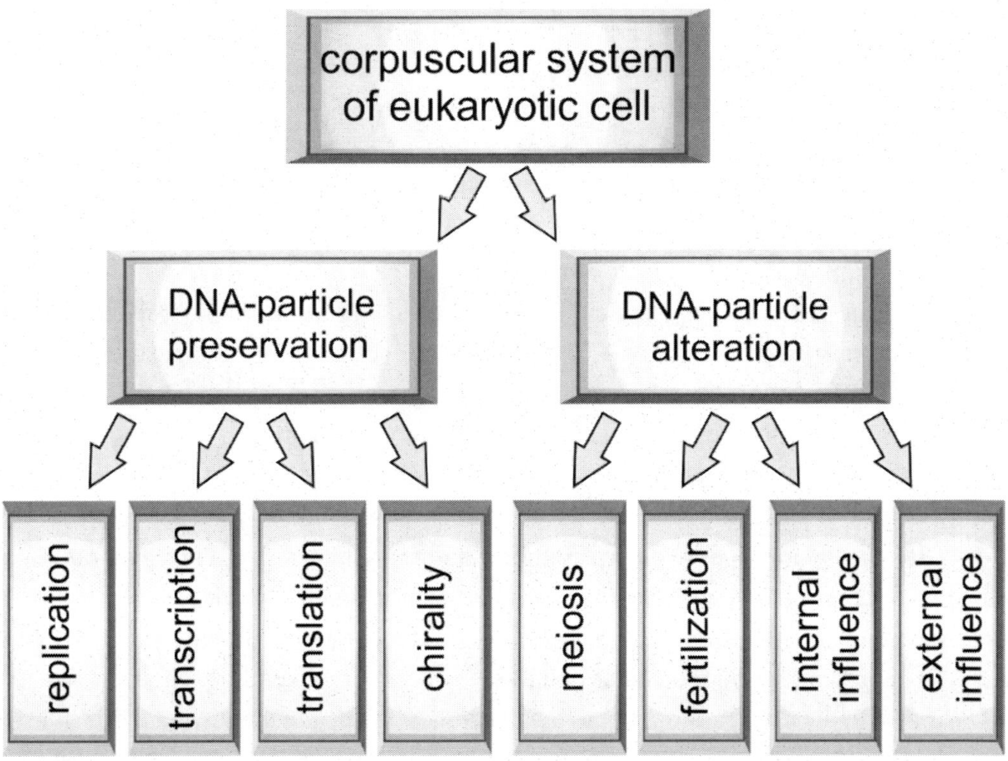

Figure 19. Corpuscular system of eukaryotic cell.

The corpuscular system of eukaryotes, like in the prokaryotes, consists of two blocks of opposing mechanisms: some of them are connected with the preservation of a DNA-particle, while the others lead to its alteration. The key mechanisms, which take part in the preservation of a DNA-particle, are: Replication, transcription, translation and chirality. Taking into account that the genetic apparatus of a eukaryotic cell has moved into the nucleus, the internal structure of a DNA-particle has become more complicated, the number of chromosomes increased, and these mechanisms have also undergone significant changes with the exception of chirality. The preservation of the homochirality of biological molecules is an inherent property of all prokaryotic and eukaryotic organisms at the molecular level of cell organization. A system for the DNA-particle alteration was more radically transformed. After sexual reproduction appearance in plants, meiosis, fertilization and cross-pollination were added to the two mechanisms of DNA-particle alteration of prokaryotes. Now, we will move into the analysis of the mechanisms by DNA-particle conservation.

Mechanisms of DNA-Particle Preservation

Chirality

Optical activity is inherent for many organic molecules: Proteins, amino acids, nucleic acids, and carbohydrates. This phenomenon does not occur in an organic compound artificially synthesized from an inorganic substance. The presence of optically active carbon compounds in ancient sedimentary rocks is proof of the existence of life at this time. The mechanism of occurrence of the chiral neatness of biological molecules is still a subject of constant debate among scientists (Rutten, 1971; Volkenshtein, 1983; Bada, 1995; Nikitin, 2016; etc.). It is believed that only a light wave with a circular polarization can cause optical activity. In the previous work, we hypothesized that the plasma membrane can give the effect of optical activity in biological molecules (Kuznetsova, 2013).

All cellular membranes are built by a common principle. These are the thin lipoprotein films, which consist of two layers of lipid molecules with the inclusions of proteins. Lipids of plasmatic membranes are mainly represented by phospholipids. These are the elongated molecules with hydrophobic tails directed inside the membrane and located parallel to each other. Therefore, with respect to light, the bilayer of the plasma membrane can act as two polaroids arranged one after the other (Figure 8). Its external lipid layer represents a thin molecular polarizing film or a typical optical linear polarizer. Un-polarized light after passing through it becomes linear-polarized light. The linear-polarized light gets on the spiral molecules of the second lipid layer of plasmatic membrane. The internal monolayer is rotated around its own axis and the circular-polarized light enters into the cell. A circular-polarized light is absorbed by certain stereo-isomers of enzymes, proteins, carbohydrates, etc. They are included into intracellular metabolism on the basis of one of its forms (L or D). As the plasma membrane (two polaroids) is an obligatory structural element of any prokaryotic and eukaryotic cells, the optical activity phenomenon can be attributed to the immanent properties of living matter.

Replication

Replication is the first step via the formation of the species-specific molecular composition of a cell and plant. DNA biosynthesis is an endergonic reaction. This is the most energetically dependent process of the cell cycle (Webster & Vant'Hof, 1969). The appearance of a eukaryotic oscillator and metaphase chromosomes significantly changed the basic intracellular processes, including replication. The processes of replication and transcription have become spatially separated. DNA biosynthesis proceeds in the S-period, and this reaction lasts less than half of a cell cycle. The DNA amount inside a nucleus is thousands of times greater than in the single ring chromosome of a prokaryotic cell. A DNA-particle has divided into a certain (a species-specific) number of chromosome-particles. All chromosome-particles occupy a certain territory inside a nucleus, and connect with its envelope at three points: A centromeric, and two telomere regions. A chromosome-particle is a complex nucleoprotein complex. Its basis is a continuous DNA molecule, which is in association with chromatin, consisting of specialized proteins of two classes – histone and non-histone proteins. Each chromosome represents a polyreplicon system.

DNA-particle doubling is a complex corpuscular-wave process. The physicochemical principle of its action in a eukaryotic and prokaryotic cell is similar. A biological harmonic oscillator initiates this reaction in the moment of transition to the lower energetic level. DNA-wave radiation occurs at the end of the G_1 period. Free energy enters into the nucleus by two streams. Two groups of DNA-polymerase molecules pass into the active state, and a negative charge is received. Within the group, a multitude of negatively charged enzyme molecules immediately are repelled from each other and distributed along their own chromosomes. As a consequence, the chromosomes are divided into a certain number of replication units – the replicons. S period begins with a given moment.

DNA is a neutral molecule, but its nucleotides have a dipolar structure (Volkenstain, 1983). The positive charges are concentrated in the region of bases, and the negative charges are in the region of phosphate groups. Two groups of the negative-charged active molecules of DNA-polymerase are attracted to the chromosomes-particles in the area of the bases and begin to move with their chromosome-waves in the opposite directions

(Figure 21). A negatively charged enzyme breaks the electrical balance in the place of enzyme-macromolecule association. As the association is a negative charge, the negatively charged phosphate groups of two DNA threads are repelled from it and form the loop, or the DNA-replication fork.

DNA biosynthesis begins in the regions of replicative forks and is performed asynchronously within one chromosome and the chromosome set. The molecules of an active DNA-polymerase, which are moved together with a DNA-wave, automatically untwine the DNA threads in the direction of their movement. As a chromosome is the standing wave, the nucleotides of the counter DNA-wave at this time pass the enzyme in the opposite direction (Figure 9). The reading of corpuscular information occurs with the counter wave according to the complementarity principle and in a 5' → 3' direction. The replication velocity is about 100 nucleotides per second. After replication, the inactivated molecules of an enzyme pass into the nucleoplasm. Electrons remain on the chromosomes and participate in the mechanisms of mitosis.

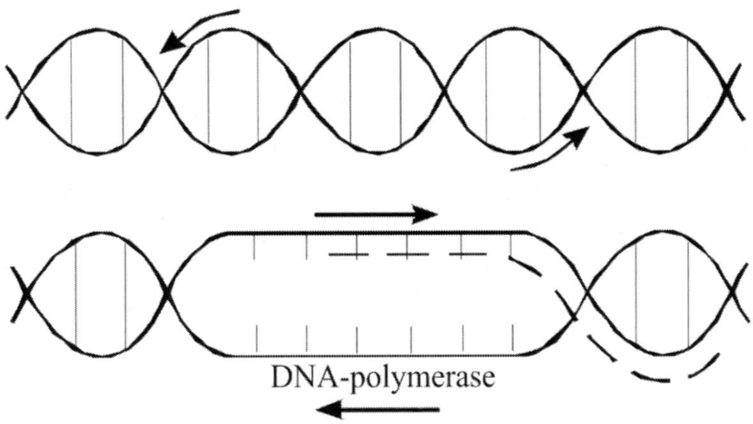

Figure 20. The scheme of the corpuscular-wave mechanism of DNA replication.

Thus, the replication is a matrix biochemical reaction and complex biophysical process, the implementation of which many molecular and structural components of the nucleus are involved. The biological harmonic oscillator is its main inductor and energy source. Its standing waves present two DNA-waves, which are moved along chromosomes in different directions with the same speed. Chromosome-waves bear the molecules of active DNA-polymerase and ensure the accuracy of reading corpuscular information from the matrix of a DNA-particle by the complementarity principle. From the enzyme position, the catalyzed reaction is a repeatedly cyclic process of joining to each nucleotide of its complementary pair. The molecules of active DNA-polymerase conduct the DNA doubling reaction within its own replicon, wastes all its energy and returns to an initial inactivated state.

Transcription

The transcription (or the matrix reaction for the transfer of corpuscular information from a DNA-particle to an RNA-particle) is the second stage in the implementation of the genetic program, laid out in the DNA nucleotide sequence into the chemical structure of proteins. Its final product is a great number of RNA molecules of three types – mRNA, rRNA and tRNA. RNA biosynthesis is an endergonic reaction, for which energy is required. An energy source is an elastic energy of strongly compressed metaphase chromosomes. The same amount of free energy enters inside a telophase cell during chromosome despiralization. A telophase cell is a closed system. In accordance with the 1st law of thermodynamics, the total energy in a closed system is preserved, and it can be converted into other forms of energy: A chemical, a kinetic, etc.

In telophase, a number of important physical events occur. Nucleotides of metaphase chromosomes during decondensation come into motion and their initial dipolar structure is restored. The motion of charges generates around two sets of untwisting chromosomes in the electromagnetic fields, which decreases as $1/r^2$. The maternal cytoplasm is attracted to the sources of electromagnetic fields, and its contents are located around them in the form of two concentric layers. At the same time, its chemical composition is sorted according to the size and electrochemical design of molecules and ions. RNA molecules of an old generation appeared in the region of the future nucleus, forming the prenucleolar bodies. Enzymes of DNA-polymerase, RNA-polymerase, nuclear proteins, actin of monomeric and polymeric forms appear in the zone of the future nucleus (Strochkova & Kazan'ev, 1975;Kazmin, 1984; Jockusch et al., 2006; Zharskaya & Zatsepina, 2007; Benken, Sabaneeva, 2011; etc.). Molecules of phospholipids move along the vector of the electromagnetic field and restore the envelope of the nucleus. Since the uncoiling chromosomes retain the V-shape, they are hit by a centromeric and two telomere sites about the envelope. A standing wave appears on the DNA, and the macromolecule will begin to exist in the form of a particle and a wave simultaneously.

The second important event of telophase is the restoration of the energetic status of the dividing cell. Free energy is released during the untwisting of the metaphase chromosomes and enters into cell, which is a closed system. In the closed system, it is preserved and passes to other types of energy through several channels: RNA-polymerase activation, ATP from ADP restitution, kinetic energy increase, biomolecule excitation, etc. (Figure 15). RNA-polymerase is located in the nuclear zone near uncoiling chromosomes and passes to the active state, and DNA-polymerase is not. Apparently, the energy of metaphase chromosomes is not enough for its activation. As is known, there is no replication in the telophase.

RNA biosynthesis begins after the nucleus envelopes the formation and DNA-wave function restoration. Transcription occurs with the participation of special enzymes of the RNA-polymerase and carried out only in certain parts of a DNA-particle, where the heterochromatin zone passes into the euchromatin ones. A DNA-particle serves as a matrix for biosynthesis of three RNA types: mRNA, rRNA and tRNA. Clusters of tandem-repeated ribosomal genes of a DNA-particle (rDNA) occupy certain regions within the chromosomes. These are nucleolar organizing regions of chromosomes (NOR).

Activated RNA-polymerase are attracted one by one to a single strand of DNA (particle, wave), forming a strong bond by the principle of induction conformity. Then they begin to passively move with the DNA-wave along the rDNA copy zone. The electrical field of RNA-polymerase is weaker than that of a DNA-polymerase. It is created by the charges' asymmetry. Therefore, an RNA-polymerase can untwist and transcribe only decondensed polyisogenic parts of a DNA-wave. The "enzyme-macromolecule" associations arise before a euchromatin segment of a DNA-wave. The transcription process begins with the certain nucleotide sequences of a promoter.

Corpuscular-wave mechanisms of RNA biosynthesis in prokaryotes and eukaryotes are rather similar. Corpuscular information is read from the counter DNA-wave. The enzyme, which is moved together with a DNA-wave in the 3' $\rightarrow$ 5' direction, automatically untwists the double thread of DNA, catalyzes the biosynthesis of new RNA molecules on a complementary basis, wastes its energy and inactivates. Hundreds of RNA molecules can synthesize simultaneously on the matrix of one transcription unit. The growing transcripts begin to bind to proteins, forming the ribonucleoproteins (RNP) with a sedimentation coefficient of 80S. When the RNA-polymerase reaches the end of the copy zone, it collides with the second heterochromatin block and the inactivated enzyme passes into the nucleus. The end result is the formation of primary transcription products: Large molecules of pre-mRNA, pre-rRNA, and pre-tRNA.

The dominant role in the creation of the intracellular corpuscular field belongs to the messenger RNA. Its portion in the total RNA pool is only 10%. Biosynthesis of mRNA occurs with the participation of enzyme RNA-polymerase II. Their molecules rapidly degrade, are very variable by the nucleotide sequence, size and average life expectancy. Then, a stream of corpuscular information created on the basis of a large number of mRNAs is carried from the nucleus to the cytoplasm.

rRNA and tRNA are synthesized on the matrix of a DNA-particle too. They are used to make proteins. The main purpose of rRNA is to ensure protein biosynthesis via the energy. Therefore, its portion in the cellular RNA is so great (50-70%). Synthesis of 70-85% of rRNA is carried out in the nucleolar organizing regions (Swanson & Webster, 1977; Lehninger, 1985; Chentsov, 2004; Alberts et al., 2002; etc.). Newly formed rRNAs are large molecules of 47-45S pre-rRNA. They initially accumulate in the nucleolus and participate in the building of ribosomes of the next generation. The ribosome formation occurs in 2 stages: An origination of ribosomal particles during the processing of pre-rRNA; and their carrying from nucleoli into the nucleus and cytoplasm. The amount of tRNA in the total mass of cellular RNA is 25%. Its molecule has a complex three-dimensional shape in the form of a "clover leaf" and a length of ≈ 30 nm (Chentsov, 2004). Its main function is the transportation of amino acids to the site of protein biosynthesis.

Therefore, the elastic energy of metaphase chromosomes is an energy source for the activation of RNA-polymerase and the biosynthesis of mRNA, rRNA, and tRNA. The DNA-wave bears the activated complex of different RNA-polymerase molecules and ensures the high accuracy of the corpuscular information reading from one thread of a DNA-particle according to the complementary principle.

Translation

Translation is the last step in the implementation of the corpuscular program, which is laid in the DNA chemical formula as a species-specific nucleotide code into the amino acid sequence of proteins. A cell possesses the unique organic composition, which has no counterparts in inanimate matter. It is represented by four types of compounds: Carbohydrates, proteins, lipids and nucleic acids. The leading role in the organization of a corpuscular field belongs to the proteins. They make up about 2/3 of the dry mass of a eukaryotic cell, and are its functional basis. All reactions of primary and secondary metabolism are catalyzed by protein enzymatic systems. Molecules of proteins after synthesis remain inside the cell, because they cannot pass through a semi permeable plasma membrane. Therefore, the intracellular corpuscular field is limited by the framework of a cell.

The biosynthesis of proteins is a complex matrix process that occurs in the specialized cellular organelles (ribosomes) during interphase (Lehninger, 1985; Anderson et al., 2002; Chentsov, 2004). Functional mRNAs give rise from large heterogeneous molecules of pre-mRNA, which are synthesized in large numbers during transcription. These initial transcripts are found only in the nucleus and are very unstable. In the nucleus, they undergo post-transcriptional processing, or the exponential disintegration of pre-mRNA, which proceeds with energy release. During processing, a large part of the initial molecule is lost. The exponential disintegration occurs in determined segments of a primary molecule: The uninformative sections (introns) of mRNA are removed and then further break up. RNA segments, which have coding information, are attached to each other and give rise to typical mRNA. mRNA have other physical and chemical properties, in 2.2-10 times shorter than the primary transcript (Chentsov, 2004). Under action of the biofield forces, the smaller molecules of mRNA come into motion and begin to go out from the nucleus through the nuclear pores into the cytoplasm.

The pre-tRNA have a length of ≈ 39 nm. They consist of two exterior functional zones and a central non-functional area. During processing, all pre-tRNAs undergo the exponential disintegration and removal of a nonfunctional zone.

In telophase, the delivery location of primary rRNA transcripts is the prenucleolar bodies, consisting of PNA molecules of an older generation. During transcription, they are reconstructed in a nucleolus. Within them, the following are accumulated: 1) Transcripts of pre-rRNA during their growth; 2) 80S RNP with 45S RNA; 3) 60S RNP – a precursors of large ribosomal subunit; and 4) 40S RNP – a precursors of small ribosomal subunit. The primary transcript of rRNA is a more stable molecule than mRNA. Nevertheless, their maturation process is also associated with the separation of a large macromolecule into smaller fragments. The amount of heavy pre-rRNA is gradually decreased, and different fragments of rRNA appear. This indicates that the exponential disintegration of RNA is a very orderly processed. In eukaryotes, there are 3 types of typical rRNA: 5.8S, 18S and 28S. Then, disintegration processes undergo both primary transcripts and mature rRNA, which take place in the nucleus and cytoplasm. 50% of typical rRNA (28S and 18S) is deconstructed without leaving a nucleus (Georgyev, 1968). The energy, which is liberated during processing, represents a centrifugal energetic stream. During interphase, it is used mainly for the energy supply of protein biosynthesis.

As was mentioned above, the translation is the third and most complex biochemical matrix reaction. It takes place in the cytoplasm and includes the same stages as in the prokaryotes. In essence, the corpuscular information, recorded as a sequence of nucleotides of mRNA is transferred from the nucleotide code to the amino acid sequence of proteins. The main organizing center for a translation is the ribosome. 20 amino acids, various enzymes and all types of RNA are involved in this process, providing the informational, transport and energetic sides of translation, in addition to the factors regulating the initiation, elongation and termination mechanisms of the protein biosynthesis.

Protein biosynthesis occurs in the eukaryotic ribosomes. Ribosomes have a spherical or ellipsoidal shape and size of ≈ 25-30 nanometers. Their large and small subunits are equal to 60S and 40S (in prokaryotes – 50S and 30S, respectively). 1500-3000 ribosomes per minute are synthesized in the intensively functioning nuclei. The energetic source is the energy of exponential disintegration of an RNA of the previous and new generations. A eukaryotic ribosome is a nucleoprotein with a ratio of protein and RNAof1:1. Ribosomal RNA is its structural carcass. It contains 4 molecules of different lengths: 5S, 5.8S, 18S and 28S. The smaller subunit is composed of one RNA molecule, which passes from the nucleus to the cytoplasm after processing. In the nucleolus, 28S, 5.8S fragments of rRNA and 5S, which was synthesized independently of them, are united into a large 60 ribosomal subunit. These subunits also go out into the cytoplasm through nuclear pores along the biofield gradient. The 40S subunit initially links to the mRNA, and then to the large 60S subunit and, as a result, a functional 80S ribosome is formed. Ribosomes in the cytoplasm can be free in the form of polyribosomes or attached to the membranes of endoplasmic reticulum.

The translation process proceeds in the ribosomes and represents the following sequence of events. Molecules of mature mRNA are attached to one or more ribosomes. The gradual exponential disintegration of rRNA molecules provides a protein biosynthesis by energy. Functional tRNAs have the configuration of a "clover leaf" with the anticodon in the center. Each type of tRNA can be attached to only a certain kind of amino acid. tRNAs deliver the amino acids attached to their CCA tails to the ribosomes. This sequence is necessary for the recognition of tRNA by enzymes. The corpuscular information from mRNA is read by the corresponding enzymes of tRNAs, and the amino acids in a certain sequence are connected to each other, forming a polypeptide chain of future protein molecule.

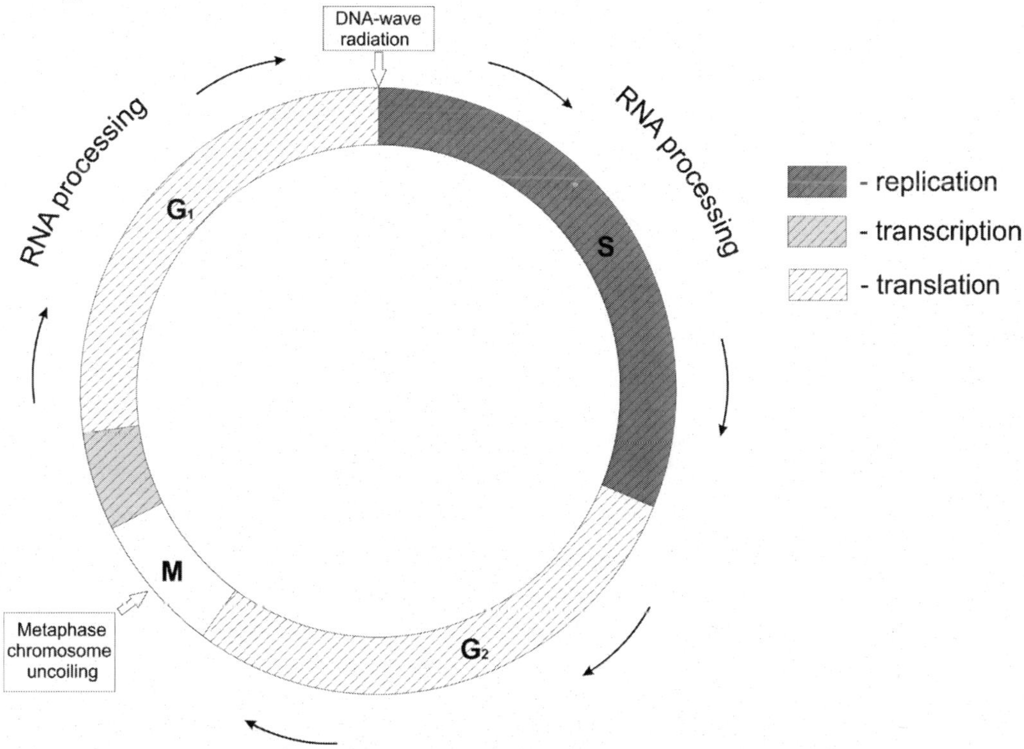

Figure 21. Algorithm division of internal energy during creation of the corpuscular system of eukaryotic cell.

We have noted that the development of a corpuscular system has a cyclic character and is fully coordinated with the movement of energy from internal sources. Figure 21 shows an algorithm of the processes of replication, transcription and translation, and the periods of energy receipts by the phases of a cell cycle.

The biological harmonic oscillator is a physical and biological body simultaneously. Its energy changes the physical situation inside the interphase cell (G_2 period end) and initiates the processes of DNA-particle doubling and 2nd internal energy source formation. Metaphase chromosomes have a constant species-specific shape and size, i.e., they carry the same amount of elastic energy. During uncoiling, the equal amount of free energy enters inside the telophase cell. This energy initiates an RNA biosynthesis and centrifugal energetic stream. During transcription, the same pool of primary pre-mRNA, pre-rRNA, pre-tRNA transcripts are formed, which undergoes a processing with the releases of free energy in the nucleus and cytoplasm. mRNA are the bearers of corpuscular information for the translation reactions. rRNA molecules are the part of the large and small subunits of ribosomes, and a main energy source for protein biosynthesis. tRNAs are a transport system and reading system for this process. The spontaneous disintegration of rRNA, mRNA and rRNA occurs by the exponential law and presents a centrifugal energetic stream, which ensures a protein biosynthesis by energy in the interphase.

Thus, during each cell cycle the protein number is doubled and has created a species-specific corpuscular composition of a cell, which depends on a degree of cell differentiation and tissue type. The energy source for the translation reaction is the exponential disintegration of different RNA molecules (mainly rRNA). Bio-building of proteins is carried out into specialized organoids – the ribosomes, where all the components necessary for this reaction (mRNA, rRNA, tRNA, amino acids, enzymes) are concentrated.

Matrix nature and energetic stability underlie the high reliability in the operation of the corpuscular system of a cell. Its preservation is ensured by the succession of three matrix processes –replication, transcription and translation. The energetics of these reactions is provided by three sources, which release the same amount of energy, respectively, during DNA-wave radiation, metaphase chromosome uncoiling and RNA molecules processing.

Mechanisms of DNA-Particle Changing

It is known that the biological systems can exist only in certain conditions (light, temperature, pressure, etc.). Independence of ontogenesis from the environment increased after the DNA-particle included a nucleus. Unicellular eukaryotic organisms have a large surface area, which is in direct contact with the environment. For survival in constantly fluctuating conditions, they use the various regulatory mechanisms and the proper responses, which ensure the constancy of the internal environment when the external changes occur. In accordance with fluctuations of external characteristics, the intracellular metabolism is able to vary within a wide range of normative reactions and adapt to them. During the evolution of multicellular organisms, the sizes of plants and animals are increased; their inner organization and specialization are complicated. With respect to the external environment, their development is carried out in two directions:1) A creation of an internal controlled environment; and 2) an emergence and improvement of functional systems, which increase the effectiveness of direct contact with them and the joint action of channels on energy utilization within the organism.

Variability, as the immanent property of living matter, is inherent in all biological systems. Quantitative signs vary in very wide limits. Owing to the variability, a structural organization of a gene pool is created, a genetic material mixing within a range is maintained and species survival in a constantly fluctuating environment is provided. It can be divided into a hereditary (genotypic, indefinite) variability and a non-hereditary (modification, definite) variability.

Modification variability reflects the changes of phenotype under the influence of different internal and external factors. This type of variability occurs within the species normative reactions, does not affect the genotype and does not require special intracellular systems for the phenotypic realization. Many modifications arise under the influence of environmental factors, and, as a rule, have the adaptive character. Probably, their occurrence is caused by the different changes of the informational component of a genome, which during the sexual process is corrected and therefore not inherited (see below).

Genotypic variability can also occur under influence of internal and environmental factors. The gene, chromosomal and genomics mutations and their combinations lie at the base. We think that this type of variability changes both a DNA-particle and DNA-information. Since mutations are associated with the corpuscular component of the genome, these changes are inherited and can lead to the origination of a new genotype. Somatic and generative mutations happen. A level of somatic variability has significantly decreased after the appearance of multicellular plants. For example, the spontaneous mutation level of the Scots pine in the center of the species range is within only 5% (Butorina et al., 2001). Combinative variability and special mechanisms, which ensure the restoration opportunities of genetic diversity and its maintenance, arise together with the sexual process. These are such fundamental processes as crossing over, cross-pollination and fertilization.

Crossing Over

Crossing over is a very complex process of exchanging segments of homologous chromosomes from the standpoint of biology and physics. This is an internal mechanism for DNA-particle changing, which occurs during meiosis in the beginning of a plant's individual life. According to our notions, this phenomenon has a quantum-mechanical nature, and occurs in certain physical conditions during meiosis. Here, we only succinctly describe its physical model.

It is shown that the chromosomes (particles, waves) retain an anaphase orientation in the interphase (Chentsov, 2004). Their telomeric ends are fixed at one end of the nuclear envelope, and the centromeres are located at the opposite end (Figure 22a). Homologous chromosomes are territorially close each other (Monachova, 1990). Each shoulder of a chromosome occupies a certain zone into the nucleus, which does not enter the zone of neighboring chromosomes. A line drawn through the telomere ends makes a system of triangles that uniformly cover the surface of the cone from the outside. In this case, the set of chromosome-particles resembles a truncated cone, which transversely lies inside the nucleus. A set of chromosome-waves represents a few harmonic oscillators, creating a biofield with a maximum intensity at the nucleus center. During the G_2 period, two positively charged centrioles make a circular drift in this zone, and carry out a conformation change of macromolecules by means of internal coiling (Kuznetsova, 2013).

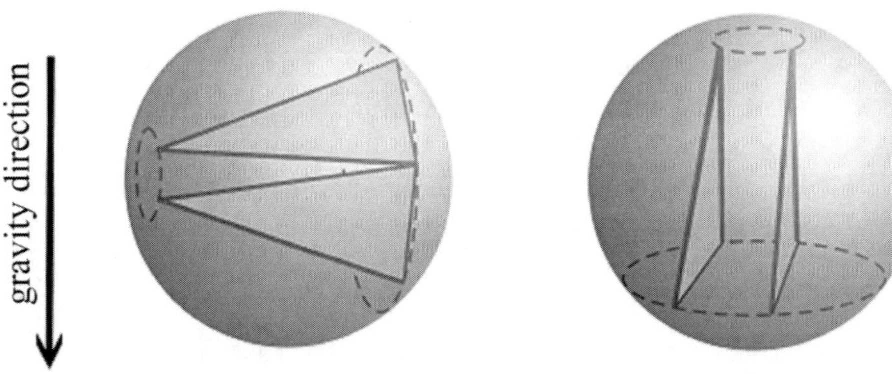

Figure 22.The spatial disposition of chromosomes-particles inside the nucleus during a beginning (a) and end (b) of meiotic prophase.

Transition from mitosis to meiosis is due to a change of physical conditions inside a megaspore mother cell. Its growth is greatly restrained by the nucellus tissue, the development of which is also limited to the ovule volume. According to the laws of physics, a division and growth of cells in a fixed volume should lead to an increase of intracellular pressure and viscosity inside the ovule. Therefore, the circular drift of centrioles occurs in a medium whose viscosity gradually increases. This makes a number of alterations in the standard scheme of mitosis. At some moment, the chromosomes begin to rotate together with the centrioles.

In a proleptotena, each chromosome lies across the gravity forces, turns around itself, and remains bound with the nucleus envelope in the regions of the centromere and two telomere. It is shown that the unusually twisted chromosomes –prochromosomes – appear in the proleptotena (Raven, 1961; Cytology and …, 1975). Chromosomes are twisted until the returning force reaches its maximum. Then, they quickly unwind and lose their contact with the nucleus envelope. The torque created by the gravity forces causes a rotation of chromosomes by 90° inside the nucleus. The change in their spatial orientation is confirmed by the appearance in the leptotene and zygotene of a figure of the so-called "bouquet," when the chromosomes are located along in the direction of gravity (Raven, 1961).

Chromosomes begin to unwind and are lengthened immediately after separation from a nucleus envelope. Their standing waves are destructed. At the same time, the torsion waves will go along the chromosome shoulders (from centromere to telomeres). They set in motion their shoulders and unwrap each chromosome inside the nucleus in such a way that the homologous chromosomes turn up opposite each other. If their set is imagined as a system of triangles, the shoulders of homologous chromosomes will be parallel (Figure 22b). After reaching the nucleus envelope, the wave function on the DNA is restored and the process of chromosome coiling is resumed.

Homologous chromosomes are the similar particles and waves by the frequency and amplitude of oscillations, shoulder length, angle between them, number of chromomeres and chromomeric pattern. Therefore, they can be considered almost identical particles, which, after turning, are in one place. One of the important consequences of quantum mechanics is that the waves of identical particles, which find themselves in the same place, must interfere with each other. Waves of homologous chromosomes begin to interfere (i.e., conjugate) with each other, creating in the zygotene a haploid set of bivalents or a connected together homologous chromosome-waves. In the pachytene, the chromosomes in bivalents become indistinguishable and behave like the separate micro-objects.

A high probability of mutual exchange between the internal parts of closely related complex micro-objects is an important consequence of wave interference. This explains the physical nature of crossing over. Really in fact, the homologous chromosomes into the bivalents are the closely related complex micro-objects, whose compound particles are nucleotides. In the zygotene, the nucleotides periodically change places in DNA segments where the resulting wave is formed. On the electronograms in this time are observed the synaptonemal complexes, which confirm the presence of nucleotide exchange. The gene recombinations arising from such exchanges give rise to new DNA-particles, which are not identical to the old ones.

Differentiation of Sexual Gametes by the Spin Direction

In the previous work, we assumed that the stability of the diploid state of the genome is ensured by the spin-spin coupling between the homologous chromosomes (Kuznetsova, 2013). It is shown that these ties can arise between nucleotides of homologous chromosomes in certain physical conditions inside the ovicell during fertilization.

Suppose it were true that nucleotides as the elementary structural units of chromosomes have their own angular momentum vector and are Fermi particles with the

½ spin. A number of consequences and phenomena follow from the fact that the nucleotides are the Fermi particles. In this case, they can enter the chromosomes in two different basic states: 50% nucleotides with the direction of the spin "up" and 50% with the spin "down." This means, during the meiosis, not only the reduction in the chromosome number can occur, but also nucleotide differentiation depending on the direction of their spin. Accordingly, the formation of haploid cells of two types is possible: 50% with spin "up" and 50% with spin "down." We think that a differentiation of gametes by the spin direction may go together with the meiotic division. Now we will consider this process in more detail in the period from leptotene to diakinesis.

As was mentioned above, the passage of mega- or microsporocytes from mitosis to meiosis occurs in leptotene after chromosomal set rotation on the 90° inside the nucleus. At this moment, the orientation of chromosomes changes according to the gravitational and magnetic field of Earth. Then their doubled DNA-particle continues to coil, shorten, and its elastic energy gradually increases until the diakinesis ends. Two positively charged centrioles, which participate in this process, are mobile structures. Therefore the electromagnetic field is generated during their circulation. However, after chromosome rotation on the 90°, the direction of the magnetic component of intracellular electromagnetic field ceases to coincide with the direction of Earth's magnetic field lines. This internal electromagnetic field interacts with the external stationary magnetic field of Earth. As a result, the intensity of total field inside the meiotic cell is constantly changing.

In the leptotene-pachytene a doubled DNA-particle undergoes the conformational changes and is shortened. At the same time, the two types of nucleotides as the Fermi particles with the "up" or "down" direction of the spin move through a non-uniform magnetic field, which is produced as the result of overlapping between the external and internal magnetic fields of the Earth and centrioles. This magnetic field interacts with each nucleotide and they are separated into two beams which depend on the spin direction. Therefore, every chromatid is formed on the basis of filterable nucleotides. As a result of nucleotide filtration and two meiotic divisions, a sexless diploid maternal cell gives rise to two types of sexual gametes with a haploid set of chromosomes: Half of them with a spin "up", the other half with a spin "down", or 50% male and 50% female gametes.

Origination of sexual chromosomes has changed the genotype and phenotype of the male and female gametes, male and female plants within the same species. This has become the basis for the further development of sexual and sex-linked signs separately in the male and female individuals of plants. The differentiation of gametes by the morphology and mobility takes place. The development of forms of the sexual reproduction was carried out from isogamy to heterogamy and then from heterogamy to oogamy.

Genotype-Environmental Interaction during Meiosis

There is an alternation of generations in the development of higher plants reproducing sexually. Meiosis provides a transition from a diploid to a haploid phase of life cycle and fertilization from a haploid generation to a diploid one. Meiosis is the unique stage of the reproductive cycle. Its corpuscular-wave nature and role in a sexual process is not fully known. A decrease in the chromosome number, gene recombination during crossing over, gamete differentiation by sex, and rejuvenation of megaspores (megasporogenesis) are associated with it. It should be noted that meiosis is the most ecologically dependent division and at the same time a very vulnerable stage of the reproductive cycle. Now, the variability of its cytogenetic characteristics (frequency and spectrum of anomalies) under the influence of unfavorable environmental conditions experimentally has been proven. Nevertheless, the generative cells during microsporogenesis and megasporogenesis must be in maximal contact with the external environment. Moreover, they must have a certain orientation with respect to the gravity direction. This is largely owing to the fact that the meiotic cell has to be capable of receiving a new cosmic information signal, the process of which will be described below.

It becomes obvious that meiosis belongs to the number of functional systems, which maintain direct contact with the environment and react sensitively to its condition. Therefore, generative shoots are located or go outside the plant crown together with reproductive structures (sporogen, microstrobiles, megastrobils, stamens, pistils, etc.) for maximal connection with the factors of the external environment. They directly affect the genetic material of the forming male and female gametes by the principle of genotype-environment interaction. Environment and crossing over essentially extend the genotypic variability within the norm reaction of species, and the genetic diversity of population gene pool is restored at the initial stage of its formation. This ensures the preservation of its norm-forming part, the survival of most stable genotypes, helping to the adaptation to the specifics of regional climate.

Meiosis is most balanced in the center of range in the conditions owed by the requirements of species biology. The stressful impact, as a rule, negatively affects on the number, structure and quality of male and female gametes, and, accordingly, the state of seed progeny. For the most part, energetically weak, unadapted genotypes are removed during natural selection of gametes, embryos and plants.

Fertilization

In lower plants, the fertilization occurs by the syngamy of two equal forms or morphologically different gametes: A male sexual cell (microgamete, spermatozoon, sperm) with a female one (macrogametes, ovicell), which lead to the formation of a diploid zygote. In higher plants, the fertilization presents two complexes of complicated corpuscular-wave processes: 1) A progamous period and 2) syngamy sexual gametes. The first period connects with the delivery of the male gametes to the female ones: From pollen grains germination in a nucellus (gymnosperm) or stigmas (angiosperms) up until the pollen tube enters into a female gametophyte and an embryonic sac, accordingly. The second period is the syngamy gametes, when a diploid genome of a future plant is formed. The quantum nature of sexual gametes and the specificity of spin-spin interaction have an important role in the mechanisms of realization of both these processes.

The isogamy nature can be explained and proceed from the quantum nature of the sexual gametes. Each isogamete can be characterized by 4 basic states (individual and species-specific) of its DNA-particle, DNA-energy, DNA-information and spin magnetic moment. In a system with four basic states, any change of the state of their genome can be described as the different combinations of these elements. But right now, we're more curious what happens when two sexual gametes have a different direction in spin.

A classic example of isogamy is the sexual process of *Chlamydomonas reinhardtii* – the singe-cell mobile round or slightly elongated little green alga. Their sexual gametes differ from the vegetative cells only by a smaller size. It is not possible to distinguish the male and female gametes. Vegetative development of *Chlamydomonas* occurs in the haploid phase, and only a zygote is the diploid. Before copulation, their cell walls are dissolved, the content of the male gamete moves into the female one and a zygote is formed. A zygote is covered with the cell wall, which contain ≈ 10% protein and glucose account for 56% of total sugar (Grief et al., 1987). The syngamy result is an increase in intracellular pressure, a DNA-wave transition at the higher energetic levels, and a zygote passes into the resting state. The next logical question is: What forces should act between isogametes, in order for the contents of two cells move into a volume of one cell. We suggest that a spin-spin interaction can be such a force. Its energy depends on the mutual orientation of the spins of gametes and the distance between them. The reproductive cycle of *Chlamydomonas* ends by meiotic division (Figure 23). Its matrix includes two reproductive stages: 1) Conjugation together with activation stage and zygote formation,  + 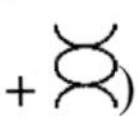) and 2) meiosis (). During meiosis, 4 haploid vegetative cells are formed: Two with spin "up" and two with spin "down".

Now let us see what happens when the gametes as the two Fermi particles have the same impulse and an identical spin direction. For this, we assume that the spin of male gametes is directed downward, and the female gametes are directed upward. Two female gametes or two male gametes cannot approach each other, because in this case the total amplitude of their DNA-wave oscillations becomes equal to zero. By the Pauli exclusionary principle these combinations of sexual crossing do not occur. A spin-spin attraction arises between gametes and their copulation becomes possible if their spins are directed in opposite directions. As a consequence, only a single combination of isogametes may end their syngamy and the zygote origination: ♂ gamete spin ↓ + ♀gamete ↑.

Generative cells are very sensitive to the direction of a spin magnetic moment. The quantum mechanical effect, which occurs between the male gamete (♂ isogamete, microspore, spermatozoid, sperm) and female gamete (♀ isogamete, megaspore, egg, ovicell) are the attraction forces. The spin-spin interaction is substantially smaller than the electric interaction of particles. The attraction forces begin to act only when two gametes are at a close distance, and then an increasing convergence towards each other is experienced.

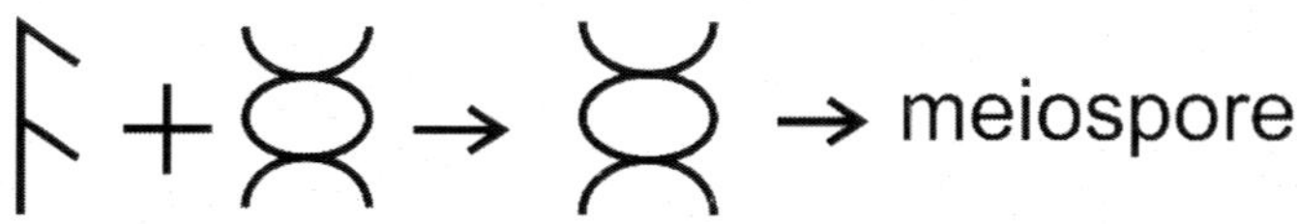

Figure 23. Scheme of reproductive cycle of the *Chlamydomonas*.

The syngamy of gametes at the early evolutional stages occurs in the beginning of the reproductive cycle and this mechanism can be conditionally called fertilization. Apparently, the attraction forces are sufficient only for the process of convergence, copulation of gametes and formation of zygotes, but not enough to connect the homologous chromosomes together and form a stable diploid nucleus. Although the intracellular physical conditions of a zygote are not able to ensure the stability of a diploid state of its genome, they satisfy the requirements (viscosity, pressure) needed for meiosis. The first division of a zygote is meiosis, and a vegetative development of most species of algae occurs in the haploid phase.

Each basic element of eukaryotic genomes and reproductive structures has undergone a certain path of independent and interconnected evolutionary development. The main stages of evolutional development of the reproductive cycle in the different systematic groups of lower and higher plants are shown in Table 1. Initially, the development of the sexual process took place in the direction from isogamy to anisogamy, and from anisogamy to oogamy. Then, the division of reproductive cycle into the male and female generative line occurs. The number of reproductive stages in each of them is increased. The formation of generative cells passes into the specialized male and female generative organs, where all necessary physical and trophic conditions are created for the realization of each stage of the reproductive cycle. Sexual cells arising into the male and female generative organs differ by size, morphology and mobility. The male gametes (microspore, spermatozoid, and sperm) are smaller and more mobile, and the female gametes (megaspore, ovicell) are larger and immobile. In the life cycle, there is an alternation of two generations: A diploid sporophyte and a haploid gametophyte.

The removal of energetically strong generative cells from a vegetative body has become a new stage of plant evolutionary development. The reproductive cycle is broken into 2 (*Cutleriales*, Polypodiophyta) or 3 (*Fucales*) parts. The reproductive process begins in the mother plant and continues on to the energetically strong tissue of prothallium. Apparently, the requisite physical conditions inside the archegonium are created, which ensure the fertilization mechanism by means of forming a spin-spin coupling between nucleotides of homologous chromosomes. The division of the life cycle into two independent biomorphs gave a new impulse to the evolutionary development of diploid sporophyte of the Polypodiophyta. Ferns are characterized by the new progressive properties, the more perfect internal specialization of vegetative and generative tissues and organs. The prothallium has a short life cycle, small sizes and a poorly developed morphological structure.

The sporophyte dominance takes place in gymnosperms and angiosperms. The reproductive process completely proceeds in its body, which includes 3 stages of creation of the hereditary program and 6 activation stages (Table 2). The reduction of the gametophyte is continued. It loses the capacity for independent existence. Fertilization occurs by the karyogamy of the sperm and ovicell, the most energetically strong generative cells by the male and female generative lines. Formation of the ovicell takes place inside the haploid (gymnosperms) or triploid endosperm (angiosperms). Therefore, the energetic potential of the endosperm inside a seed is lower than in the embryo, but higher than in the mother plant. The delivery of a male gamete to the female gamete is a complex multi-stage process. The diploid tissue of the nucellus (gymnosperms) or a column (angiosperms) is located between an ovicell and a pollen grain. The wide spectrum of crossbred and inbred genotypes of pollen passes through it, ensuring the success of the most energetically strong, adapted and quality components. In cross-pollinated species, the tissue of the nucellus and column has come to be regarded as a special system, which promotes cross-pollination (Crowe, 1964, Stephenson & Bertin, 1983; Kuznetsova, 2003).

Formation of the stable diploid genome and corpuscular matrix of the embryo is carried out by means of spin-spin coupling between the nucleotides of homologous chromosomes. In fertilization, the ovicells of the mother tree take part, the DNA-particle of which underwent changes in the result of crossing over. Sperm can have either a DNA-particle of the same tree changed during meiosis (self-pollination) or a specific DNA-particle of another tree (cross-pollination). The spin-spin coupling unites the male and female homologous chromosomes, but does not change their nucleotide composition. In gymnosperms, the fertilization process is stretched in a time. In detail, it is described in Part II, Chapter 3concerning the example of *Pinus sylvestris.* In angiosperms, the formation of spin-spin coupling between homologous chromosomes occurs inside a zygote during the process of karyogamy. In the early and late embryogenesis of higher plants, the energetic matrix and informational matrix of a seed are successively formed. A seed becomes the main reproductive organ. The level of development of the morphophysiological and behavioral sexual dimorphism reaches its maximum.

Seed Reproduction Systems

In angiosperm, the type of seed reproduction system is realized at the species level. It is considered that cross-pollination was their ancient system (Crowe, 1964; Nettencourt, 1977; Aghajanyan, 1992). Cross-pollination provides higher movement rates of corpuscular information by the territory of a species range. At present, 3 groups of species, differing by the fertility level, meet: A self-incompatible (SI), a self-fertile (SF) and a self-compatible (SC).

Self-sterile species in the course of evolution acquired a complex of mechanisms that promote cross-pollination (de Nettencourt, 1977, 1999; Surikov, 1991; Aghajanyan, 1992; Franklin-Tong, 2008; etc.). They are divided into: 1) Morphological devices of the flower; 2) temporal intervals of development of the male and female generative spheres; and 3) physiological incompatibility. Mechanisms promoting cross-pollination rather than self-pollination in the hermaphroditic flowers are also divided into two groups: 1) Spatial barriers or herkogamy (heterostyly, dimorphism by column length, enantiostyly); and 2) temporary barriers or a dichogamy (protogynia, proterandria).About half of the flowering plants are characterized by the physiological systems of self-incompatibility. Self-incompatibility via the pollen reaction to inbreeding is divided up into two types – a gametophytic and a sporophytic. This is the most important mechanism, which prevents the closely related intraspecific relationships in higher plants. As a consequence, the self-fertilization in SI species, due to the inhibition of germination or growth of inbred pollen tubes, is impossible. In SF species, a ratio between self-pollination and cross-pollination is labile and varies in accordance with changes of environmental factors (weather conditions, insect-pollinators).

SC species are the typical self-pollinating plants. It is believed that they originated from self-incompatible ancestors. However, the general direction of the evolutional development of angiosperms reproduction systems is the transition from the self-incompatibility to the self-compatibility.

Gymnosperms are a cross-pollinated species. Their seed reproduction system has the number of distinctive features (Kuznetsova, 1996; 2012). Differentiation by the seed reproduction systems is realized within a population as the intra-population variability of trees by a self-fertility level. Morphological devices preventing self-pollination are absent. Mechanisms of physiological incompatibility are at a lower stage of evolutionary development. Therefore, in the progamous period, the intensity of incompatibility reactions between the nucellus and an inbred pollen tube is insufficient to stop its growth. Inbreeding depression is realized during early embryogenesis. In optimal years, more than 90% of inbred embryos are removed from the reproductive process. In abnormal years (a cold, a dry), the proportion of inbred seeds are more resistant to weather stress, and it is increased in proportion to the weather. Consequently, this more primitive type of seed reproduction fully performs the regulation of closely related intraspecific relationships. The self-fertility system, firstly, promotes the mixing of corpuscular information within the species range due to the selective advantage and survival of crossbred embryos and the removal of inbred genotypes. Secondly, the ecological dependence of incompatibility reactions adapts the gene pool of surviving seeds to the weather conditions of each year.

To summarize the above, it should be noted that the corpuscular and energetic systems have traveled a long (an independent and an interconnected) path of evolutional development. The results of this analysis with the schemes of species reproductive cycles are presented in Tables 1 and 2. Initially, the haploid asexual type of the corpuscular system arose. It originated from the prokaryotic genome. Then, the various forms of the sexual process appeared, when a vegetative haploid plant (unicellular, multicellular) is formed from a meiospore. The diploid type of corpuscular system emerged later in the Mesozoic era. Diatoms (*Diatomeae*) are probably the first diploid species. Presumably, the diploid corpuscular system and first diploid species originated from unicellular haploid species of the Desmidiaceae, whose sexual process includes two reproductive stages: 1) Syngamy gametes by the conjugation type and activation stage ( + 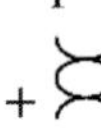) and 2) meiosis (). The reproductive cycle of *Diatomeae* has three stages, which are arranged in reverse order (meiosis, ; activation stage, ; gamete syngamy, ), so the last generative cell is the zygote.

The next step of change of the haploid corpuscular system was an alternation of haploid and diploid generations in the plant life cycle. Two different lines of its organization can be clearly traced: 1) A species with the domination of the haploid phase (gametophyte) and 2) a species with a more developed diploid generation (sporophyte). Then, the diploid generation of lower plants and haploid generation of seed plants completely lost their independence and began to exist on the plant vegetative body, respectively, of gametophytes (Bryophyta) and sporophytes (seed plants). After that, the speciation process on the basis of the haploid corpuscular line ceased. Higher plants for continuation of evolutional development by a diploid corpuscular line moved to a next more reliable type of energetic protection (double protection). On this basis, they generated a great number of gymnospermae and angiospermae species and created a new era in the history of the plant world. Now, there is a transition of angiosperm to a higher level of life cycle organization and energetic protection system –apomixes, or an asexual type of plant reproduction.

The following is a short description of the 4 types of sexual reproduction, the changes of which give an idea of the main trends in the evolutional development of haploid corpuscular and energetic systems of lower plants (Table 1).

Apogamous Type

An apogamous type of asexual reproduction occurs in some unicellular algae. The *Chlorella* reproductive cycle consists of only one activation stage, and as a rule, 4-8 (up to 32) spores are formed. Species are characterized by a primitive prokaryotic technique of the DNA-wave transition on the higher energetic levels. This is the pressure increasing during cell divisions inside a fixed volume of the mother cell. Meiosis and sexual gametes are absent. Energy protection systems are also absent.

Sexual Type without Alternation of Generations

The main distinctive features are the obligatory existence of meiosis, sex gametes and zygotes. This type of sexual reproduction is inherent in most systematic groups of unicellular and multicellular algae. The reproductive process is completely carried out on a haploid plant. The minimum number of reproductive stages is 3 and the maximum is7 (Table 1, reproductive cycles: a) *Chlamydomonas* and b) *Volvox*). The number of activation stages is 1-2. The last reproductive stage is meiosis, and the first vegetative cell is the meiospore. Sexual gametes are the same or unequal by morphology, size and mobility. The types of sexual processes include hologamy, isogamy, anisogamy, and oogamy. The energy protection mechanism is the haplont way.

Sexual Type with Alteration and Independent Existence of Generations

This type of sexual reproduction occurs only in several orders of Phaeophyceae. In their life cycle, there is an alternation of haploid and diploid generations. Two species-specific lifeforms (biomorphs) exist independently of each other. They are the isomorphic or heteromorphic types. The sexual process is initiated on a haploid vegetative plant and ends on a diploid gametophyte. The reproductive cycle is extended, and separates into 2 (*Cutleriales*) or 3(*Fucales*) parts. The number of activation stages is increased to 4. The main organ of propagation is an energetically strong haploid zoospores. The energy protection mechanism is the bimorph way.

Sexual Type with Alteration of Generations

Table 1. Scheme of evolution development of the haploid corpuscular and energetic systems of lower plants on the basis of the idea of corpuscular-wave nature of sexual reproduction process

№	Type reproduction	Scheme of reproductive cycle and energy protection mechanism	Systematic group of lower plants
1	Apogamous type	- spore Energy protection mechanisms - absence	Claucophyta, *Chlorella,* etc.
2	Sexual type without alternation of generations	 a Energy protection mechanism – haplont way 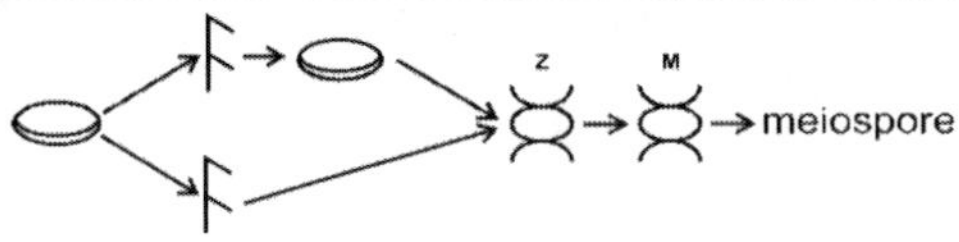 b Energy protection mechanism – haplont way	All divisions of haploid lower plants, reproductive cycles of which end the meiotic division, except for Rhodophyta, Phaeophyceae, Bryophyta
3	Sexual type with alternation and independent existence of generations	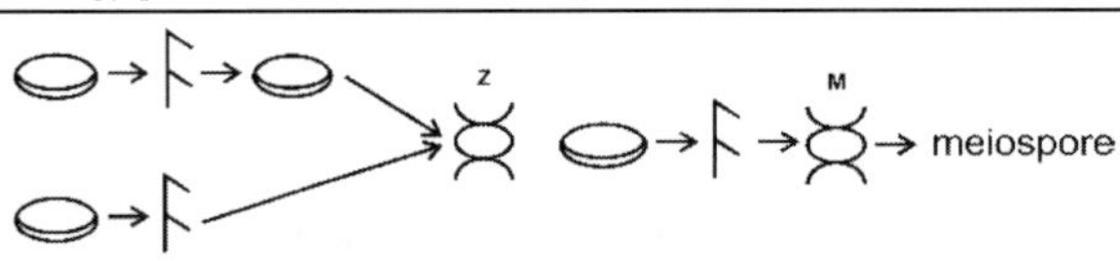 Energy protection mechanism – bimorph way	Phaeophyceae (*Cutleriales, Fucales*)
4	Sexual type with alternation of generations	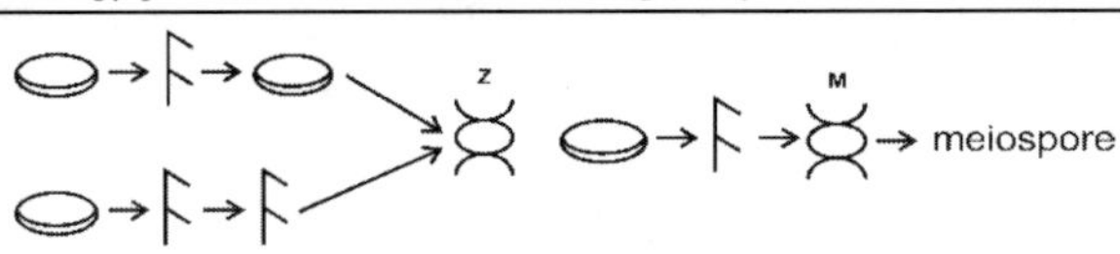 Energy protection mechanism – haplont way	Rhodophyta, Bryophyta

Z – zygote; M – meiosis; - activation stage; - stages of the emergence of a new reproductive initial; - stages of the creation of a hereditary program

Multicellular red algae and mosses are the last representatives of this evolutional line. There is an alternation of generations within gametophyte dominance. The meiospore is a generative cell, which gives rise to a grown-up plant. The reproductive process completely carries out on the gametophyte and exists at its expense. The new generative organ – the archegonium – appears, or the 4th activation stage in the moss reproductive cycle (Table 1). Further growth of energetic advantage over inanimate matter is one of the prerequisites for sushi mastering. The haploid way is again an energy protection mechanism. A return to the old way, which exhausted its possibilities probably became one of the reasons for the end of plant evolution on the basis of the haploid corpuscular line.

Table 2. Scheme of evolution development of the diploid corpuscular and energetic systems of lower and higher plants on the basis of the idea of corpuscular-wave nature of sexual reproduction process

№	Type reproduction	Scheme of reproductive cycle and energy protection mechanism	Systematic group of lower and higher plants
1	Sexual type without alternation of generations	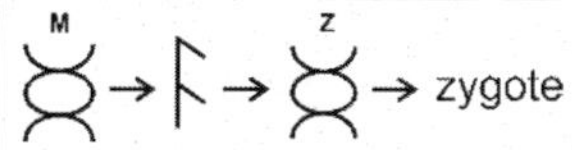 Energy protection mechanisms - absence	Bacillariophyta
2	Sexual type with alternation and independent existence of generations	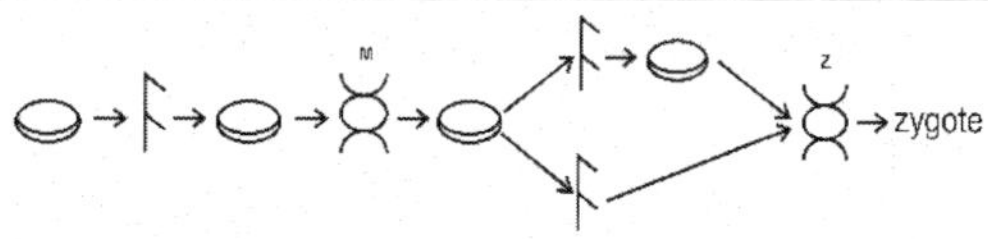 Energy protection mechanism – bimorph way	Diploid orders of Phaeophyceae (*Chordariales, Desmarestiales, Laminariales*, etc.)
3	Sexual type with alternation of generations	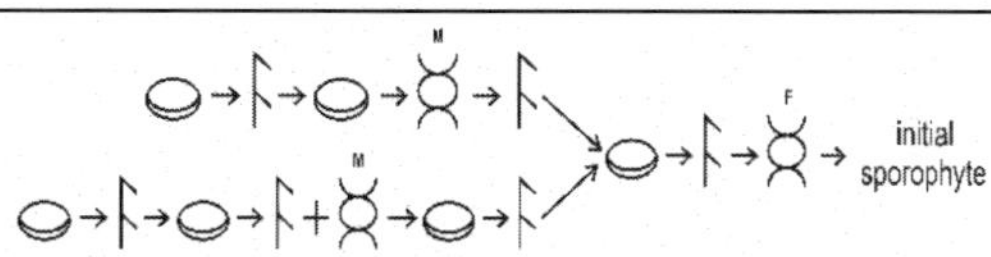 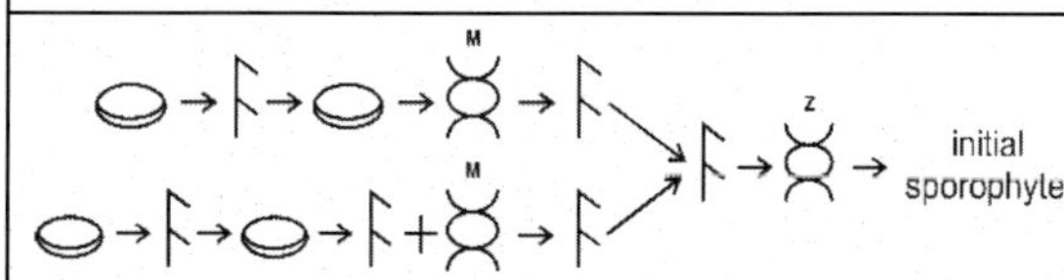 Energy protection mechanism – double protection way	Gymnospermae, Angiospermae
4	Apomictic type	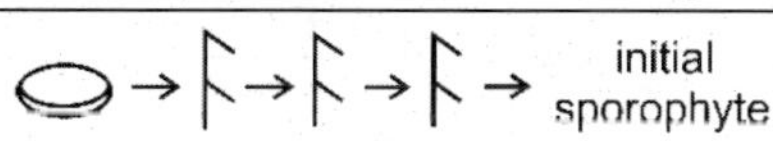 Energy protection mechanism – double protection way	*Asterales, Compositae*, etc.

Z – zygote; M – meiosis; - activation stage; - stages of the emergence of a new reproductive initial; - stages of the creation of a hereditary program

Speciation on the basis of the diploid corpuscular line begins with the origination of unicellular diatoms (Bacillariophyta) and includes 4 stages of the evolutionary development of lower and higher plants (Table 2).

Sexual Type without Alternation of Generations

The diatoms reproductive cycle begins with meiosis and has three stages: Meiosis, activation stage and fertilization. The increased intracellular pressure required is to realize these reproductive stages (including formation of a spin-spin coupling between homologous chromosomes). It is achieved with the help of a solid bivalvous silica shell and grinding individuals during their vegetative development. The main type of the sexual process is a conjugation of the following: Isogamy, anisogamy, and oogamy. There is also an autogamy. An energy protection system is absent.

Sexual Type with Alteration and Independent Existence of Generations

This type includes ferns (Polypodiophyta) and the diploid orders of brown algae (Phaeophyceae), the life cycle of which is characterized by the alternation of two equal or morphological different generations. The energy protection mechanism is the bimorph way. The reproductive cycle of brown algae includes 3 activation stages. The transition from the gametophyte dominance to the sporophyte predominance and, accordingly, from the haploid to the diploid corpuscular system occurs within one class of Phaeophyceae. A gametophyte dominance occurs in *Cutleriales*; a thallus of a gametophyte and sporophyte has the same sizes in the orders of *Dictyotales* and *Ectocarpales*; a sporophyte of *Desmarestiales* and *Laminariales* is dominated, and their gametophytes present a small prothallium. Its energetically strong tissue provides all necessary conditions for the formation of a stable diploid genome.

The structure of reproductive cycles of the equisporous ferns and mosses are similar enough (Table 1, 2). The number of activation stages is increased to 4-5. The main exception is the different sexual process direction: a last generative cell of ferns is a diploid zygote, and in the mosses is a haploid spore. The gametophyte of ferns is strongly reduced. Formation of antheridium and archegonium and fertilization occur on it. On the contrary, the sporophyte is sufficiently developed. It has initial roots, but no rhizoids. Stem is more differentiated, and the leaf plate combines the functions of photosynthesis and sporulation. The bimorph way of protection and more high energetic potential allowed the Polypodiophyta to master many different ecological niches of the land at the beginning of the Mesozoic era.

Sexual Type with Alternation of Generations

This type is represented by two large systematic groups of higher plants: Gymnosperms (more ancient and now extinct) and angiosperms, which reached a peak in the Cenozoic era. The reproductive process is wholly carried out ina sporophyte. It is divided into two parts: Gametogenesis and embryogenesis. During gametogenesis, a formation of the energetic and corpuscular programs of the future plant takes place. In embryogenesis, an energetic, informational and corpuscular matrix of space organization of a plant body is created. The seed is the main organ of propagation.

The alternation of generations with sporophyte dominance takes place. The number of activation stages is increased to 6. Fertilization occurs in two steps: The progamous period (up to the pollen tube entry into a female gametophyte or embryo sac) and the syngamy gametes (proembryogenesis in gymnosperms and double fertilization in angiosperms). In angiosperms, the ovule is enclosed inside the ovary cavity and all reproductive processes by the female generative line are strictly oriented with respect to gravity. Types of pollination are cross-pollination and self-pollination. All gymnosperms are a cross-pollinated species. The evolution of angiosperms is directed from self-incompatibility to self-compatibility. An energy protection mechanism is the double protection, which is created on the basis of the matrix reproductive cycle. In gymnosperms, this is the only part of the female generative line. In angiosperms, it encompasses all of the reproductive cycle up to fertilization. In the first case is a haploid endosperm, in the second, a more energetically powerful triploid endosperm (after double fertilization) is formed. In angiosperms, the number of ⬭ stages, which is connected with the energy losses, decreases (Table 2). During their evolution, the growth of biological energy was also due to the unequal distribution of biological energy between the vegetative and generative sphere in favor of the latter. Therefore, the species of woody plants were replaced by the shrub, and then grass.

Apomictic type. Apomixis is the next stage in developing in our time principally new type of asexual reproduction on the base of diploid corpuscular line. Transition to the apomixis occurs in the most highly organized families of Angiospermae (*Rosaceae, Rutaceae, Orchidaceae, Solanaceaes, Asteralceae, Poaceae, Compositae,* etc.) (Asker & Jerling, 1992; Kashin & Kuprijanov, 1993; Savidan, 2000; etc.). It is a continuation of gametophyte reduction; a weakening of the male generative line; a passing from cross pollination to self-pollination, and from self-pollination to reproduction without pollination. The replacement of typical sexual reproduction is realized through a speciation process with different intermediate forms (pollination presence, pollen germination on the stigma without fertilization, etc.). The apomictic type of asexual reproduction guarantees a high degree of viability and polymorphism of seed progeny, a high and stable reproduction rate and the ability to master new ecological niches.

An embryo formation without pollination, meiosis and fertilization are the main traits of apomixes. The reproductive process is fully realized inside the ovary cavity. The male generative line is absent. Activating meiosis is replaced by an activating mitosis. All cells of the embryo sac are diploid. The reproductive cycle includes 3 activation stages (Table 2), which are directly interlinked with each other and have the same orientation with respect to gravity. During embryogenesis, only energetic and informational matrices are created. The energy protection system is entirely formed on the basis of the reproductive cycle. The mechanism of energy protection is double protection. The genetically determined energetic gradient is between the initial cells of the embryo (ovicell) and endosperm and, respectively, between the tissues of the embryo, endosperm and sporophyte. The organ of propagation is the seed.

Summary

If we represent the haploid and diploid corpuscular lines as a common evolutionary process, then the following main trends of plant development are revealed.

- The evolution of sexual reproduction begins as an asexual mechanism of haploid unicellular algae and is now moving into a highly organized system of asexual reproduction of diploid multicellular plants (apomixis).
- Progressive evolution has occurred and is taking place due to the continuous growth of the biological energy of species. This gives them an energetic advantage over the inanimate matter and the more highly organized species over the low-organized ones. The prokaryotic technique is the first way of transitioning into the biological harmonic oscillator to higher energetic levels and cell rejuvenation. The physical essence of this technique is to increase intracellular pressure in a fixed cellular volume during cell division. This way was replaced by activating mitosis, during which an uneven distribution of energy between sets of daughter chromosomes occurs.
- The speciation process was initially associated with the increasing of the stage number in the reproductive cycles of lower and higher plants (Table 1, 2): 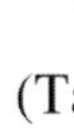 - from 1 to 6 stages (Gymnospermae, Angiospermae);  - from 0 to 6 (*Selaginellaceae*);  - from 0 to 3 (*Selaginellaceae,* Gymnospermae, Angiospermae). The reproductive cycle is also complicated by means of separation of its matrix into female and male generative lines. This led to an increasing of the complexity of life forms and species diversity on the Earth, and the plant mastering of various ecological niches.
- Plant evolution was carried out from a gametophyte advantage to sporophyte dominance; from isogamy to anisogamy, from anisogamy to oogamy; from syngamy gametes to the fertilization mechanism with the formation of stable spin-spin coupling between homologous chromosomes. Perhaps, the transition from a haploid to a diploid corpuscular genome became possible as a result of the energy potential increase, the emergence of specialized male and female generative organs, and the creation of the bimorph way of energy protection.
- The plant world is organized according to the principle of bipolarity. Embryogenesis appears in the seed plants, during which a corpuscular, energetic and informational plan of plant space organization is formed.
- Further improvement of reproductive cycle structure begins from the gymnosperms. Stages with energy losses () first decrease in number and then are removed. In conifers, the amount of  stages is 6, in angiosperms 4, and in the family of *Compositae* 1. In angiosperms, the energy potential of species

are increased due to an unequal distribution of DNA-energy between the apical and floral meristems in favor of the latter, which leads to the change intheir lifeforms: The transition from the woody to the shrubby and then to the grassy plants. The reproductive cycle of an apomictic type begins from the highest level and represents 3 closely related stages, when the last generative cell of one stage becomes the first cell of the next.

- An energy protection system has great importance in the organization of their life cycle. In the speciation course, this system underwent its evolutionary path of development: From its absence to the mechanism of double protection, which is completely formed on the basis of the reproductive cycle. The change in the system type of sexual reproduction is usually caused with a replacement of the way of energy protection. The return of the sexual process of the red algae and mosses from a bimorphous, more progressive protection way to a less progressive haplont way has led to an impasse in the speciation on the basis of the haploid corpuscular line.

INFORMATIONAL COMPONENT OF GENOME (DNA-INFORMATION)

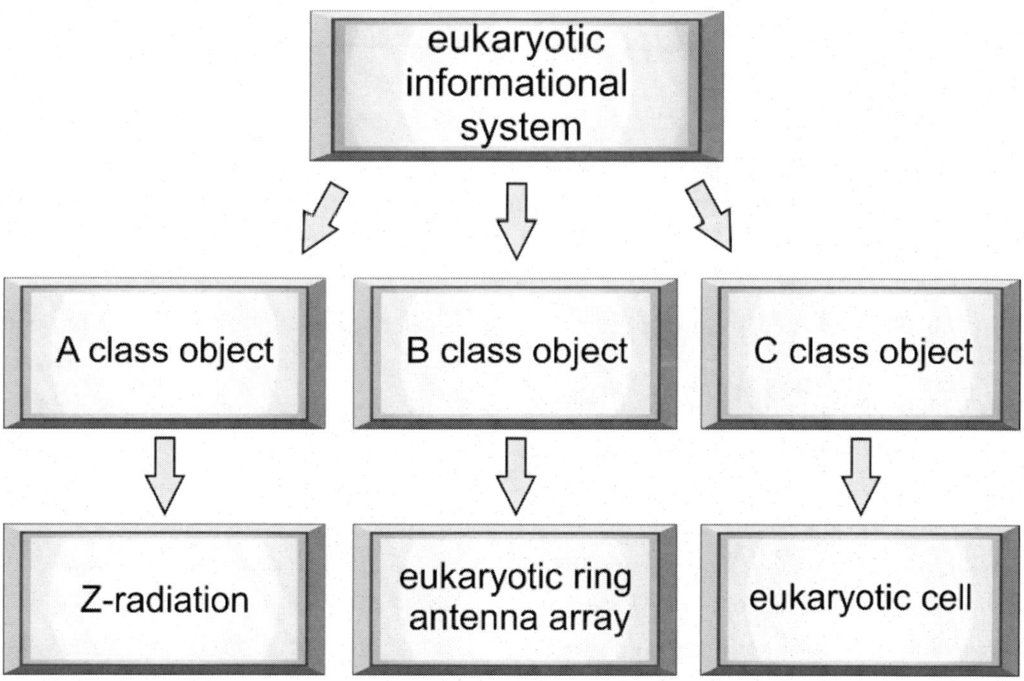

Figure 24. Informational system of eukaryotic cell.

DNA-information is the third main component of the genome. It has the terrestrial and cosmic origin in our three-plane corpuscular-wave model. Its functional role is the creation of intraspecific and interspecific heterogeneity in the alternation of generations, as well as the plant adaptability to the different environmental conditions according to the principle of "genotype-environment" interaction. The eukaryotic informational system, in comparison with the prokaryotic system, has a more complex structure. And just like in prokaryotes, it consists of the same three basic elements: Objects of class A, class B, and class C (Figure 24).

A Class Object

A class object of prokaryotic and eukaryotic systems is the Z-radiation or the cosmic bioactive radiation that's continuously coming to the Earth together with cosmic streams of particles and electromagnetic radiations (Chizhevsky, 1972). Space information, as an objective unshakable substance of our world, exists in a variety of a set of species informational signals. In order to catch their own species informational signals, it is necessary that a material species-specific substance and favorable environmental conditions by this time on Earth have already been created. Therefore, DNA-information is inextricably linked with the energetic and corpuscular components of a genome. In unicellular species, it is a eukaryotic cell and in multicellular plants it is a generative cell with the species-specific DNA-particle and DNA-energy. DNA-information is a new type of antenna for receiving and transforming of space species information into cellular DNA-information of plants of certain species.

At this time, we can only make assumptions of when, where and how the first eukaryotic cell arose about 1 billion years ago, what transformations occurred and in what sequence it underwent along its path. Obviously, however, there was an oxygen-rich atmosphere created by prokaryotic organisms on Earth at this time. As mentioned above, in the eukaryotic cell there had been many changes from the position of DNA-energy and DNA-particle in connection with the origination of a nucleus or a biological harmonic oscillator of a more effective and reliable type. From this, we can conclude that the informational component of the genome in the transition from a prokaryotic to eukaryotic type of cell organization has been stronger than the corpuscular and energetic component. This is based on the fact that the appearance of a nucleus and the displacement of DNA (particle and wave) into its cavity caused the disappearance of a ring chromosome in the eukaryotic cell. A ring chromosome in the informational system of prokaryotes is an object of the class B or a prokaryotic antenna of the CTHE type.

B Class Object

We approached the problem of studying the organization of the informational system of eukaryotic cells and multicellular organisms from the position of the trinity principle of energy, matter and information and the corpuscular-wave nature. Two theoretical provisions were the foundation of our working hypothesis. Firstly, the energetic, corpuscular and informational organization of a genome, which is laid on the basis of a DNA-molecule (particle and wave) and species-specific reproductive cycles. Secondly, the three-plane internal structure in the form of an energetic, corpuscular and informational matrix is created during fertilization and embryogenesis of higher plants. These processes are described in detail in Part II, Chapter 3.

The search for a new type of receiving antenna began with mitosis, the physical model of which was described in detail in our previous work (Kuznetsova, 2013). It found out that the eukaryotic cell, in contradiction to the prokaryotic cell, cannot be a receiver of species cosmic information during mitotic divisions. Next is the consideration of the sexual process. For this analysis, it's taken us the following data on the spatial arrangement of chromosomes inside the nucleus in the different types of cell divisions.

In the reproductive cycle of plants, the following types of cell divisions occur: Mitosis, asymmetric mitosis, activating mitosis, meiosis, free-nuclear mitosis, endomitosis, and cleavage. The concept that the chromosomes retain the ana-telophase orientation during interphase was first expressed by C. Rabl in 1885. At present, it has been experimentally proved that the chromosomes in the interphase have the ana-telophase orientation by Rabl and are connected to the nuclear envelope by their telomeric and centromere regions (Monakhova, 1990; Chentsov, 2004). Each shoulder of the chromosome inside the interphase nucleus occupies a certain zone, which does not enter the volume of the neighboring chromosomes. If their ends are to connect lines, then a set of triangles or chromosomes-particles-waves in the form of a truncated cone lying across the force of gravity will be obtained (Figure 22a). Centromeres by the perimeter of the upper base and telomeres along the lower base of the cone are distributed. In the further analysis, this spatial localization of chromosomes inside the interphase nucleus will be considered as the non-working state of the receiving eukaryotic antenna.

In order for the generative cell as the B class object, to operate, it shall satisfy three requirements: 1) As is known, the nucleus of a eukaryotic cell is destroyed during mitosis. The wave function on the DNA disappears, the biological harmonic oscillator is destroyed, and the cell passes from the open system into a closed system. We think that the set of chromosomes can serve as a receiving intracellular antenna only where there is a DNA-wave. In this case, the antenna will be able to convert the Z-radiation energy into the electromagnetic oscillations of a DNA-wave. 2) In the prokaryotic cell, the transition of eukaryotic antenna in working conditions and reception of the cosmic species signal also must occur during cell division, when the spatial orientation of chromosomes changes within the nucleus. 3) Eukaryotic antenna for effective reception of space species signal should be located vertically upwards and have a direct exit into the atmosphere with a minimum of the mechanical interference.

After analyzing the above types of cell divisions, we stopped at meiosis. This choice is not accidental. As a B class object, meiosis and meiotic chromosomes satisfy all physical requirements.

Really, the destruction of the nucleus envelope and the formation of the cell division spindle begin only at the stage of metaphase I. The wave function on DNA remains in the prophase of meiosis (from leptotene to diakinesis). The cell is an open system. As was already mentioned, the change of the chromosomes' spatial position relative to the gravity direction occurs during proleptotene (Figure 22b). The truncated cone "gets up" on the lower base, and the chromosomes take an upright position. Elongation of the apex stem with sporogone in mosses, mega- and microstrobilus in gymnosperms, the growing point of a flower escape in angiosperms, like masts, lift up the generative structures before meiosis, which confirms the important role of gravity in the processes of plant development (Merkis, 1990). Therefore, male and female generative organs, male and female gametes always have the same orientation with respect to gravity direction.

Let's consider the physical essence of cell transition from mitotic to meiotic division on the example of Bacillariophyceae. Diatoms are the unicellular algae with a rigid, non-stretchable shell, consisting of silicon dioxide. All cells of the first generation have approximately the same sizes. A particularity of their vegetative reproduction is the grinding of individuals in the alternation of generations into 2 or more times. Nevertheless, these different generations contain approximately the same number of biomolecules, since the number of proteins, RNA, lipids, etc. in the end of each cell cycle is doubled, and their intracellular concentration due to the reduction in cell size will gradually increase. According to the gas law, this leads to a growth of intracellular pressure and an increase of nucleoplasm viscosity. We consider that these two physical factors cause a change in the type of cell division and the transition from mitosis to meiosis. Reduction division begins after the cell reaches the minimum dimensions, the threshold level of intracellular pressure and viscosity.

We think that an increased intracellular pressure and viscosity into the microsporocytes and megasporocytes of higher plants also causes the transition of these generative cells from mitosis to meiosis. Initially, their cell cycle occurs by the type of mitosis. At the interphase beginning, their genome is 1 DNA-particle – 1 DNA-wave. DNA-wave radiation at the end of the G_1 period and the entry into the nucleus of two quanta of free energy initiates the process of new cell formation (DNA-polymerase activation, DNA replication, and centrioles creation). All this time, a set of chromosome-waves create a biofield with a maximum intensity in the central part of the nucleus. The state of microsporocytes genome after S period is 2 DNA-particles – 1 DNA-wave.

During the G_2period, the centrioles are held by the biofield forces inside the nucleus, where the process of chromosomes coiling begins by a one-side nucleotide turn.

It was established that the sporogenous tissue of archesporium is characterized by the increased mitotic activity in the pre-meiotic period. Microsporangium (an entirely formed anther) by this time already reaches a species-specific size and its volume no longer changes. According to the laws of physics, a further increase of the number and sizes of sporogenous cells inside a fixed volume of archesporium leads to an increase in the density and viscosity of the nucleoplasm. The consequence of a physical situation changing is the cessation of divisions of sporogenous cells and the beginning of microsporocytes formation.

The transition from a mitotic to the meiotic path of development begins with the appearance in proleptotene of "pro-chromosomes," or the unusually twisted chromosomes (Cytology and ..., 1975). Since the rotation of centrioles around a set of chromosomes occurs in a more viscous medium of nucleoplasm, the chromosomes at some point "give in" to the rotation and begin to intertwine around themselves, remaining connected by means of centromeric and telomeric ends with the envelope of the nucleus. After the restoring force reaches a maximum, they begin to quickly unwind in the opposite direction and lose contact with the nucleus for a while. A set of chromosomes, like any physical body lying across the gravity forces, undergoes the effect of torque, which is created by gravitational attraction. Therefore the cone unfolds inside the nucleus and "stands" on the lower base (Figure 22b). At the same time as the lifting and uncoiling chromosomes will be flow of torsion waves along their shoulders. Each chromosome is rotated into the nucleus so that the homologous chromosomes turn out to be located opposite each other. If we now represent their set the set of homologous triangles, then the surfaces and sides (shoulders of homologous chromosomes) will be parallel. After a turning, the uncoiling and elongated chromosomes soon reach the envelope of the nucleus. And the new spatial position of DNA (particle and wave) at this time is fixed by its volume.

Homologous chromosomes are equal particles. They are close along the length of shoulders, the angle between them, the frequency and amplitude of oscillations, the number of chromomeres and the chromometric pattern, and after the turn (early leptotene) they closely approach each other. Therefore, the waves of homologous chromosomes, as two identical particles and coherent waves, begin to interfere between themselves. Interference occurs only in the zones of chromomeres (or regions of the internal coiling of doubled chromosomes). Conjugation of homologous chromosomes begins with several points in the zygotene. The volume and number of chromomeres increases in the course of the further process of chromosomes spiralization. The formation of a resulting wave along an entire length leads to the origination of bivalents. At this time, the mutual exchange of nucleotides (crossing over) between homologous chromosomes is carried out. In pachytene, the chromosome-particles in bivalents become indistinguishable.

Now, we will consider at what stage of meiotic division and how the reception of species cosmic information can takes place. To do this, let's once again look at the above-described process from the position of the B class object. We are interested in what component of the information system can perform the function of a receiving antenna. Suppose that the active antenna is the set of chromosome-waves after their turning in the leptotene. In this case, each pair of homologues, at the moment of their lifting and rapprochement, becomes a separate antenna element. And all sets of chromosomes represent the species-specific ring antenna array for a reception of certain species cosmic signal.

The lower base of the truncated cone, where the telomeric ends of homologous chromosomes are attached to the nucleus envelope, is the base of the ring antenna array with a radius of **r**. Antenna elements, or the set of homologous chromosomes, are located around the lower base. This array construction provides a high degree of symmetry of the antenna elements (Figure 25a). The centromeric ends of chromosomes in the upper base of the cone are also attached to the nucleus envelope, which provides a fixed spatial position of the antenna and its elements during the reception of a new cosmic informational signal (Figure 25b). This individual ring antenna array we will further refer to as a receiving antenna of eukaryotic type (or eukaryotic antenna array).

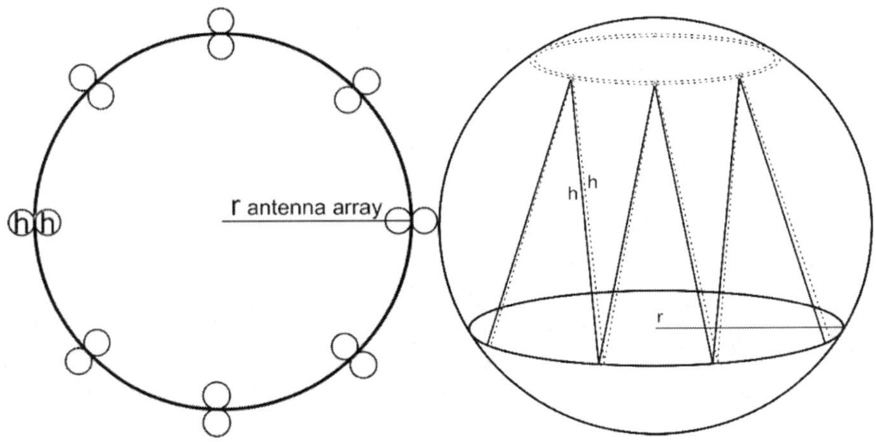

a b

Figure 25. Eukaryotic antenna array during meiosis: a – top view; b – side view.

The antenna system with the ring antenna array is a widespread class of antennas, which is characterized by a more complex construction than the Contrawound Toroidal Helical Antenna. It consists of a set of separate vertical antennas and is used as an underground and surface antenna with the full azimuth coverage. This type of antenna has high reliability, vibration resistance and minimal overall dimensions due to the combination of the class antennas of curtain array and the distribution system of its elements. These qualities allow for its use in communication systems placed on the mobile objects. The advantage of this construction is the high gain of each element of the antenna and the effective spatial summation of the power of low-power transmitter emitters. Moreover, the alignment of elements in the antenna array is almost the same as the alignment of its solitary element.

Now, we will focus on one more feature of the function of the informational system of multicellular organisms. The thing is that the DNA-wave of meristematic somatic cells, which gives rise to the first generative cells, carries an informational signal of the mother plant. It was received by the male and female generative cells during meiosis and underwent a transformation process in the course of reproductive and ontogenetic development of the plant.

This is because the DNA-information, as any information, is inherent in the property of frailty, resulting from various internal and external noise and interference. In prokaryotes, its restoration occurs during each division by means of superposition of two waves – a new (unchanging) cosmic signal and the changed cellular DNA-wave of the previous generation. In unicellular and multicellular eukaryotes within the sexual process, the reception of cosmic species signal occurs only during meiosis. Their life cycle is divided into a period of sexual reproduction and ontogenesis and lasts much longer. Therefore, the orderly ontogenetic transformations of informational parameters, distortions and losses, which undergo the original DNA-information (formed on the basis of the reproductive cycle) in the channel of ontogenesis with noises, are very significant and irreversible. It is shown that the degree of chromatin condensation in the undifferentiated embryonic cells is insignificant and has a diffused structure. In specialized and apoptotic cells, it is represented in the form of large blocks of heterochromatin. These changes of DNA-information cannot be restored prokaryotically – by means of the superposition (imposition) of two modulated DNA-waves.

The effect of ontogenetic transformations of the DNA-wave and irreversible distortion of species DNA-information are manifested as the age-related morphogenetic changes of phenotypes and genotypes. It seems obvious that the old informational signal of previous plant generation must be removed or strongly weakened before receiving a new species cosmic signal. The experiences by chemical mutagenesis confirmed that the exchange of DNA-information between the adjacent sexual generations is weaker. The gene pool of pine seeds, which was broken by an inhibition of doses of chemical mutagens, in the next generation almost completely returns to the species norm (Kuznetsova, Mashkina, 2011). The first induced progeny was characterized by a large number of mutant, morphologically altered, dwarf plants. A large portion of empty and underdeveloped seeds in their cones takes place. In the second induced progeny, the stress response manifests only as the residual variability, as a "memory" of the stress experienced.

We assume that the demodulation of the DNA-wave occurs in the beginning of reproductive cycle separately for the male and female generative line – during microsporogenesis and megasporangiogenesis. Reproductive initial of microsporangiogenesis is one or a few surface vegetative cells of microsporophyll primordium, the polar axis of which coincides with the gravity direction. During microsporangiogenesis, the initial cell undergoes three activating mitosis. The end result is the origination of an apical energetically strong cell, which gives rise to a sporogenous tissue. Three cells of the basal pole, which lost the energy, participate in the formation of microsporangium wall layers. The cells of the apical pole are then moved inside the primordium tissue, push apart its nearby cells, and feel the pressure from the side of the cells of the microsporangium. At this time, the sizes and shape of generative cells do not change and they are difficult to identify. Perhaps, when the generative cells are compressed, the old changed informational signals are strongly weakened and the mother DNA-information is converted into harmonic oscillations of the carrier DNA-wave (Figure26a). Then the initial sporogenous cell begins to grow. Its nucleus is significantly increased, the nucleoplasmic ratio changes, the proportion of diffused chromatin increases, which also leads to the restoration of the characteristics of the carrier DNA-wave before the reception of a new cosmic informational signal.

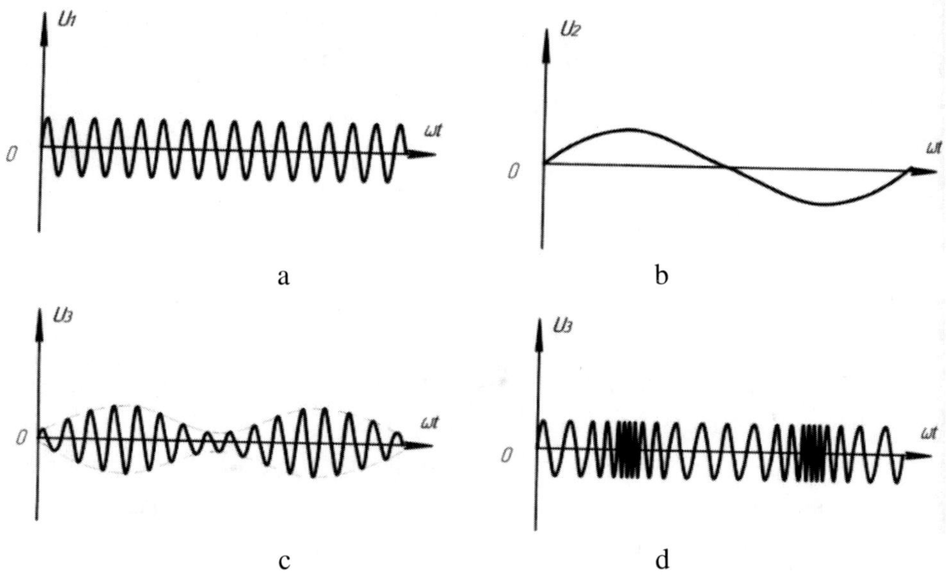

Figure 26. The carrier DNA-wave of microsporocyte before meiosis (a); cosmic species signal (b); amplitude (c) and frequency (d) modulation of DNA-wave.

The DNA-information after the removal of the old information signal is only a carrier DNA-wave. Perhaps, this is a new eukaryotic way of preservation of the species information from degeneration in the alternation of generations. The necessity of this procedure is due to the complication, differentiation and specialization of somatic cells and tissues.

Electron microscopic study of sporogenous cells showed that they in different species of higher plants are very similar. They differ from other cells of microsporangium by the shape, size and ultrastructure of the cytoplasm and nucleus. These are the cells of polygonal and almost isodiametric shape with very thin walls and an altered nucleoplasmic ratio. The nucleus is large, round and without invaginations. Chromatin is mainly represented by the euchromatin, and the heterochromatin proportion is significantly reduced. Perhaps the increase of the nucleus sizes and the low degree of chromatin condensation indicate that their DNA-wave is a carrier standing wave, either without an information signal or with a strongly attenuated signal.

Thus, the eukaryotic antenna array consists of a haploid set of chromosomes or **n** groups of antenna elements. Each antenna element contains two parallel arranged homologous chromosomes (or 4 equal DNA-particles), which are fixed on the nucleus envelope in the three points, arranged uniformly by a circle and symmetrically about its center. The species signal in two homologous antenna elements has a constant phase difference. Such waves are coherent. When the coherent waves add up, the interference of DNA-waves occurs. The conjugation of homologous chromosomes leads to the formation of the resulting DNA-wave on the bivalents and amplification of species cosmic informational signal takes place.

The eukaryotic antenna array has high reception characteristics in the species-specific frequency ranges of bioactive radiation. In a constructive plane, this is due to the fact that each plant has the individual and species-specific karyotype or the haploid (**n**) set of chromosomes. Species differ from each other by the chromosomes number, the size of one or more chromosomes, their structure and form. Therefore, the eukaryotic antenna array represents the individual receiving antenna with the species-specific construction, which is characterized by high sensitivity, enabling a cell to receive only its own species signal. This is the first DNA-information preamplifier, which acts in the system of a class Bobject.

A weak cosmic species signal is captured and amplified by a receiving eukaryotic antenna array (Figure 26b).The carrier DNA-wave of the microsporocyte and megasporocyte in the leptotene undergoes a transformation (modulation process) of one or several of its informational parameters: A frequency or an amplitude (Figure 26c, d). This process in all multicellular biological systems occurs at the beginning of the individual life (during male and female meiosis). DNA-information after fixation of the species cosmic informational signal on the material carriers (DNA-wave of microsporocyte and megasporocyte) changes constitutive once more in the process of formation of the resulting wave during fertilization. The ontogenesis, with respect to the DNA-information, can be viewed as a closed communication system within which their individual morphogenetic information is realized into the species-specific plants and animals.

The uniqueness of meiosis lies in the fact that the several important events occur simultaneously. During male meiosis, three basic states of the genome of a microsporocyte change: a DNA-particle, a DNA-information, and a spin direction. 1) The exchange of corpuscular information (or the nucleotides as the internal paired segments of DNA-particle) between non-sister chromatids takes place during interference of homologous chromosome-waves (or two closely related complex objects). We assumed that the mechanism of crossing-over has a quantum nature. The final result of meiotic division is the appearance of 4 haploid DNA-particles that are not identical by the nucleotide composition. 2) There is a reception of a species cosmic informational signal, which is fixed on the DNA-wave as on the intracellular material carrier. The initial DNA-information and its transformations during life cycle of species-specific plants is the basis of their morphogenetic differences and ontogenetic changes. 3) The differentiation of generative cells by sex occurs. This phenomenon is also of a quantum nature. Filtration of gametes by the spin direction is the consequence of interaction of an inhomogeneous intracellular magnetic field with two types of nucleotides as with Fermi particles. The end result is the formation of male gametes: 50% with the spin down and 50% with the spin up. The combination of three different corpuscular-wave processes in the mechanisms of the male meiosis correspond only one ☋ reproductive stage in the scheme of the species reproductive cycle (Table 1, 2). In the previous work (Kuznetsova, 2013) we designated this stage by two symbols as ☋ + ⚥ (DNA-particle + spin direction). So, if we say that during male meiosis the "hereditary basis" of the generative cell changes, this means an alteration of three base states of the male gametes: A DNA-particle, a DNA-information, and a spin direction.

Let us continue the analysis of the B class object organization on the example of the cell cycle of a microsporocyte from the moment when a weak species cosmic signal is captured by a eukaryotic antenna array. This signal modulates the oscillations of DNA-wave with the species-specific frequency and/or amplitude. Initially, the received new informational signal is amplified on two elements of the antenna array (conjugation of homologous chromosomes-waves and formation of the resulting wave on bivalents). Then this strengthened informational signal also requires amplification for transferring information from a B class object (modulated DNA-wave) to a C class object (a cell, and then into a plant phenotype). In this regard, the eukaryotic cell is very similar to a radio receiver. The operational amplifier underlies the work of any radio signal receiver. It is used wherever the signal amplification and/or its conversion are required.

The presence of an operational amplifier in the prokaryotic cell is not important as in eukaryotic cells. Prokaryotes are very small unicellular organisms. The nucleus is absent, and a DNA-wave carrying the informational signal is in direct contact with the cytoplasm. Reception of new cosmic species signal happens every division after 20-30 minutes. The amplification of the information signal occurs by the superposition of two waves: A new cosmic signal on the cellular DNA-information of the previous generation. Then, this informational signal during the cell cycle is realized in the phenotype of a species specific unicellular organism.

The informational system of eukaryotes has a more complex organization. This is due to the fact that the eukaryotic cell is much larger than the prokaryotic cell. Most of eukaryotic species are multicellular organisms. The genetic apparatus is located inside the nucleus and DNA-information does not have direct contact with the cytoplasm. Therefore, the weak cosmic signal should not only be amplified by means of obtaining a powerful copy of the input signal, but also by subsequent amplification of this copy after each cell division of ontogenesis. This is necessary to create an informational field and energetic heterogeneity within a cell, tissue and plant as a whole. In addition, it is required to prevent informational losses, and to suppress noises that occur in the process of development of a multicellular organism. This significantly increases the reliability of the intercellular transmission and capacity of the information signal. According to Kaznacheyev & Michailova (1981), the informational interactions are carried out at the cellular level. It is established that the cells constantly exchange regulatory information, even when they are separated by quartz glass.

The mechanism of action of this additional operational amplifier (terminal amplifier) of the informational signal and intracellular structures providing the strengthening effect is not known. However, this does not mean that it is absent. Below, we are trying to understand which intracellular structures and biological molecules can be involved in the implementation of this function inside a nucleus of a eukaryotic cell.

Since the operational amplifier, as the main part of the B class object, appears only in eukaryotes, we again turned to the analysis of internal differences in the organization of a genome of prokaryotic and eukaryotic cells. After examining their genetic apparatus, we paid attention to the peculiarity of the chromosome structure. Eukaryotic chromosomes present a complex of DNA and histone proteins. These proteins are never found in the ring DNA molecule of a prokaryotic cell. Specific nuclear histones account from 40 to 80% of all proteins concerning the chromatin of the cellular nucleus (Chentsov, 2004). These are the highly alkaline proteins, which are present in huge numbers in all eukaryotic cells ($\approx$ 60 million copies of each type of histone per nucleus). Histones are highly conserved proteins that underline, apparently, their important role in the biology of eukaryotes.

We were interested in the following information, in terms of the possible participation of histones as an intracellular operational amplifier (Alberts et al., 2002; Zhimulev, 2003; Chentsov, 2004; Teif et al., 2011). Histones are intranuclear proteins, which are connected with the DNA, and form the filamentous complex molecules of deoxyribonucleoproteins (DNP). In eukaryotes, they are represented by only 5-7 types of molecules. The ratio of DNA and histone in the nucleic-histone complexes is 1:1. The species and tissues specificity of these molecules is weakly expressed. Histone proteins have a small molecular mass and are characterized by the pronounced basic properties. Positively charged amino groups of lysine and arginine at their N- and C-ends determine the positive charge of these proteins. A positive charge helps the histones connect directly with the negatively charged phosphate groups of DNA. Their synthesis in the cytoplasm and DNA in the nucleus occurs simultaneously. Histone molecules after inclusion in DNP are characterized by a low replacement rate. DNA in the nucleus is packed in the nucleosomes, and their sequence forms a nucleosome filament. Each nucleosome consists of an octamer of histones, around which the DNA makes 1.67 turns of a double helix. Nucleosomes are joined together by a segment of linker DNA. The wide distribution of this group of proteins, the obligatory presence in the composition of chromosomes, and the interspecific similarity of their primary molecular structure testify to the important role of histones in the defined vital functions of eukaryotic cells.

It is considered that the main function of histones is the participation in the packaging of DNA and the regulation of chromatin activity during transcription. We have assumed in terms of the information system that the histones are used in the channels of preliminary analog signal processing for providing of amplification of the species informational signal.

Let's pretend that the nucleosomes, distributed along the DNP strand in the chromosomes is a system of bipolar transistors in combination with the isolation of its individual elements. The differential inputs of each amplifier element (i.e., nucleosome) consist of two inputs: V_+ and V_-. In each nucleosome, the V_+ inputs are created by the positively charged histone molecules, and V_- – by the negatively charged phosphate groups of DNA. In this situation, the informational signal at the input is fed to the positive non-inverted chromosome output (+ charged lysine and arginine groups). Since the DNA molecule is characterized by a high resistance, only the amplitude of species signal changes at the output. Isolation of nucleosomes (or separate bipolar amplifiers) within each chromosome is provided by the DNA linker sites. In this case, a lot of nucleosomes can be represented as a system of bipolar transistors, each of which has a high gain.

Thus, a B class object of eukaryotes has two types of functional amplifiers. 1)The preamplifier is the structural element of the eukaryotic antenna array. It is located in the immediate area to the signal source, which is needed for the transmission of information without deformations and noises. The preamplifier is intended to strengthen a weak cosmic signal to the level that for the normal operation of the terminal amplifier is necessary. It acts in the moment of fixation of the new informational signal on the DNA-wave, once in the life cycle during meiosis. 2) A terminal amplifier is the typical power amplifier of the informational signal that is a main structural chemical element of all eukaryotic chromosome-particle. The terminal amplifier is intended to strengthen the oscillation power of a modulated DNA-wave (DNA-information with a narrow spectrum of frequencies) during each cell division and throughout all ontogenesis of the plant.

C Class Object

The plant world are characterized by a great number of species. The attempts explain the species heterogeneity by means of information theory have a long history (Shannon, 1949; Ashby, 1950; Wiener, 1983; Korogodin & Korogodina, 2000; Thilms. 2012; etc.). In our model, the informational system has the terrestrial and cosmic origin and consists of three basic elements – the objects of the A, B and C classes. We examined two of them. It should be noted that A and B class objects of eukaryotic cells and radio engineering systems are rather similar, despite the known specificity and differences.

Class C objects are the objects for which this species cosmic information is needed. It is the eukaryotic cell and eukaryotic organism (unicellular, multicellular), where this information is decoded and realized in the form of species-specific plants and animals. The search for mechanisms of decoding information begin from the hypothesis about embryonic fields.

The concept of the biological field and its own field of cells was first proposed by Gurvich (1944). He theoretically substantiated that the source of the biofield must be in the nucleus. It has a vector character, a radial form, and a finite attenuation decrement. The cellular field, which decreases as $1/r^2$, acts both inside the cell and outside of it. Its forces exert an effect on the intracellular structures and molecules, giving them a certain mutual arrangement. The fields of cells are superimposed according to the rules of vector addition, causing the movement of nearby cells.

If we proceed from the two-plane corpuscular-wave nature of a cell, then the source of the biofield is also inside the nucleus (Kuznetsova, 2013). It is a molecule of DNA (particle and wave) whose nucleotides have a dipolar structure. The charge motion during nucleotides oscillation together with a DNA-wave inside the nucleus generates an electromagnetic field, an intensity of which drops off at a rate of $1/r^2$. In turn, the intensity and extent of the biofield depends on the energetic potential of a biological harmonic oscillator. Its maximal species-specific level is in the generative cells (meiospore, zygote, initial sporophyte), and the minimal level - in the differentiated aging and apoptotic cells.

In the three-plane model, the harmonic oscillations of a carrier DNA-wave (DNA-energy), with a fixed frequency and amplitude of oscillations (Figure 26a) creates a biofield, which is divided on the close-range zone inside the cell (changing as $1/r^2$, $1/r^3$, etc.) and far-range zone beyond it (changing as $1/r$) and is homogenous within its concentric layers. The modulated DNA-wave (DNA-information), with a different frequency and amplitude of oscillations (Figure 26c, d), generates a heterogeneous informational field inside and outside the cell. These two fields are, according to Weiss's terminology, morphogenetic fields. This field dropping with a distance proportional to $1/r^n$, $1/r^{n-1}$ ... $1/r$ and distribution of the electromagnetic field in the zone of action is heterogeneous. Its distribution has a certain genetically determined order, embedded in the DNA-information of male and female gametes, and then in the informational matrix of a differentiated embryo and then a plant. In turn, the initial orderliness of the intracellular structure and functions originates from the information encoded in a species cosmic signal.

The morphogenetic field is an information and energetic matrix together, in accordance with which the internal space of each cell is organized. It has an electromagnetic nature and interacts with charged ions, and molecules of a dipolar structure. In accordance with the size and charge, their spatial position inside the nucleus and cytoplasm, or outside the cell are determined. After each division, the DNA-information and DNA-energy of each daughter cell create again intracellular ordering in accordance with the energy potential and actual information embedded in their genome.

As you know, the higher the degree of order and complexity of systems, the more energy is needed to achieve and maintain. The eukaryotic cell has a very complex organization. The energetic, corpuscular and informational components of the genome are interrelated with each other by the trinity principle of energy, matter and information. The origination of a new type of receiving antenna (eukaryotic ring antenna array) and an increase in the capacity of a biological harmonic oscillator substantially changed the corpuscular, energetic and informational substance of eukaryotic cells. The corpuscular body of a eukaryotic cell differs from prokaryotic ones by size, shape, and internal organization. The shape varies depending on the function and the place that the cell occupies into the plant. They by the shape can be very diverse: Rounded, elongated, isodiametric, fusiform, etc. A eukaryotic cell depending on the energetic state can be in different vital states: a proliferating, differentiated, equilibrium and dead cell. The informational component determines the morphological specificity of a meristematic, parenchyma, conductive and other cells and tissues.

Since the prokaryotic and eukaryotic cells differ by the degree of complexity and functions, this affects their ultrastructure and the connection of intracellular processes with intracellular energetic channels. All eukaryotes are aerobic organisms. A hereditary apparatus (DNA-particle, DNA-energy, DNA-information) of a eukaryotic cell is located inside the nucleus, which is separated from the cytoplasm by a double membrane. The cytoplasm is divided by a system of intracellular membranes into compartments, which are also spatially structured. These membrane systems allow them to unite and to separate the energetic channels and metabolic reactions into the multi-stage conjugate processes. The cytoplasm of the plant cell contains two types of specialized organelles (mitochondria, chloroplasts), where complex oxidation-reduction reactions are carried out to convert the energy of light and energy of organic substances into the chemical energy of ATP. Endoplasmic reticulum, Golgi apparatus, lysosomes, and vacuoles, which are separated from the cytoplasm by elementary membranes take part in the processes of intracellular synthesis, decay and transport.

The objects of classes B and C during the evolution of plant information systems have passed 3 stages of development.

One-Step Mechanism during Meiosis

As mentioned above, the DNA-wave transition to higher energetic levels takes place during sexual reproduction, and loss of biological energy occurs during ontogenesis. The last stage of the reproductive cycle of lower plants with a haplontic life cycle is meiosis (Table 1). The reception of the species cosmic signal happens in the last stage of the reproductive cycle, in the time of the reduction division of diploid zygote. Accordingly, all reproductive structures with generative cells before meiosis are carried vertically upward, and should be in direct contact with the external environment. The height of the antenna installation has a great influence on the efficiency of its operation.

Meiosis is a very important stage in the life cycle. This is a unique process from the standpoint of biology and physics. The start time and its duration are closely connected with the biology of the species and the habitat conditions. The meiotic division is the most ecological sensitive stage of the reproductive cycle, because the generative cells for better reception of the informational signal must be in direct contact with the environment. Genotype-environment selection of the most adaptive genotypes takes place simultaneously with meiotic division. Any fluctuations of environmental factors lead to an increase of the number of chromosomal aberrations, different changes of the informational signal, loss of viability and mass death of gametes. Therefore, the DNA-energy, DNA-particle and DNA-information of the surviving male and female gametes carry the information about the current state of external environment. This question will be discussed in more detail in Part II, Chapter 2.

The haploid generation predominates in the lower plants with the haplontic life cycle. The single diploid cell in their life cycle is the zygote. In the mosses, the diploid generation exists, but it is more poorly developed (sporangium with spores). Genome of the first vegetative cell (zoospore, microspore, megaspore, etc.), which was formed during meiosis, can be characterized with the 4 basic states: DNA-particle, DNA-energy, DNA-information and spin direction. During ontogenesis, all vegetative cells of algae and mosses receive the same haploid DNA-particle. DNA-energy in accordance with the 2nd law of thermodynamics decreases from the maximal energetic level of meiospore to zero. DNA-information, as the information of the whole, passes regular age changes and is realized in species-specific plants of algae and mosses.

Two-Step Mechanism during Meiosis and Fertilization

The last generative cell in higher plants with a diplontic life cycle is the initial sporophyte. This is the zygote (Phaeophyceae, Angiospermae) or 4 cells of an embryonic tier of secondary proembryos (Gymnospermae). Male and female meiosis and such phenomena as the change of a DNA-particle, the reception of a new cosmic signal and the gamete differentiation by the spin direction take place in the first half of a reproductive cycle. Then, the male and female gametes go to the further formation of a genotype of the future plant. The development of microspores and megaspores proceeds by the different paths up to fertilization. Therefore, the formation of the corpuscular, informational and spin vector components of the diploid genome occurs in two stages – during meiosis and fertilization. A supplement of biological harmonic oscillator by the energy by means of the transition of a DNA-wave to higher levels takes place on the 3-5 activation stages by the male and female generative line.

After meiosis, each microspore and each megaspore are the bearers of individual genetic information, one of the partners in the fertilization process, and act within the haploid intact cell system. In the course of microgame to phytogenesis inside an anther, their energetic state is again increased. The pollen grain during maturation acquires a species-specific shape and size in accordance with the DNA-information embedded in its genome. There are no genetically identical gametes among them. The pollen grains from different systematic groups of higher plants differ in size, morphology, pollen envelope structure (pores, furrows, slits, etc.), propagation manner, and mechanisms of intraspecific and interspecific incompatibility (Knox, 1990). 4 haploid microspores originating from one microsporocyte after male meiosis have the same DNA-energy, a same DNA-information, a different DNA-particle, two gametes with spin up, and two gametes with spin down directions. The population gene pool of microspores, in the formation of which many different trees participate, distinguishes by the DNA-energy, DNA-particle, DNA-information, and spin direction (with gametes ratio 1: 1).

As mentioned above all higher plants reproducing sexually receive a specific cosmic signal only once in a life cycle – during meiosis. Then, the development of microspores and megaspores is continuing by the male and female generative line and their initial informational signals are transformed. The degree of its disturbance depends on the violation level of genetic material of the mother plants, and the adverse weather conditions during meiosis, pollination and fertilization. Most of the genetic violations are removed before fertilization. Other gametes carry an informational signal, which is something differed from the initial cosmic.

The syngamy of male and female gametes occurs during fertilization. This is a complex corpuscular-wave process, which leads to a change in the four basic states of the diploid initial cell of the future plant: DNA-energy, DNA-particle, DNA-information, and spin direction. The stability of the diploid state of a genome is ensured by a very firm spin-spin coupling that forms between the homologous chromosomes of the sperm and ovicell with opposite spins in the certain physical conditions inside the ovicell (Part II, Chapter 3).

The fertilization between two sperms or between two ovicells does not exist. The reason is that their combination, by the Pauli exclusionary principle, does not give a bound state. Conversely, the forces acting between the chromosomal sets of an ovicell and sperm, when the spins are opposite, lead to a strong mutual attraction that connects their homologous chromosomes together. The process of spin-spin coupling formation requires a certain amount of time. Therefore, the period between syngamy gametes and first mitotic division of initial sporophytes (zygotes or 4 embryonic cells of secondary proembryo) has an unequal duration in different species. This is the time when a veritable diploid genome arises. This is the most stable and reliable type of organization of the genetic material of biological systems.

Three-Step Mechanism during Meiosis, Fertilization and Embryogenesis

At this stage of evolutionary development of the information system, the creation of an information matrix is added to the above-described mechanisms by the formation of DNA-information.

The seed plants have a well-developed body, which is characterized by the same tissue systems (Ezau, 1980). The seed is the main organ of sexual reproduction. The seed development occurs on the mother plant during embryogenesis. The formation of a three-plane matrix or the corpuscular, energetic and informational plan of the internal space organization of the body of a future plant takes place in the process of fertilization and embryogenesis.

In accordance with the three-plane matrix, all somatic cells of each plant have the same diploid DNA-particle. It is formed either during fertilization (Phaeophyceae, Angiospermae) or during pro-embryogenesis (Gymnospermae) by forming spin-spin coupling between homologous chromosomes in the first case in the nucleus of the zygote, and in the second in the cells of the embryonic tier of a 16-cellular proembryo in the period of their dormancy. The species-specific corpuscular matrix of a plant does not change during ontogenesis.

All plants have the same energetic matrix or the order (plan) of internal energy distribution within the plant body, which does not change during ontogenesis. These are three zones of energetically strong cells (apical shoot meristem, apical root meristem and lateral stem meristem). The energetic matrix is superimposed on the corpuscular matrix, and is formed during early embryogenesis. The process of creating an energetic matrix is described in detail in Part II, Chapter 3. The essence of it includes the first two divisions of embryogenesis, which are described as activating mitosis. As a consequence, the daughter cells that are different by the level of biological energy arise. An apical cell on the halazal pole of a future embryo gives rise to a multicellular embryo, within which two complexes of energetically strong cells are located – a shoot apical meristem and a procambium. The basal cell on a proximal pole of an embryonic complex continues to divide by means of activating mitosis, forming the third pole of energetically strong cells, or an apical root meristem.

The informational matrix is superimposed on the corpuscular and energetic matrix. Probably, the information signal of a zygote, which was formed after fertilization, is transformed during late embryogenesis. This process affects the groups of embryonic cells differently, depending on their location within the undifferentiated embryo. This is due to the fact that the embryonic chamber, where it develops, has an elongated form, and its volume has been fixed by the seed size. The transverse mechanical compression of an embryo leads to a change in the cell shape, the direction of their division, the order of location, and causes the same type sustainable alterations of the information signal in the same groups of cells. The central cells of an embryo acquire a narrow elongated shape, forming a hypocotyl-root axis. One or more cotyledons develop at its upper pole. In the mature embryo, the location of tissues and the cell cytological characteristics of embryonic complex are similar to the main tissues of seedling.

The totipotency loss by the specialized cells, the formation of the same tissue system and the strict sequence of histogenesis processes after the transition from the embryo to the adult plant indicate that the informational signal of the DNA-wave irreversibly changes during informational matrix formation. Hence it follows that the informational matrix is a cast of the location of tissues in the body of a future plant. It is laid in the form of a certain set of sustainable modificational changes of the DNA-wave and is anchoring certain distribution of their action zones within an embryonic complex (Part II, Chapter 3).

All species of plants are adapted to the specificity of their ecological niche according to the principle of "species-environment" and "genotype-environment" interaction. Their morphogenesis is inherently similar. An adult plant presents a complexly organized whole, consisting of different interrelated and interdependent organs and tissues. Moreover, each organ and tissue in accordance with the informational matrix occupies its own, strictly defined place in its body and has the species-specific morphology and organization of the internal structure, which are fully adapted to the immobile lifestyle of plants. An important role belongs to the meristems, a cell of which preserves the totipotency and functional activity throughout its life. Meristematic zones, the energy potential of which decreases more slowly than the whole plant, allow us to add new branches, roots, leaves, increase the volume of trunk in proportion to volume of the crown on the basis of individual and species-specific information, laid into the energetic, corpuscular and informational program of genome and a three-plane matrix.

The three-plane corpuscular-wave matrix is a progressive evolutionary trait of higher plants. In the Cenozoic era, the seed plants have achieved very high species diversity, the level of internal and external differentiation, and specialization of a sporophyte. They are characterized by the highest degree of organization. Their body is divided into different vegetative organs – a root, a stem and a leaf, which are inherent to the same tissue systems. In turn, these organs reveal the unity and interdependence in the course of plant development and functioning. The organization degree of any system is determined by the distribution and redistribution of corpuscular-energetic streams, which are provided by the informational streams.

Two-Step Informational System without Meiosis and Fertilization

This type of reproductive process is inherent in the evolutionary progressive species of angiosperms, and it is known as the apomixes phenomenon. The formation of the informational system is currently taking place and it is possible to trace the stages of given evolutionary transition. Their main traits are as follows: Meiosis and fertilization are absent. A diplontic type of life cycle with the asexual reproduction and ontogenesis, in which only diploid generative cells and plants participate occurs. Reproductive development occurs on the sporophyte. The main organ of reproduction is the seed. The energy protection system is the double protection way. The genome of all cellular systems is the diploid asexual ones are included. Within the individual life of a plant, it has only three basic states: The ordered variable DNA-energy, DNA-particle and DNA-information. The creation of energetic, corpuscular and informational programs of plant development takes place on the basis of a species reproductive cycle. The formation of an energetic and informational matrix of the plant internal organization is carried out during embryogenesis.

The reproductive cycle of apomikty consists only of four stages (Table 2). Three of them are activation stages (ᛚ); they track each other and are strictly oriented with respect to the gravity direction. The DNA-energy transition to the higher energetic level occurs on these stages without energetic losses. DNA-information is formed in the first two activation stages. In megasporangiogenesis, the old informational signal is erased or significantly weakened (1stᛚ). During megasporogenesis, the eukaryotic antenna array is received as a species cosmic signal in the time of activating mitosis (2ndᛚ instead meiosis) and a new DNA-particle is formed during mitotic crossing-over. An energetic matrix is created in the early embryogenesis; an informational matrix develops after it occurs during the course of late embryogenesis.

Chromosomes during activating mitosis and the first reduction division of female meiosis have the identical orientation with respect to the gravity direction. Nevertheless, the internal arrangement of antenna elements (homologous chromosomes) in them on the eukaryotic antenna array is different. In the activating mitosis standing waves are not destroyed and the subsequent processes, which lead to the formation of bivalents do not occur. In both cases, homologous chromosomes are located nearby. No less important is the capacity of the biological harmonic oscillator. Based on the general evolutionary trend, it is higher in the apomicts species than in the angiosperms. This increases the sensitivity of their eukaryotic antenna array to the field. Probably, it allows the receiving antenna to catch the new species signals from the stream of bioactive Z-radiation. Secondly, in the time of activating mitosis, a reception, amplification of informational signal and mitotic crossing-over or an internal exchange of nucleotides and chromatid sites occurs between the homologous chromosomes (as the same elements of ring antenna array and the same particles). At present, the phenomenon of mitotic crossing-over is found in many species of plants and animals and it is widespread in nature (Carlson, 1974; Inge-Vechtomov, 2010; Hartl & Ruvolo, 2012, etc.).

Summary

In this work, we tried to reduce the diversity of phenomena and processes to their internal unity, as the forms of manifestation of the general laws. Plants can be characterized as open and closed three-plane quantum-mechanical systems, in which the DNA-particle, DNA-energy, DNA-information, spin vector and all corpuscular-wave processes on their basis are ordered and organized in space and in time processes, which are implemented in accordance with the fundamental laws of physics. If during ontogenesis the DNA-particle does not change, then the DNA-energy and DNA-information represent a set of successive age-related states, with corresponding external and internal forms of their manifestations. Energetic, corpuscular and informational components of the genome are linked together by the principle of trinity of energy, matter and information. DNA-particle and DNA-energy have an earthly origin, and DNA-information is an earthly and cosmic nature. Life as a cosmic organization, as a cosmic (a general) and an earthly (a private) phenomenon has been substantiated earlier in the works of Vernadsky (1967), Yugay (1976), etc.

To summarize the above, it should be noted that in Part I, the theoretical aspects of the three-plane corpuscular-wave concept of the organization of cells and plants are described. So that the question is logical, how do we know that a DNA-wave exists? The direct evidence of a DNA-wave nature is lacking. Therefore, we used a widespread scientific approach as arguments to prove the contrary. It turned out that all the factual material of cell biology and plant development biology entirely fits into the three-plane corpuscular-wave concept, explains the nature of many phenomena, and does not contradict the fundamental laws of nature. Proceeding from this method of evidence, all biological phenomena and processes turn out to be as we suspected and what they should be if the DNA-wave, biological harmonic oscillator, and DNA-information really existed. In Part II, the three-plane corpuscular-wave model of a plant life cycle is described in detail on the example of forest tree plants.

Part II. Three-Plane Corpuscular-Wave Model of Life Cycle of the Forest Tree Species

Chapter 3

INITIATION OF REPRODUCTIVE PROCESS AND INCEPTION OF MALE AND FEMALE GENERATIVE ORGANS

The occurrence of bipolar sexual differentiation in the ontogenesis, stages and mechanisms of gamete development by the male and female generative lines, which leads to the formation of a gene pool of the ovicells and pollen grains, are still actual problems for the developmental biology of animals and plants. The data of classic embryology, cytoembryology and cytogenetics of coniferous and *Pinus sylvestris* was described in the works of Schnarf (1933), Chamberlain (1935), Singh (1978), Kozubov et al. (1982), Nekrasova (1983); Trenin (1986, 1988); Tretyakova (1990), Owens & Bruns (2000); Belousov (2005); Williams (2009), etc. Reproductive development occurs on the basis of the species-specific matrix of pine reproductive cycle, a construction of which is the genetically determined and has a double corpuscular-wave nature (Kuznetsova, 2002, 2013).

The process of "biological assemblage" of the genotypes of future plants is ends with the creation of three individual species-specific genetic programs (the corpuscular on the basis of the DNA-particle, the energetic on the basis of the set of energy levels of the DNA-wave and the informational on the basis of the modulated DNA-wave). The success of reproductive development depends on many internal and external factors such as weather conditions, genetic anomalies, pollen concentration in the air, etc. (Sarvas, 1962; Kozubov, 1974; Necrasova, 1983; Williams, 2009; Efimov, 2010; Svintsova at al., 2014; Kuznetsova, 2015; etc.).

Initiation of Reproductive Development

The transition of a plant from the vegetative to the generative path of development is associated with the activity of the apical shoot meristem. This is a complex genetically determined intrarenal process of transformation of vegetative cells into generative cells. Every year, the shoot apex of a tree performs the same cycle of processes as differentiation, histogenesis and organogenesis. This cycle involves the successive formation of metamers of the new central bud, inside of which the rudiments of a future shoot with a leafy primordium are developed from the apical meristem. Needle primordiums have the same size and are arranged in the right geometric order.

A comparative study of vegetative shoot apexes of coniferous plants revealed the similarity of their inner structure (shape, size and ultrastructure of cells), the types of cytological zonality (rhythm and direction of mitoses) and the localization of zone boundaries (Singh, 1978, Owens, & Molder 1979, Kozubov et al., 1982, Skupchenko, 2004). The layer of apical initial cells is located in the terminal part of the apex. The shoot apical meristem is always protected from the environment by the covering scales of buds. At the beginning of the vegetative season, it is located inside the bud of the previous generation, and after its budbreak, the apical initials are under the covers of the new bud. The main type of division of the apical initial cells is asymmetric mitosis and, once in the vegetative season, usual mitosis occurs. This mitosis leads to an increase of meristematic tissue volume and a decrease in the DNA-energy of apical initial cells by one energetic level. Asymmetric mitosis ends with the formation of one daughter cell, which is identical with the mother cell by the DNA-energy, DNA-particle and DNA-information. After division, it grows and becomes an apical initial. The second daughter cell, whose energetic level has become lower by two levels, develops along the path of differentiation and specialization. It becomes an initial cell of the 2nd order, giving rise to the zone of the central maternal cells.

In the apical bud, the zone of the central maternal cells is located under a layer of initial cells. It consists of 2-6 layers of large meristematic cells, with large rounded nuclei. The development of these cells follows on the descending energy vector of several cell lines associated with different processes of differentiation, histogenesis and organogenesis. The central part of the zone of mother cells, the polar axis of which coincides with the gravity direction, forms the core meristem and the core of the future shoot. The central part of the zone of mother cells, the polar axis of which coincides with the gravity direction, forms the meristematic tissue of the core and the core of the future shoot. In the peripheral zone of central maternal cells, where, apparently, there exists a spatial instability, the several groups of cells originate, the polar axis of which deviates from the gravity direction. They begin to divide in different directions and form the primordium, or the several relatively independent local complexes of totipotent cells within the peripheral meristem. Primordium grows quickly to a certain size. Their development leads to the formation of lateral outgrowths or the rudiments of covering scales of future vegetative (generative) buds.

Lateral vegetative and generative buds are initially laid in the form of several hemispherical protuberances along the sides of the new central bud. Their development begins with the emergence of the primordium of the future lateral branch, micro- or megastrobile. Vegetative primordium from generative ones at this stage cannot be distinguished. It is necessary to take into account that the process of its formation is associated with a decrease in the energetic potential of the meristematic cells of each cellular generation by the number of mitotic divisions. Hence, it follows that the DNA-energy level of apical initials of the lateral branches is always lower than at the apical meristem of the central shoot.

Changes in cytological zones in the transition from the vegetative apex to generative ones are similar in different species (Kozubov et al., 1982). Therefore, their development during the formation of lateral branches, microstrobiles and megastrobiles can be described as the next intrarenal processes.

The undifferentiated meristematic cells of generative primordium have the same DNA-energy, DNA-particle, DNA-information and the great potential for division and growth. However, their development takes place inside the buds of the previous generation, the volume of which is fixed. Its dense scales initially restrain, and then stop the primordium growth in width and in height. The definition of the further path of its development is apparently due to a number of reasons. Firstly, the sizes of generative primordium during undifferentiated growth increases more rapidly in width than in height and its top becomes more flat. Secondly, the initials of the future vegetative buds become the group of meristematic cells; and the initials of the generative buds are only a few cells (1-4), the polar axis and mitotic spindle of which coincide with the direction of gravity.

As mentioned above, the apical initials are divided only by anticlinal walls, i.e., by means of asymmetric mitosis. The group of initials of a less flat vegetative primordium can be represented as a new layer of meristematic cells consisting of identically oriented totipotent mother cells. With their division begins the process formation of a new element of the tree crown – the lateral branches.

During division in a fixed volume of vegetative buds, the layer of apical initials, as one whole begin to be pushed out by the underlying primordial cells from the inside, and squeezed by the adjacent cells from the sides. Therefore, after asymmetric mitosis, a group of daughter cells of the apical pole moves outward and the daughter cells of the basal pole remain on the surface layer of the primordium. This ensures the active uptake of the entire group of apical initials. These cells become the apical meristem of new lateral branch. They will conserve the energetic status and totipotent state of meristematic cells of their primordia and the growth of the bud upward during asymmetric mitosis. Soon after leaving the buds, a stem begins to lengthen, leaves unfold and a new lateral branch is formed.

Only 1-4 vegetative cells of a more flat surface layer of generative primordium, the polar axis of which coincides with the gravity direction, become the reproductive initials. In order to become a reproductive initial, they must satisfy one more requirement. Each reproductive initial with respect to the next lower layer of the primordium should be located at the junction of several of its cells. During division in an unchanged volume of the generative bud, single initial cells undergo pressure from the surrounding primordial cells, which fix their position within the surface layer. The first division of reproductive initial is axial transverse division. After the first division of the reproductive initial, the basal daughter cell remains in the surface layer, and the apical cell is pushing the underlying cells and passes between them inside the primordial tissue.

Generative primordium has an oval shape. It consists of undifferentiated, weakly vacuolated and closely pressed to other meristematic cells. Primordial cells are totipotent, i.e., contain the same DNA-energy, DNA-particle and DNA-information. In relation to the apical meristem, their genome has a DNA-particle and DNA-information of the mother plant, and the level of DNA-energy is lower. The polar axes of primordial cells are at different angles towards the gravity direction.

The reproductive initial is a vegetative cell. Its apical pole in connection with the transverse division and cell growth should move only inward towards the primordial tissue. Therefore, the dividing initials are wedged and move apart the underlying cells. In this case, this part of the cell is strongly compressed. A significant gradient of intracellular pressure between the apical and basal pole along the polar axis of the mother cell arises at this moment. As a consequence, the necessary physical conditions for the activating mitosis are created. The daughter cell of the apical pole in the course of this mitosis receives a portion of energy, and the cell of the basal pole loses it. Then, the apical cell is pushed inside the primordial tissue. Vegetative cells after transition to higher energy levels rejuvenate, acquire an energy advantage over nearby cells and become the first generative cells.

As already mentioned (Part I, Chapter 2), the DNA-information of the apical meristem of a tree combines noises with the species-specific signal arising from external and internal disturbances, which distort it during long ontogenesis. Probably, at the beginning of each reproductive cycle there is one more important event–a decrease in the level of noises of the reproductive initials. Reduction of chromatin compaction level in the nucleus of microsporocytes and megasporocytes testifies to this (Konar & Moitra, 1980; Trenin, 1988). Apparently, the attenuation of a species signal and the level of ontogenetic distortions at the transverse compression of the nucleus and the increase of its volume during and after division take place. Transformation of maternal DNA-information in the genome of microsporocytes and megasporocytes is carried out during three and two activating mitosis, respectively.

This is the turning point from vegetative to generative development, and the beginning of the development of a population gene pool and life cycle of plants of the next generation. The life cycle of all forest tree plants includes four periods. During the first three periods of sexual reproduction (initiation, gametogenesis and embryogenesis), the energetic, corpuscular and informational programs of tree development and a three component plan of its internal organization are laid. The last period is the longest and is associated with their realization in the species-specific plants and population ecosystems during ontogenesis. The main events of the 1st period are the initiation of reproductive process, the transition DNA-energy from the descending to the ascending principle of motion and the change of energetic and informational basic states of a genome of generative cells.

SPACE ORGANIZATION AND AGE-RELATED CHANGES OF TREE CROWN

The 1st period of the life cycle included the processes of inception of the male and female generative organs and intrarenal stages of initiation and development of generative cells of a new generation. We think that the main task of this period is also the emergence of a great number male and female generative buds on the tree, as the material basis for the future restoration of the genetic diversity of the future population gene pool. Sex differentiation of male and female generative organs begins from the construction of the tree crown, which has two types of generative apexes – the male and the female. It is considered that the sex differences are inherent in the nature of their meristematic cells, and are manifested in the morphology of branches.

As was mentioned above a reproductive cycle of all species of plants has the species-specific structure, and is divided between the male and female generative lines. Origination and formation of male and female gametes is a strictly genetically determined process. Furthermore, the spatial position of stages and generative cells relative to each other and the direction of planes of cellular divisions have a certain orientation with respect to gravity. The accidental cells and cell (or nuclear) divisions in the reproductive cycle are absent. The rhythms of the development of the male and female generative sphere do not coincide with each other. Asynchrony is manifested in the distribution of the generative organs in the tree crown at the time of their occurrence, morphology and duration of development. It is a species-specific process, which we will describe in an example of conifers and above all one species – *Pinus sylvestris* L.

Scots pine comes into a phase of sex reproduction in 9-10 years. Initially the young plants form the only female generative organs and after 1-2 years begin to appear male reproductive structures. After fruiting beginning, the seed production of trees continues until old age. Usually three types of shoots – the male, the female and the growth (or the vegetative, asexual) are met in the tree crown of all monoecious bisexual plants. Their asexual development begins with the inception of the terminal and lateral buds. The next year, the growth of terminal and lateral shoots and the formation of terminal and lateral buds occur. This cycle of branch development is reproduced annually. After fruiting beginning, the active growth of the trees in height continues (Vorobiev et al., 1989). With age, the intensity of their growth gradually decreases. Between their own age and the growth of branches of the 1st order, the feedback is also confirmed (Powell, 1977).

Doak (1935) gave a detailed description of the morphogenesis of shoots in the pine. He showed that the formation of two types of shoots – the elongated and the shortened – is a hallmark of the beginning of the sex specialization and studied an unusual phenomenon in the Scot pine – the morphogenesis of the shoots of a mixed type, i.e., the shoots, which form the reproductive organs of both sexes. In bisexual shortened shoots, microstrobiles occupy the lower part of the axis; a zone with the needles is located is then located, and megastrobiles have a distal position before the shoot growing point. Later, this scheme was defined more precisely (Minina & Larionova, 1979). They described the morphogenesis of the pine for the asexual, female and male types of branches.

Shoots of different sexual types are also definitely distributed in the crown of the trees. A crown, depending on the shoot placement, is divided into four tiers: Growth, female, mixed and male. The growth shoots deprived reproductive organs are in the uppermost part of the crown. The female generative organs are located closer to the upper tier, the mixed in the central part, and the male in the lower tier of the tree crown. All newly formed branches of the 1st order are initially asexual. It is important to note that the female buds have the same spatial position, so the megastrobiles carrying the shoots grow only vertically upwards. Unlike female buds, the male buds occupy the different spatial position inside the tree crown (vertical, horizontal, oblique). They can be located in the same direction with the bearing branch, perpendicular, or at an angle to it (Figure 27-a-f). The polar axis of the bud and its microstrobile have the same disposition, which often do not coincide with the gravity direction.

With the age of the tree, the length and mass of female shoots decrease (Vorobiev et al., 1989). The percentage of female shoots in its crown is becoming smaller, and the stabilization of sexual processes takes place. For example, in a 40-year-old cedar tree, their number was 97.8%, and for a 200-year-old tree only 37.5%. Preservation of cones during their development in young trees is less than 50%and more than 75% in older trees (Nekrasova, 1983).

In the male bud, the zone with the primordium of microstrobiles is in its foundation, and afterward a primordium of brachyblasts arises (Nekrasova, 1983). The following spring from these buds will grow new shoots, in the ground of which are located the microstrobiles, and in the upper part are shortened shoots with needles. Microstrobiles are laid on one-year shoots of the current year on the branches of the 2nd and 3rd order. The maximal intensity of their growth is observed at the beginning of their development, and then the rate of growth of branches decreases (Vorobiev et al., 1989). Usually, male shoots on the branches are formed annually. With the age of the tree, the level of male sexualization reaches a maximum. For the last 2-4 years sexual identity can be identified by the bare areas on the branch (Figure 27a).

We think that the differentiation in primordium of growth shoots, megastobiles and microstrobiles are determined at a different level of biological energy (Kuznetsova, 2013) before their morphogenesis. Apical meristems of the top growth buds of asexual shoots has the highest energy potential. Further along the gradient, there should be a tier of female floral meristems with the female buds on the shoots of the 1st order, and even lower is a tier of the male floral meristem with the male buds on the shortened shoots of the 2nd and 3rd order.

a-b c-d e-f

Figure 27. Vertical (a, b) and horizontal (c) disposition of the buds and microstrobiles of Pinus sylvestris.

Characteristic features of sexual reproduction are: Firstly, the reproductive processes have a certain orientation with respect to gravity; secondly, the connected and independent development of the generative (male and female) and vegetative sphere with each other; and thirdly, their relationship with habitat conditions. In July and August, the initiation of reproductive development begins from the inception of the trees a large number of male and female generative buds. Formation of megastrobiles in the apex is observed later than the microstrobiles (Minina & Larionova, 1979). The difference between the dates of occurrence of these reproductive structures is a week or longer, depending on the habitat and weather conditions.

In pines, a process of formation of the male and female gametes occurs on two generative lines – a male and a female, and continues during 3 vegetative seasons. The following occurs on the male generative lines: 1) From microstrobile inception to pollination,≈ 10 months (July - May) is necessary; and 2) from pollen germination to fertilization – ≈ 11 months (May - June). Female cones are characterized by a two-year development cycles. Therefore, concerning the female generative lines: From megastrobile inception to fertilization, it takes ≈ 22 months (August – June next year). All process of seed formation (from inception of primary generative cells up to seed fly out) lasts about 2.8-2.9 months.

According to literature data, there is a direct correlation between the number of male microstrobiles and weather conditions: The higher the rainfall, the lower the relative humidity and the air temperature from July – August, the greater the amount of microstrobiles will be formed in the next year (Efimov & Chertov, 1976; Efimov, 2010). It is shown that the correlation coefficient between the intensity of seed-bearing and the moisture deficitis0.690 (Girgidov, 1973). However, these data were not confirmed by our research (Svintsova et al., 2014).

In 2010, the inception of the pine reproductive organs occurred under the conditions of severe drought in the history of meteorological observations. In July, less than half of the norm of rainfall was experienced and the average monthly temperature rose to 26.4°C; that is 6.7°C higher than regional norms. Within 4 weeks of August, the amount of precipitation lowered to 7% of the monthly norm; divergence of the average daily temperature was7.8°C (the drought ended on August 28th). However, in the spring of the next year, the number and morphology of microstrobiles and the pollen abundance and its viability in the Voronezh region did not visually differ from optimal years. The viability of pollen by the results of the germination at the standard temperature of 27°C is high and varies within the limits of 82.6 – 96.1%. In the absence of significant differences between the trees, the low intra-population variability of the number of germinated pollen grains is revealed, and the coefficient of the variation is4.1%. Intensive growth of pollen tubes is observed in almost all trees.

Inception of generative organs and the initiation of reproductive development occur within the closed buds. Therefore, the generative organs and reproductive structures during intrarenal development are separated from the external environment by the dense scales. Outside, the buds are covered with a thin layer of white resin. Apparently, the stability of the genetic material is partially provided by a needles bunch, which creates shade and protects the generative buds from excessive heat and mechanical damage. Therefore, the inception and intrarenal development of male and female generative organs proceeds in the relatively normalized microclimatic conditions. Figure 27-a shows the needle orientation on shortened male shoots during initiation of reproductive processes (the end of July– the beginning of August).

As known, a generative apex of all conifers arises on the basis of the vegetative apex as a result of the transition of the vegetative meristem to the floral once (Singh, 1978; Kozubov et al., 1982). The early stages of the development of generative apexes are under strict genetic control and at the same time an environmentally-dependent process. The apexes, which will give rise to the microstrobiles from the apexes of the future megastrobiles, are very difficult to distinguish. If we proceed from the corpuscular-wave nature of DNA, the ontogeny of the crown can be represented as an ordered process of biological energy change – from the level of an apical meristem of the axial top bud to the level of the meristem of the lower branches according to the downstream energy vector. Apparently, in the life cycle of each plant there are two sets of DNA-energy levels: From the one set of levels, a transition of the vegetative cell to the reproductive development in the male generative line is possible, and from the second the reproductive development in the female generative line is possible. Taking into account the different spatial localization of generative organs in the tree crown, the female generative sphere requires a higher energy level, and the male a lower one. Inception of generative organs for the survival of a species is one of the most important and least studied periods of plant life. A mother tree creates the favorable external and internal conditions for the conversion of generative buds and meristematic cells in the two types of local complexes, which consist of the totipotent by DNA-energy, DNA-particle and DNA-information vegetative cells.

COMMON CHARACTERISTIC OF REPRODUCTIVE CYCLE OF PINUS SYLVESTRIS L

Initiation of the reproductive process is preceded by the process of development of the two types of generative primordium in the male and female generative buds. The male primordium gives rise to the microstrobiles and the megastrobiles are associated with female primordium. Formation of these two complexes of undifferentiated meristematic cells, which on the cut have the form of a circle, is the stage by the emergence of a new reproductive initial () (Kuznetsova, 2002, 2013). A reproductive cycle of all species of high plants, including the *Pinus sylvestris*, by the male generative line and female generative line begins from this stage (Figure 28).

Table 3. Scheme of activation stages in the reproductive cycle *Pinus sylvestris* L

Reproductive stage name; reproductive initial	N division	Cells of apical pole	Cells basal pole	Number divisions of basal cells	Scheme divisions
Male generative line					
Microsporangio- genesis microsporangium initial[a]	1 2 3	Microsporangiuminitial Microsporangiuminitial Primary sporogenous cell – gives rise archesporium	Primaryepidermal cell Primarysubepidermal cell Primarytapetal cell	7-8 8-9 7-8	
microgametophytogenesis Microspore[a]	1 2 3 4 5	Antheridial initial Antheridial initial Generative cell Spermatogenous cell Functional sperm – participates in fertilization	1st prothallial cell 2nd prothallial cell Siphonogenous cell Sterile cell Nonfunctional sperm	0 0 0 0 0	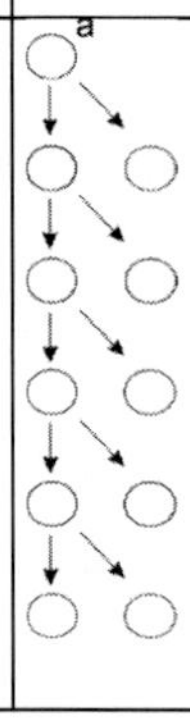
Female generative line					
Megasporangio- Genesis megasporangium initial[a]	1 2	Megasporangium Initial Primary sporogenous cell – gives rise ovule	Integument primary cell Nucellus primary cell	Unknown Unknown	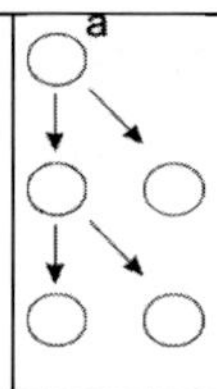
Megasporogenesis Megasporocyte[a]	1 2	Micropilar cell Functional megaspore – gives rise female gametophyte	Chalazal cell Nonfunctional megaspore	1 0	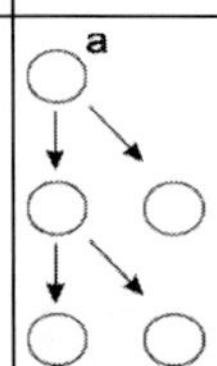
Oogenesis Archegonial initial[a]	1 2	Central cell Ovicell – participates in fertilization	primary neck cell Ventral canal cell	2 0	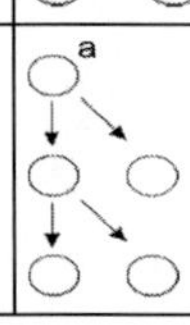
Proembryogenesis					
Proembryo-genesis 4-nuclear proembryo[a]	1 2	primary embryo tier embryo tier of 16-cellular proembryo – gives rise embryo	Primary proximal tier Suspensor tier	1 0	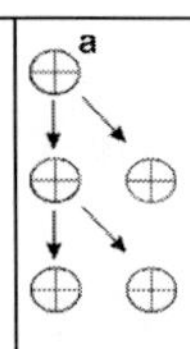

a – reproductive initial, or the first cell, from which begins the activation stage.

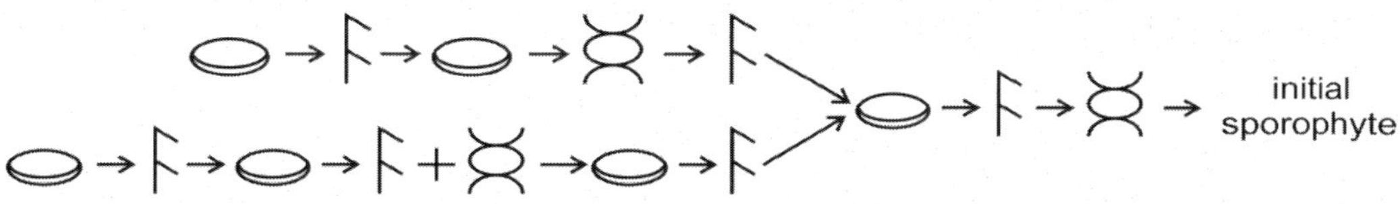

Figure 28. The scheme of reproductive cycle of *Pinus sylvestris* L.

The matrix of the reproductive cycle of *Pinus sylvestris* L. consists of the male and female generative lines, and the 14 stages of three types: 6 stages by the creation of energetic program (activation stages, which we mark on the scheme as the symbol; Table3); 3 stages by the formation of the hereditary programs () and 6 stages by the emergence of a new reproductive initial () (Figure 28). 12 stages in the matrix of Scots pine participate in the formation of the energetic program of the future plant (6 activation stages of and 6 stages by the emergence of a new reproductive initial (). Three basic states of generative cells (corpuscular + informational + spin direction) change during meiosis and fertilization. In the previous work (Kuznetsova, 2013), we designated this stage on the schemes as + ⚥. Now, for simplicity, we will use only one symbol:.

Most multifunctional and least studied stages of the reproductive cycle are still meiosis (male, female) and fertilization. They become the basis for the formation of DNA-particle, DNA-information and gamete differentiation by sex. The nature and mechanisms of their actions open up new possibilities for the study of the regularities of the genotype formation of future plants.

GENOTYPE-ENVIRONMENTAL INTERACTIONS

The primordium development occurs in July and August, when the metabolic activity related with the shoot growth is stopped, but cold weather has not come yet. The reproductive processes proceed within the closed buds covered with a resin layer, which protect them from the sun, extreme temperature changes, wind and mechanical damage. All types of meristems (apical, lateral, floral, etc.) of one mother tree are characterized by the identical DNA-particle and physiological homeostasis – constancy of osmotic pressure, ionic and protein composition, etc. (Nobel, 1993; Usmanov et al, 2001). Therefore, the beginning of the formation of generative organs occurs at the same (stable) substrate (Table 4). The generative buds are laid mainly on the lighted part of the crown, as light renders the stimulating action of the reproductive process. Perhaps four main factors – the conditions of homeostatic substrate, normalized microclimate, certain orientation with respect to gravity, and optimal illumination – are necessary for initiation and efficiency of the mechanisms of sexual reproduction.

Table 4. Genetic and environment components of population gene pool variability in forest trees during different periods of its formation

Period of development/Type of variability	I period inception of generative organs	II period gametogenesis	III period embryogenesis	IV period ontogenesis
GENOTYPE				
genotypic	–	+	+	–
modify	–	+	–	+
ENVIRONMENT				
substrate	stable	stable	changeable	changeable
climate	stable	changeable	stable	changeable

The first reproductive structures are diploid generative cells, and their development by the male and female generative lines occur independently of each other. Therefore, we will analyze the processes of their initiation and formation up to meiosis.

MALE GENERATIVE LINE

The development of microsporangium and their walls in all species of *Pinaceae* is characterized by constancy. Gymnosperms by the type of micro- and megasporangium are eusporangiate plants, which produce many spores and have a large sporangium developing from a few initial cells (Kozubov et al., 1982). In *Pinus sylvestris,* microsporangium initials become several superficial cells of the primordium, the polar axis of symmetry of which coincides with the direction of gravity (Brunkener, 1973; Kozubov et al., 1982; Bhatnagar & Moitra, 1996, etc.). At the time of microsporangium formation, the initial undergoes three activating mitosis, during which a biological harmonic oscillator in the apical daughter cells moves to the higher energy levels, and the energetic potential of other parts of the daughter cells (in accordance with the law of energy conservation) falls into the same number of levels lower. In the course of these three divisions, the apical cells move deeper into the primordial tissue. This leads to a change in the informational component of its genome. The last cell of microsporangiogenesis is the primary diploid sporogenous cell (1st stage activation, ⊦) (Figure 27, Table 3).It is the first generative cell and the most energetically strong cell, which is identical with the mother plant via the DNA-particle and differs from it through the DNA-energy and DNA-information. Figure 29 shows the part of male the generative line during the 1st period of the life cycle, which ends with the formation of a new reproductive initial – a microsporocyte.

If we exclude the winter, the time from microsporangium inception up to pollination is approximately 4 months. Microsporangium protects the male gametes from the external environment and a large part of its cells provides their trophic function. After the pollen departure, it is completely destroyed.

As was mentioned above, the process of microsporangiogenesis includes 3 activating mitosis, during which three energetically weak cells of the basal pole are formed. After activating mitosis, each energetically weak cell lost only a part of its biological energy. Accordingly, the duration of its life is limited. Then, their biological harmonic oscillators, in accordance with the 2nd law of thermodynamics, go down up until the zero energy level and initiate the new cell divisions. The number produced from these derivatives is genetically determined. All three derivatives of the cells of the basal pole, as the energetically weak cells, are attracted to the energetically strong cells of the archesporium and form around the three concentric layers. The cells of the outer layer – derivatives divisions of the 1st basal cells – give rise to the epidermis of the microsporangium wall. The subepidermal layer originates from the 2nd cell of the basal pole. In pines, it consists of the 3-4 layers. Tapetum occurs from the last basal cell. At the beginning a cell, all layers have an ultrastructure of the meristematic type (Trenin, 1982). Only the outer layer of cells differs from the internal by the degree of vacuolization, which indicates a higher level of their differentiation. The formation of the wall of the microsporangium and pollen ends in the spring of the next year.

Primary sporogenous cells receive the energy during microsporangiogenesis, divide and begin to form the sporogenous tissue in the center of the microsporangium. Its cells are significantly differed from the cells of the wall. This include larger, similar in ultrastructure, undifferentiated, izodimetric shaped and weakly vacuolated cells. In the course of mitotic divisions, the number of sporogenous cells is rapidly increased. As their periclinal divisions occur in the different directions, the archesporium on the cut has the shape of an oval. This suggests that the local complex of generative cells inside the microsporangium presents the 2nd stage  in the male generative line (Figure 2, the upper line of the scheme). All sporogenous cells are totipotent generative cells by the corpuscular-wave nature, i.e., have the same DNA-particle, DNA-energy and DNA-information. At this stage, microstrobiles survive the winter (Trenin, 1988). Megastrobiles survive the winter at the primordium stage (Kozubov et al., 1982; Trenin, 1988).

FEMALE GENERATIVE LINE

Another situation takes place in the female generative line. In pines, the development of megastomias begins somewhat later, and their development is carried out more slowly. In the inception year (year^{n}), only an anlage of the main axis exists with spirally arranged protuberances, which present3-4 layers of the undifferentiated meristematic cells of the rimordial tissue (Hagerup, 1933; Brunkenener, 1973; Singh, 1978; Kozubov et al., 1982, Trenin, 1988; Williams, 2009; etc.). These cells are characterized by a dense cytoplasm and a large nucleus.

Inception of megastrobiles occurs in July and continues acropetally. In the autumn, the degree of their differentiation may be different within the  stage. In the spring of the next year (year^{n+1}),only primordium or the anlage protuberances of a coating and seed scales are observed. The cells of the peripheral layer are densely pressed to each other and represent a radial row. As a rule, the two (rarely 1 or 3) superficial cells of the primordium, the longitudinal axis of which coincides with the direction of gravity, become the ovule initial (Brunkenener, 1973). Therefore, a female generative line, such as the male generative line, begins from the stage of the emergence of a new reproductive initial (1st stage – , Figure 29). However, if the male generative cells to the end of the 1st vegetative season pass the most part of the reproductive development by the male generative line, the female generative sphere to this moment is at the beginning of reproductive development.

Megasporangiogenesis, or the process of ovule formation, begins in the spring. Hagerup (1933) described in detail the process of development of the nucellus and integument tissues. In May, the median and two lateral protuberances occur in the base of the seed scale. Lateral protuberances grow more quickly and give rise to the integument, and the median gives rise to the nucellus.

We suggested that the main principle of ovule formation should be similar to the process of microsporangiogenesis. If we proceed from the ovule structure, the megasporangiogenesis presents the activation stage, during which an ovule initial undergoes the two activating mitosis. This is the 1st  stage on the scheme of the female generative line (Figure 29; Table 4). The distinctive feature of activating mitosis is the unequal distribution of biological energy between daughter cells. The apical cell receives more energy, and the basal cell is less. This in many respects determines their destination, life expectancy and participation in the reproductive process. During divisions, the derivatives of the first cell of the basal pole form an integument, and from the second basal cell a nucellus arises. The last apical cell, which has received the biological energy, becomes a primary sporogenous cell and it as the energetically strong generative cell continues the reproductive process. During activating mitosis the 1st and then the 2nd apical cells are moved into the primordial tissue and continue to further develop inside it. The initial informational signal with the noises on the DNA-wave of the primary sporogenous cell is significantly weakened after two activating mitosis of the ovule initial and passage of its derivatives between the primordial cells.

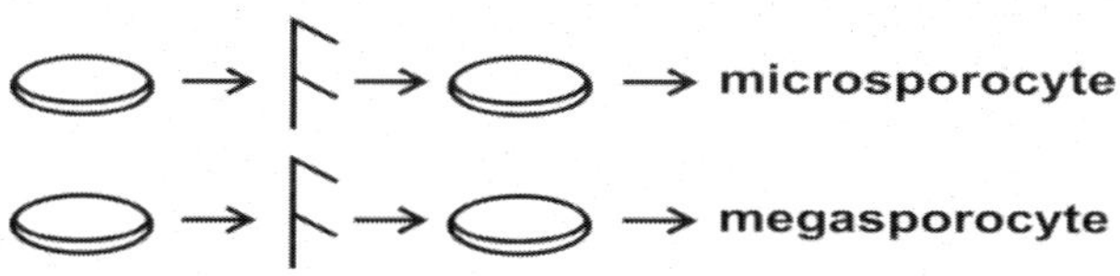

Figure 29. Scheme of diploid part of reproductive cycle by male and female generative line of the *Pinus sylvestris* L.

During pollination, the ovuleisa multicellular differentiated formation. Morphologically, it represents an important structural formation of a higher plant – the megasporangium. The ovule is intended for the protection and providing of trophic function of such important processes as meiosis, fertilization, gametophytogenesis and embryogenesis. The ovule of the pine refers to a type of the upright structure (orthotropic), the massiveness, large reserves of nutrients and the certain orientation towards the gravity is characterized. It consists of a central part – the nucellus, surrounded by the integument. The ovule is located on the upper part of the seed scale, in its foundation. It is oriented down to the axis of the strobile, which makes it difficult to penetrate the pollen into the micropyle. The integument at the top of the ovule is not fused, forming a passage into the micropile. A nucellus in all *Pinacea* is quite massive and crassinucellar. During ovule development, its structure is changed. In the spring, nucellar cells are actively divided and have a similar ultrastructure of the meristematic type. The upper part of the nucellus consists of several rows of cells, oriented in the longitudinal direction. Its tip before pollination changes and forms the micropyle. Cells of the integument and nucellus are joined in the basal part of the ovule. The primary sporogenous cell is situated deep inside the nucellar tissue. Around it, the energetically strong cell forms several layers of the nucellar tapetum.

We think that the first energetically strong generative cell is the primary sporogenous cell, having received a portion of biological energy during megasporangiogenesis. It appears inside the nucellus closer to the chalaza. A primary sporogenous cell, as the energetically strong cell, begins to attract the nearby cells of the nucellus, forming the concentric layers of the nucellar tapetum. After that, a primary sporogenous cell begins to increase in size, undergoes 2-3 mitotic divisions and gives rise to 4-8 sporogenous cells – megasporocytes (it is the second stage – , Figure 29), of which only one will become the megaspore mother cell. All sporogenous cells are the totipotent by a DNA-particle, DNA-energy and DNA-information. For example, they are identical to a maternal plant with a diploid DNA-particle and differ from it by DNA-energy, which become higher during the 1st stage of activation; this is known as megasporangiogenesis. It follows that only the preparatory processes for the formation of the female gametes take place in the analyzed period. Accordingly, a primary sporogenous cell is just the first generative cell and an intermediate reproductive structure, which is separated from the ovicell by a number of cell generations.

Thus, the main purpose of this period of the reproductive cycle is the inception of a large number of male and female buds with a large number of primordium of the microsporophylls and megasporophylls –future male and female reproduction structures as the material basis for the reconstitution of genetic diversity in the beginning of sexual reproduction. Genetic diversity provides the adaptation for the population gene pool to the different habitats on the territory of a species range according to the principle of "genotype-environment" interaction. The primordium of the microsporophyll and primordium of megasporophyll originate from vegetative cells of the floral meristem, which are identical with a mother plant by the DNA-particle and DNA-information and have the two certain energetic levels for transition from vegetative to generative path development. The male initial cells are characterized by lower levels of DNA-energy, and female initial cells are higher. Taking into account the fact that the diploid phase of the pine reproductive cycle is the same by the structure (Figure 29), the formation of the microsporocytes and megasporocytes represent the different processes, which proceed independently of each other. They are characterized by the different timing of initiation, various structures of male and female generative organs, the different number and ultrastructure of generative cells.

Chapter 4

FORMATION OF ENERGETIC, CORPUSCULAR AND INFORMATIONAL PROGRAMME DURING GAMETOGENESIS

The peculiarity of the gametogenesis period is that its beginning is associated with the changes of all 4 basic states of DNA (particle and wave): DNA-energy, DNA-particle, DNA-information and spin direction. This period coincides with the formation of new energetic, corpuscular and informational programs of development of future plants and the population gene pool of the next generation. Therefore, the main task of the reproductive process within this time interval is the restoration of genetic diversity and the population gene pool up to the level, which can ensure the survival of species within the range territory. For this purpose, the original (initial) cell of the population gene pool of male and female gametes should satisfy the two main requirements. Firstly, it should be a very diverse and balanced by a genotypic composition and, secondly, superfluous in the number of reproduction structures. This is necessary for ensuring the selection quality during competition of genotypes. In the beginning of reproductive development, these requirements are achieved owing to a genetic diversity of mother trees, a huge number of microstrobiles and megastorbili, microsporangium and ovules, microsporocytes and megasporocytes, which were formed during the previous period of sexual reproduction.

MICROSPOROGENESIS AND MEGASPOROGENESIS

The gametogenesis period begins with meiotic division by the male and female generative line. These are two complex corpuscular-wave mechanisms, which lead to the formation of four gametes with a reduced number of chromosomes, a different DNA-particle (crossing-over) and DNA-energy (only female meiosis). Also, gametes during meiosis receive a species cosmic signal, and their differentiation by the sex takes place (Part I, Chapter 2). Therefore, male (microsporogenesis) and female (megasporogenesis) meiosis are regarded as separate stages of gametogenesis.

Gametogenesis includes 6 main processes by the male and female generative lines: 1) Microsporogenesis and megasporogenesis; 2) male and female meiosis; 3) pollination; 4) microgame to phytogenesis and megagametophytogenesis; 5) oogenesis; and 6) fertilization. Microsporogenesis and megasporogenesis are differed according to the algorithm of development of male and female gametes. Both these processes and particularity meiotic division are very sensitive to the environment, when the genotype-environmental connections are formed.

In April, when the average temperature begins to exceed +5 ºC, the wave function on the DNA molecule is resumed. The transition of a biological harmonic oscillator to a lower energy level and entering into a cell of the quantum of free energy initiates its division. Thus, in the new vegetative season development of a plant in accordance with a program laid down in its DNA-energy, DNA-particle and DNA-information is continued by the male and female generative lines.

The reproductive process is well adapted to the regional climatic conditions. Sarvas (1973) has studied the annual cycle dynamics of the generative sphere of the Finland coniferous and found that phenological rhythms of species are docked with average multiyear sum of effective temperatures. The average daily air temperature above 5°C is the threshold value for its calculation. For example, a Scots pine for the beginning of meiosis requires the amount of accumulated heat, which corresponds to 5.6% of the average annual temperature; for pollination - 17.3%; and for fertilization - 31.0%. A pine in the forest-steppe zone of Russia needs approximately the same amount of heat as the one in Finland (Efimov, 2010; Kuznetsova, 2012). Its value in different years varies and is equal for meiosis, pollination and fertilization, respectively, 5.0-6.0%, 14.5-16.5%, 27.0-29.0%.

In the spring, the development of male gametes by the male generative line continues (up to the pollen departure) and is carried inside a microsporangium. Before meiosis, all its tissues are formed (Rowley & Walles, 1982). The wall consists from the epidermis, 3-4 middle layers and tapetum. All these tissues fulfill a trophic function at different stages of pollen formation. Epidermal cells have already a sign of specialization. Cytoplasm density decreases and increases its vacuolization. By the end of meiosis, the cells of epidermis develops a large vacuole, which is filled with electron-dense tannino-like inclusions. The nucleus is shifted toward one of the walls. Cell walls are thickened greatly. In the early stages of male gamete development, middle and tapetal layers are differentiated by the cell form and have an ultrastructure of the meristematic type. A characteristic feature of subepidermal cells is a large number of chloroplasts. They are concentrated around the nucleus, located in the center of the cell. Differentiation of tapetal cells occurs later than in the other layers of the microsporangium wall. During meiosis, they increase in size and become the cells of the secretory type by the ultrastructure. At the stage of microspore tetrads, the tapetum begins to break down and the lysis of intracellular contents takes place, which is completed during microspore free stage. By the time of departure, only the epidermis remains from the layers of the microsporangium wall. Microsporangium of a longitudinal crack is opened.

Sporogenous cells are located in the central part of the microsporangium. Initially, they have the isodiametric shape, slim straight cell envelopes, and a large nucleus with several nucleoli. In spring, the division of sporogenous cells begins in the closed male buds, i.e., into a fixed volume. This leads to an increase of intracellular pressure and viscosity of nucleoplasm and cytoplasm. This changes the physical situation inside the microsporangium and sporogenous cells. The change of intracellular viscosity leads to a reduction of the biofield, the cessation of divisions and the separation of sporogenous cells from each other. The sizes of sporogenous cells are increased and change the morphology of cytoplasm and the nucleus during their development in the microsporocytes.

Further divisions and growth of all vegetative and generative tissues of microstroliles inside the buds are continued. The bud scales are moved apart under their pressure. Microstrobiles go outside and start to increase rapidly in length and thickness. Their further development from meiosis up to pollination is in close contact with the external environment. Before meiosis, microsporocytes are a rather large round-shaped cells. The nucleus has large sizes and chromatin is represented mainly by the euchromatin. The subsequent growth of intracellular pressure initiates a particular type of cell division – meiosis.

Cell differentiation in microsporangium at different stages of meiosis is described in detail in the works of Rowley & Walles (1985, 1987, 1988, 1993). In this period, the development of microsporangium is mainly under the control of endogenous regulation (frequency of mitosis, rejuvenation of sporogenous cells and degrees of differentiation of the 1st, 2nd and 3rd cells of the basal pole, etc.) and smaller external factors. The coexistence of tissues with the different energy potential suggests the occurrence of certain relationships between them. Therefore, the energetically weak cells of the microsporangium wall are attracted to the archesporium. A nearby layer of tapetum begins to provide its trophic function. Since the sporogenous and tapetal cells have the identical DNA-particle, then the abscission layer between them is not formed. In the process of microsporangium formation, the shape and size of generative buds changes. By the end of the vegetative season, it becomes cone-shaped, reaches a length of 1.5 cm and the outside is covered with dense scales and a thin layer of resin (Figure 27b).

At this time, the reproductive process by the female generative line is at an earlier stage of development primordium of megasporophyll. In the spring, several cells of the surface layer of primordium becomes the ovule initial (Brunkener, 1973) and goes to the megasporangiogenesis stage ( - the 1st stage activation on the female generative line; Figure 29, lower line; Table 3). At this stage, the ovule initial undergoes the two activating mitosis. The last apical cell of these divisions is the primary sporogenous cell. This sporogenous cell, which has received a portion of biological energy, becomes the first energetically strong generative cell on the female generative line. Before pollination it undergoes 2-3 mitosis and gives rise to 4-8 sporogenous cells (the 2nd stage on the female generative line, Figure 29). At the same time, the basal cells, which during activating mitosis have lost this portion of biological energy, form two layers of the wall of the megasporangium – the integument and nucellus. A nucellus is located in the central part of the megasporangium. This is a rather massive multicellular tissue with the same type by the ultrastructure undifferentiated cell. In spring, a large number of amyloplasts is in its cells.

In Scots pine, male and female meiosis begins and ends within the spring. Its beginnings depend on the weather conditions of that year. In the Voronezh region of Russia, the beginning of male meiosis occurs in the 2nd half of April –the beginning of May; pollination begins on the 2-3rd decade of May; and female meiosis begins in the 2nd half of May –the beginning of June. The total duration of microsporogenesis varies from 20 to 25 days (Khromova, 1972; Kozubov et al., 1982). Since the first prophase ends and to free microspores formation passes an average of 12 days (Kozubov, 1974). In April-May, the temperature in the research region varies widely: Often, frost, abnormal heat and atmospheric drought occur. At this time, microstrobiles and megastrobili have small sizes and the previous year's needles are situated horizontally; therefore, male generative buds during microsporogenesis and megastrobili during megasporogenesis are in direct contact with the external environment. As a consequence, male meiosis occurs in the changeable climatic conditions (Table 4).

The processes of natural selection are accompanied by mass genotype death.

Therefore, their primary spectrum should be the capacious and very various. In genetic diversity, the internal mechanisms, superfluous reproduction and environmental factors (temperature, solar isolation, etc.) are taken into consideration.

The formation of the population gene pool of a new generation begins from the expansion of genetic diversity, which is formed on the basis of a great number of microstrobiles, megastrobili and generative cells. The population gene pool after meiosis presents the spectrum of the haploid male and female gametes. Their environment is the soma (body) of mother trees, which carry out the trophic, protective and selective functions. Before fertilization, their genotypes present for the unicellular system (ovicell), or the sperm-nucleus. Male and female gametes qualitatively (size, mobility and internal organization) and quantitatively are differed from each other, and against mother trees, from which they originate.

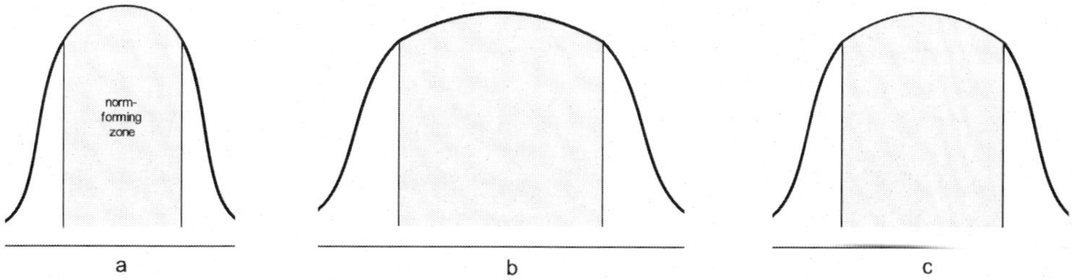

Figure 30. The spectrum of the genotypic variability of gene pool of maternal population (a), male and female gametes (b), seeds (c) by the DNA-particle.

Genotypic composition of forests pass through long and strict selection, after which they become the balanced and stable enough ecosystems to the main set of regional biotic and abiotic stressors. Lerner (1954) described the population's ability to maintain the balance of their genetic structure and to withstand external influences as the genetic homeostasis. The genotypic structure of equilibrium populations has a normal distribution with the modal meaning in the zone of norm-forming genotypes that corresponds to the equilibrium of their state (Figure 30a). In the balance, i.e., in a state of dynamic equilibrium with the environment, these forest ecosystems are able to exist a long time.

Genotypes of male and female gametes are differed from the genotype of the mother tree by the four basic states. Their genotypic spectrum is much wider than a spectrum of initial maternal population (Figure 30b). The number of rare genotypes in the variation curve is increased, and their norm-forming zone becomes wider (Zhivotovsky & Duharev, 1985). The mode in the normally fluctuating environment is situated in the center of the norm-forming zone. Hence, it follows that after microsporogenesis and megasporogenesis the connection of population gene pools of the newly formed male and female gametes with the genetic material of maternal plants becomes considerably weaker.

As already mentioned, the population gene pool of pollen at this time has the largest genetic capacity (genetic heterogeneity + superfluous multiplication). The possibilities of recombination variability of their DNA-particles are very high. Schwartz (1980) gives the following arguments: If a hereditary apparatus consists of only 1,000 genes, then each of may be manifested by the 10 alleles. In this case, the recombination variability originates $10^{1000}$of non-identical genotypes that outnumber the electron number in the visible parts of our planet. The possibilities of genotype multiplication are also large. During pollination, the pollen abundance in a pine forest is 200-250 pollen grains per 1 mm^2 (Nekrasova, 1978). So far, as the population gene pool of each tree has a large genetic capacity, in this case it should have the great potentialities for the reproduction. In the nineteenth century, a preservation of small number trees with high seed production on the clear-cutting areas is one of manner of forest restoration (Redko, Redko, 2004).

Male Meiosis

In the reproductive cycle, a meiotic division precedes the formation of haploid male and female gametes. From the standpoint of physics and biology, this is two very complex and multistep processes. This is the 1st stage by the formation of new corpuscular and informational programs on the basis of DNA-particle and DNA-information ( stage; Figure 28) by the male and female generative line.

Meiosis consists of two successive divisions, the result of which is a qualitative and quantitative change in the genome of gametes on the basis of DNA-energy, DNA-particle, DNA-information and spin direction. During male meiosis, from the one mother generative cell (microsporocyte) the 4 microspores with the same DNA-energy and DNA-information, different DNA-particle, two gametes with spin-up and two gametes with spin-down are formed. A megasporocyte gives rise to the four megaspores with the same DNA-information, the different DNA-particle, one energetically strong and three energetically weak gametes, two gametes with spin-up and two gametes with spin-down cycles.

Meiosis is one of the most sensitive processes and weather exposure affects the stage. In the spring, branches of needles are arranged horizontally. During microsporogenesis and megasporogenesis, genetic material of developing male generative buds and megsstrobili are contacted directly with the external environment. In the plantation the most part of microstrobiles and megastrobiles locate in the top canopy, edges, clearings and outer border of forest (Figure 31). Here, they are under the powerful influence of wind, frost and sun. Accordingly, the effect of weather conditions and the spatial location of the generative organs on meiosis are very large. Therefore, meiosis in the microsporocytes occurs asynchronously in different trees, in different buds of a single tree, in separate microsporophylls of one microstrobile, and inside a microsporangium (Trenin, 1988).

The obligatoriness for such a space arrangement of reproduction structures in the tree crown is due to the technical requirements that they must satisfy, as the part of an antenna complex, for receiving a species cosmic signal during meiosis. Generative buds, megastrobili and chromosomes should be directed vertically upwards to receive a new signal during leptotene-pachytene. If possible, the generative organs should be outside the crown, so that the genome of generative cells has maximum access to bioactive Z-radiation. This ensures a high efficiency of receiving the informational signal. In addition, this arrangement of microstrobiles and megastrobili contributes to a better distribution of pollen and increases the capability of cross-pollination during free pollination.

a b c

Figure 31. Megastrobili (a, b) and microstrobile (c) during pollination.

During meiosis, the process of the direct selection of more resistance to weather stress genotypes occurs. In close contact with the environment, gametes are able to catch all diapason of change and specifics of regional climate and lay down this information in the form of a frequency distribution of the genotypic composition of surviving gametes. Under optimal weather conditions, the initial spectrum of genotypic variability of pollen and ovicells obey the law of normal distribution. In the course of microsporogenesis and megasporogenesis, a genotypic variability of gametes on the basis of DNA-particle and DNA-energy are extended (Figure 30b), and on the other, their variability is decreased, as a result of direct selection, or the selective survival of certain genotypes during meiosis and open pollination. During direct selection, a variation curve is narrowed uniformly, maintaining a normal genotype distribution of a seed gene pool (Figure 30c).

As it has been shown in numerous works, the impact of different adverse factors on meiosis causes a modification variability of generative cells (Borodin, 1987; Dyuzhikova, 1997; Inge-Vechtomov, 2007, 2010, etc.) (Table 2). In Scots pine, stress modifies the character of passage of the post meiotic stage, expanding the spectrum of genotypic variability (Figure 30b). Genetic effects, which are registered on the population-ontogenetic level, are manifested as the bigger variability of seed progeny, the increasing of the proportion of empty seeds and their weight (Kalchenko et al., 1993, 2001). Apparently, the modification variability of gametes affects the informational component of a genome without changing the DNA-particle. It was established that widening the spectrum of modification diversity of the male gametes under the influence of environmental factors occurs during male meiosis. The most susceptible stages, exposed to effect a high positive and low temperatures, is metaphase I and anaphase I and metaphase II and anaphase II, which leads to an increase of a univalent number in a diakinesis - metaphase I, to the formation of dyads, triads, pentads and hexads (Andersson, 1980; Kozubov et al, 1982; Noskov & Tretyakova, 2006; etc.).

Thus, in the course of the gametogenesis period, during microsporogenesis megasporogenesis the diversity of the population gene pool of the male and female gametes are increased due to a genotypic variability, with the participation of different corpuscular-wave internal mechanisms and a non-hereditary or modification variability under the influence of biogenic and abiogenous environmental factors. As all cells of the mother tree, including the wall tissues of microsporangium and megasporangium, are characterized by the physiological homeostasis, the processes of microsporogenesis and megasporogenesis within the mother tree occur at the one and the same (stable) substrate (Table 4).

MICROGAMETOPHYTOGENESIS

Some Questions of Biology Gametogenesis in the Year of Pollination

During microsporogenesis and megasporogenesis, a population gene pool of pollen and ovicells is formed. These are the highly differentiated generative cells with a species-specific shape and size, having a number of devices to put into execution of the chief mission of sexual reproduction, known as syngamy gametes. The main difference of sex gametes in comparison with the somatic cells are as follows: Firstly, a genome of microspores and megaspores has the haploid set of chromosomes. Secondly, during gamete maturation the cytoplasm of pollen grains and ovicells are filled by the nutrients, which provide the growth of pollen tubes and embryos in the early stage of their development. Thirdly, these generative cells are characterized by a strongly changing nucleocytoplasmic ratio. Fourthly, during maturation, pollen grains and ovicells transition into a state of metabolic resting. Fifthly, the pollen, which has not gotten into the pollen chamber and unfertilized ovicells are not able to develop and die shortly thereafter.

Pollen grains of different taxonomic groups of higher plants are differentiated by the mechanisms of pollen development, morphology, structure of the envelope, and the relationships that are formed between the pollen and the surrounding cells (Schoen & Cheliak, 1987; Knox, 1990). According to Knox (1990), the main adaptive changes of pollen are:

- Reduction of chromosome number and recombination of genetic material, which takes place during formation of microspores (meiosis);
- Development of barriers, protecting the microspores from contact with the genetically distinctive form from them in maternal cells (callose envelope);

- Formation of systems, protecting the gametes from a possible mechanical damage and environmental stressors during pollination (envelope of pollen grain);
- Development of devices, ensuring the directional transfer of pollen, its attachment to the stigma surface in flowering plants or to the nucellus in gymnosperms (pollen grain form, with different degrees of sculpting of its external envelope);
- Development of mechanisms of pollen recognition in the zone of contact with the stigma (or nucellus) on the basis of proteins and glycoproteins of pollen grain envelopes;
- Creation of a complex system, which provides pollen germination and pollen tube growth in order to continue fertilization (a participation in this process of a number of specific enzymes and the phenomenon of the heterotrophy of pollen tubes).

Explaining the nature of the majority of the above adaptive systems and mechanisms is not possible from the position of corpuscular biology. We have found that the life of a prokaryotic cell may be represented as a combination of three basic states of DNA molecule: DNA-particle, DNA-energy and DNA-information (Kuznetsova, 2016). In eukaryotic cells, these three base states include one more state – the spin direction of haploid gametes.

At present, the DNA molecule as a bearer of species-specific genetic information, is mostly studied. It is proved that a change of nucleotide composition of the species-specific DNA-particle takes place only during meiosis. Then, this basic state of macromolecules is maintained throughout a plant's entire life. The main mechanisms of modifying its chemical formula are two complex quantum-mechanical processes – conjugation and crossing-over of homologous chromosomes during zygotena and pachytena. As a result, four viable male gametes are formed in the course of meiosis, characterized by a haploid set of chromosomes and a difference of DNA-particles.

The function of DNA (as a bearer of species-specific energetic information) is associated with its wave nature and the continuous action into a cell of a special physical system – a biological harmonic oscillator or a nucleus together with an oscillating DNA (Kuznetsova, 2013). Life of all plants can exist and develop only under the continued support and improvement of an inner energetic system and the relationships between an organism and the environment by means of outer energy sources. According to our corpuscular-wave model, the energetics of a eukaryotic cell is provided by 4 independent sources: The inner energy of 1) a biological harmonic oscillator and 2) metaphase chromosomes; and the outer energy of 3) photosynthesis and 4) respiration. The first source of inner energy in the general energy system belongs to a special place. Energy of a biological harmonic oscillator (DNA-energy) is the basis of the nature of many life processes and phenomena: Intraspecific and interspecific competition, the stability and living states of plants, cellular differentiation, apoptosis, direction of evolutionary development, change of dominants during the evolution of plants and animals, etc.

In a previous work, we have assumed and shown that during sexual reproduction on the basis of the matrix of the *Pinus sylvestris* reproductive cycle, not only individual DNA-particles of the future plant are formed, but a genetically determined process of the rejuvenation of their genome also occurs. The essence of this process consists in the passage of a DNA-wave to the higher energy levels. This is a complex and surprisingly perfect mechanism of energy storage, which occurs in the lines of generative cells – from the cells of a floral meristem up to the initial cells of a sporophyte on the basis of a DNA molecule (a particle and a wave).

Life is a continuous process of energy consumption and its transformation. It links together all the phenomena in a living and non-living nature. The energetic component of a genome will be described as biological energy (or DNA-energy). The harmonic oscillator with regards to a multicellular organism could be compared with the "energetic reservoir," which is filled with the energy up to a species-specific level during sexual reproduction and is released from it during ontogenesis. Therefore, the genotypes of gametes differ not only in the nucleotide compositions (DNA-particle), but also in the energy state of their genome.

It is known that the pollen of each species is characterized by the species-specific phenotype: Size, shape, presence or absence of air bags, etc. If a genome had only two basic states (on the basis DNA-particle and DNA-energy), the pollen grains of all plants would be the same spherical shape without any distinctive morphological features. Therefore, we think that the phenotype of the plants, as well as the morphology of microorganisms, defines the DNA-information, which occurs in each gamete during meiosis. In this case, all the species diversity, represented in the world of plants, animals and microorganisms, is a consequence of the manifestations of the information, which is recorded on a DNA-wave in the form of a species-specific information signal.

Lastly, another basic state of the genome is connected with the presence of the spin of the chromosomes. In a previous work (Kuznetsova, 2013), we have suggested that during meiosis tetrads of microspores, the chromosomes of which are Fermi particles and characterized by the different spin direction are formed: 2 microspores with the spin upwards (↑) and 2 microspores with the spin downwards (↓). Taking into account the quantum mechanism of this process, the gamete differentiation by the spin may take place during the leptotene-pachytene of meiosis, as the result of the spatial orientation change of chromosomes in leptotene that interact with the non-uniform magnetic field while the nucleotides of doubled chromosomes pass through it. The non-uniform magnetic field is created due to the applying of a magnetic field by the Earth on the intracellular field, generated by two moving positively charged centrioles. This causes the separation of nucleotides into two beams depending on the direction of their spin.

Therefore, the genotype of each pollen grain is characterized by a haploid species-specific DNA-particle, a species-specific level of DNA-energy and a definite direction of spin (50% pollen grains has the spin↑ and 50% the spin ↓).

Experimental data of pollen grain differentiation by the sex are not numerous. In these works, it is shown that the sizes of the pollen with a certain sex correlates (Joyet-Lavergne, 1931; Minina, 1984). Pollen grains are characteristic of the physiological differences. At the time of electrophoresis, the part of male gametes moves towards the cathode and the other to the anode (Ostapenko, 1969).Histochemical studies have shown that the pollen grains differ by their redox properties. Some of them are characterized by the oxidizing properties and the other by the reducing ones (Minina &Tretyakova, 1983; Minina, 1984).

Ecological genetics as a science is closely related to the stability of biological systems, a phenomenon which has also a corpuscular-wave nature. Analysis has shown that living matter at all its hierarchical levels (DNA, cell, plant, species) is inherent both in a corpuscular and a wave nature by the principle of "all in all" (Kuznetsova, 2013). This conception expands our understanding of the function of adaptive mechanisms and devices by ensuring the preservation of a population gene pool in all levels of its organization in the constantly fluctuating environment.

The separation of the gene pool of pollen, ovicells, embryos and plants on the genotypes of a species norm, outsiders and leaders is the general biological phenomenon that is governed by the law of normal distribution. Their biological energy as any quantitative trait varies within a certain range. In a normally fluctuating environment, its variability presents a set of continuous variational values and has the appearance of a normal curve with a mode in its central part(Figure 32). This explains the differentiation of plants in height. For forest plant species, this phenomenon is usually described as the plus, normal and minus trees or as the different "individual strength of trees" (Eytingen, 1962). It is shown that the "individual strength tree" appears early, at relatively the free (8-15 years) growth of cultures, when the competition between trees is still weak (Maslakov et al., 2001), confirming the genetic conditionality of this phenomenon.

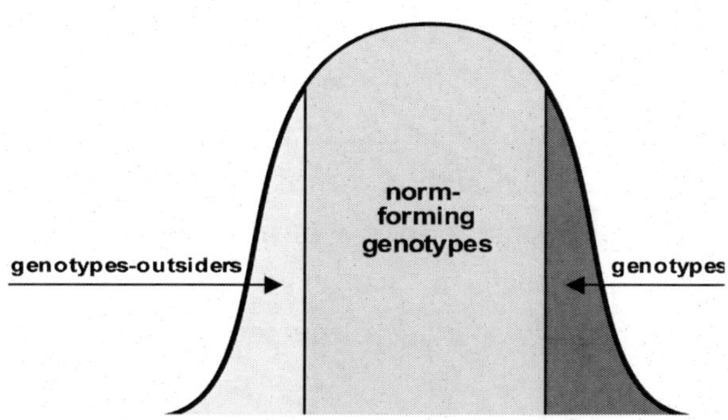

Figure 32. Genotypic variability of the initial population gene pool of pollen, ovicells and embryos by the level of biological energy (DNA-energy).

Male Generative Line

As was mentioned above, genotypes of male gametes are formed during meiosis and distinguished by the corpuscular (nucleotide composition of DNA-particles), energetic (a level number of DNA biological energy), and informational (modulated DNA-wave) components of the genome and spin direction. The equilibrium is the most stable state of any genetic system. It contributes to the survival of male gametes in case of sharp temperature fluctuation in spring, when the contact of reproductive structures with the environment is closest. In a normal fluctuating environment population gene pool of microspores and pollen, the gene pool of maternal population increases many times over and is differentiated into three zones by the level of DNA-energy, as shown in Figure 32. During pollination, the proportion of pollen, fallen into the pollen chamber, is not large. Most of it remains on the ground or in the tree crown and dies. The variation curve of genotypic variability is narrowed evenly. However, after pollination the genotypic composition of male gametes by the DNA-energy and spin direction, if we proceed from the probability of random events, fully reflects the character of pollen distribution prior to pollination.

A "bio-build" of biological energy is a strictly genetically determined process. The male and female energetic line on the example of the *Pinus sylvestris* reproductive cycle has been described in detail in our previous paper (Kuznetsova, 2013). Here, we only briefly discuss its structure. The scheme of the male generative line and the number and disposition of its stages are shown in Figure 33. The last generative cell, which is formed on the male generative line, is the functional sperm nucleus and the last reproductive structure, which develops on the mother tree, is the pollen grain.

Two activation stages () on the scheme are the processes of microsporangiumgenesis and microgametophytogenesis (Table 1), when the DNA-wave of a generative cell moves to higher energetic levels. Analysis showed that the activation stages are made up of activation divisions, the number of which can vary from 1 to 5, and this depends on the species and reproductive stage. In the course of activating mitosis, a process of energy transfer from one set of chromosomes to the other takes place, which strictly undergoes in accordance with the law of energy conservation, and only in specific conditions of sexual reproduction. The stage  is a male meiosis (or meiosis under microsporogenesis). The functional purpose of this stage is well known. Firstly, during meiosis four haploid microspores are formed, which participate in fertilization. Secondly, by means of a crossing over mechanism, changes in the nucleotide composition of DNA, or the chemical formula of a DNA-particle of microspores are experienced. Thirdly, the differentiation of the gametes by the sex occurs. On the 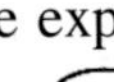 stages (generative primordium, archesporium, etc.) a choice of one, two or more of new reproductive initials are produced, the symmetry axis of which coincides with the gravitational direction.

Stage activations can only begin from the division of initial cells. On the male generative line, it is the initial of microsporangium and microspore.

In embryology, the nature and functional purpose of activation stages, and such phenomena as the apoptosis, which accompany a reproductive process of any species, is not known. The main feature of all these stages is that the end of each stage in the reproductive process only goes one (or several) of generative cells and the others (after passing a fixed number of divisions, or without divisions) die. These phenomena discuss in more detail the example of a microgametophytogenesis (Figure 34; Table 3).

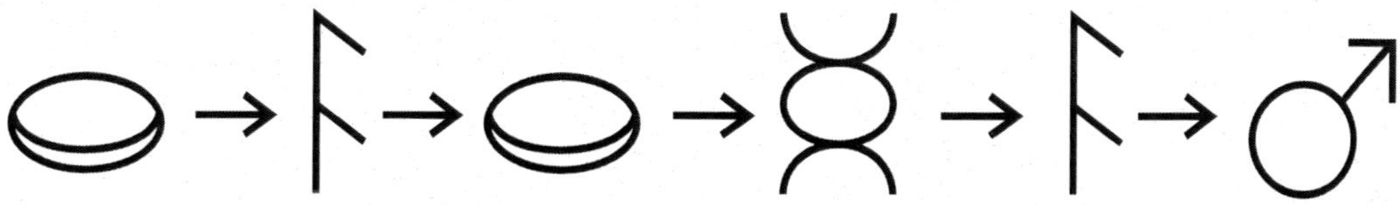

Figure 33. Scheme of male generative line of the *Pinus sylvestris* reproductive cycle.

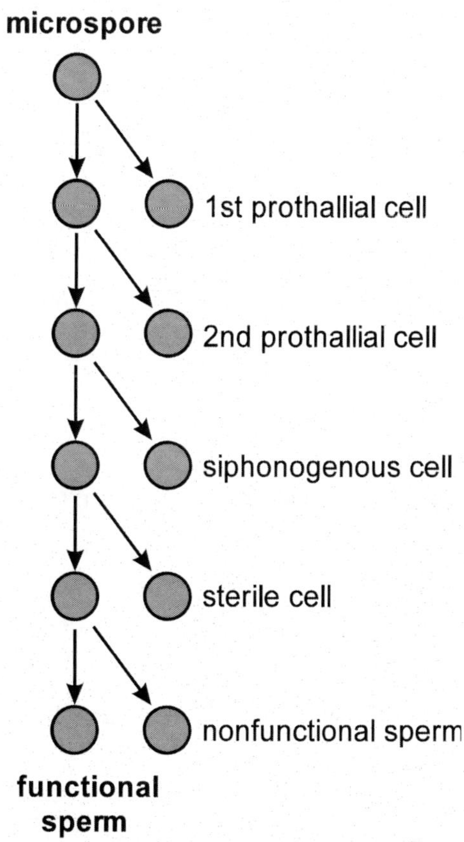

Figure 34. Scheme of activating mitosis during microgametophytogenesis.

Microgametophytogenesis begins from a division of the microspore, which is the initial cell of the 2nd stage on the male generative line. At the output of this activation stage is the nucleus of functional sperm, which participates in the process of fertilization. The specificity of activating mitosisis that the spatial relationship of the polar axis of the initial generative cells and the spindle of their divisions coincide with the gravity direction. The spatial position is one of the main reasons, which determines the direction of movement of energy within the stage activation and the different destiny of daughter cells. The daughter cell that received the portion of biological energy is the cell of the apical pole and energetically strong cell. During microgametophytogenesis, the energetically strong cells are: The antheridial initial, generative cell, spermatogenous cell and functional sperm. All the other daughter cells which have lost energy are the cells of the basal pole and energetically weak cells. These are: The 1st prothallial cell, 2nd prothallial cell, siphonogenous cell, sterile cell and nonfunctional sperm. After a while, these generative cells necessarily die and go out of the reproductive process. Apparently, the main functional purpose of apoptosis as a biological phenomenon is as follows. The group of apoptotic cells, which lost a greater part or all their energy, gave life to a certain number of energetically strong generative cells, taking part in the process of rejuvenating the genome in accordance with the law of energy conservation. Many cells of the basal pole before their death undergo a few divisions (primary neck cell – two divisions; chalazal cell – one division) and can carry a certain functional load. For example, a siphonogenous cell forms a pollen tube and delivers the male gametes to the female ones.

The formation of microspores is a turning point during the development of male gametes. A microspore is the first generative cell of the reduced male gametophyte – microgametophyte, which has its own corpuscular-wave genome – the individual haploid species-specific DNA-particle, the certain spin direction and the individual and species-specific level of DNA-energy, the energy potential of which is higher than that of the mother plant.

Microspores as the energetically strong generative cells are separated from each other and from the mother tree by a thick callose envelope (Figure 35a). The callose formation is the result of incompatibility reactions, which arise between a microsporangium and each male gamete after meiosis. Since the DNA-particles of microspores have changed after crossing-over and the relationship between them has the character of the interaction of two heterogeneous systems. The complex of necessary substances for the microspores as the energetically strong cells are received from the microsporangium tissue. Accordingly, they are characterized now by a heterotrophic type of nutrition.

In the spring, nearby tapetal cells of the microsporangium, which have gotten close to the zone of the stronger bio-field of sporogenous cells and in the conditions of higher pressure inside the closed buds, pass into the regime of high metabolic activity.

Therefore, they are transformed into the cells of the secretory type. The secretory cells are characterized by a specific ultrastructure and high synthetic activity (Kozubov, 1974, 1982). In this period, they often contain 2 nuclei with 2-3 nucleoli. The cytoplasm is characterized by a developed network of the endoplasmic reticulum, sometimes twisted into spirals or rings. Then, the cytoplasm of tapetal cells is compacted. The cell envelope in the contact zone with sporogenous cells forms folds, then it begins to disintegrate. The cytoplasm after the disintegration of the inner wall intensively produces the Ubich bodies, which have an irregular oval shape with the outgrowths and participate in the sporoderm formation (Rowley & Walles, 1987). The processes of the tapetum cell degradation are completed at the free microspore stage.

Nutrients move to the archesporium by the vector of the stronger bio-field. On the boundary of the two heterogeneous systems is located a semi-permeable envelope. Therefore, after meiosis a part of the biomolecules, moving together with a stream of nutrients to the male gametes, does not pass through it and is deposited around each microspore. Under the impact of strong biofield forces, they are polymerized and a callose layer is formed. Callose is the insoluble polysaccharide of β-1,3-glucan (Xiao Shi et al., 2016). At the same time, between a plasmatic membrane and callose layer a new envelope begins to develop – the sporoderm. Incompatibility processes lead to the disintegration and isolation of microspores inside a tetrad. Microspores in tetrads are in the protected locations of the callose case. We think that due to their isolation, the spin-spin interaction between the microscopes with the opposite spin direction becomes impossible.

Every male gamete receives two portions of free energy – after the 1st and 2nd meiotic divisions when the metaphase chromosomes uncoil. Therefore, microspores have the great potential for future growth and development. The callose layer is able to control and restrict their growth only to a certain limit. The stability of this layer is not so great and the microspore tetrads are destroyed quickly. Then, the microspores grow a sporoderm rapidly and simultaneously, and the new, more durable envelope is formed. According to Nekrasova (1983) at the stage of tetrad break-up, their average size is 32.5 x 25.5 μm; after the 1st prothallial division, they are increased by 1/3 - 43.4 x 33.6 μm.

Sporoderm limits the growth of microspores that cause the growth of intracellular pressure. Therefore, the necessary physical conditions into male generative cells are created and they pass to the next activation stage by the male generative line. This stage is known in literature as the microgametophytogenesis, which includes 5 activating mitosis (Figure 34; Table 4). During this stage, a reproductive process is continued inside each microspore. Before maturation, the two air sacs in the pollen grains are formed (Figure 35b). They are laid early, as widened cavities between a plasmalemma and a callose coat.

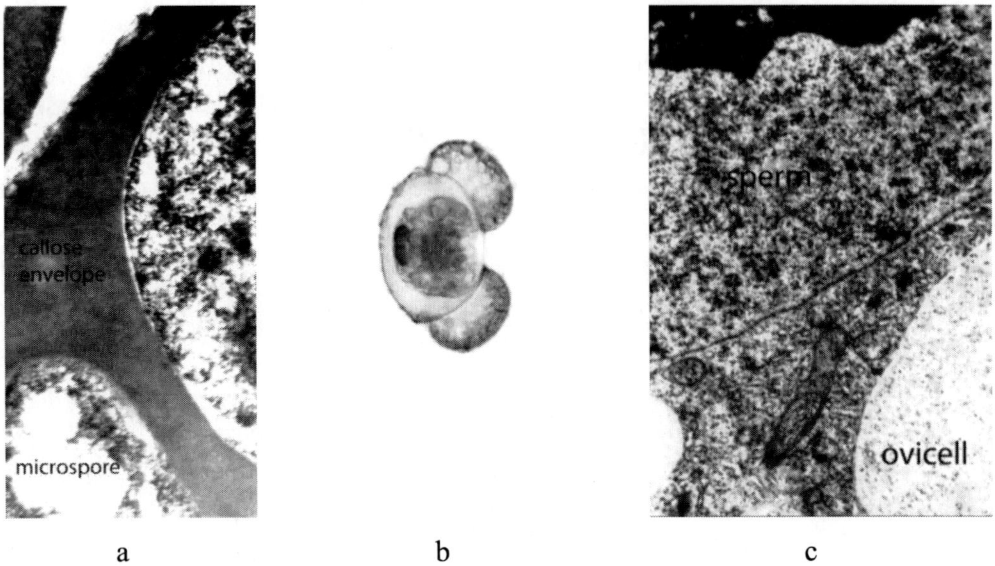

Figure 35. The main reproductive structures on male generative line: microspore tetrad (a), pollen grain (b) and sperm nucleus (c).

The first three activating mitosis of microgametophytogenesis occur within the microspore. 1st activating mitosis of future male gametophytes occurs within 5-7 days after the stage of the tetrad microspores break-up (Kozubov, 1974). It ends with the formation of the antheridial cell and non-viable 1st prothallial cells (Figure 34; Table 4). At the time of the next prothallial division, the celldies and shrinks. 2nd activating mitosis occurs1-2 days after the first and gives rise to an energetically strong antheridial cell and an energetically weak 2nd prothallial cell. Microstrobiles begin to turn yellow. 3rd activating mitosis takes place before the pollen departure from a microsporangium. As a result of this division, a large energetically weak siphonogenous and a small energetically strong generative cells are formed. It should be noted that the interaction of the archesporium with the tissues of the microsporangium is carried out at a high level of orderliness. The life expectancy and the number of cell walls of the microsporangium are such that they completely provide a trophic function of male gametes, from the primary sporogenous cell origination up to pollen departure.

The pollen grain is an independent, relatively long-lived, microscopically small, physiologically active vessel that exists without chlorophyll formation. Male gametes are characterized by their own genome – the haploid set of chromosomes, individual DNA-energy, DNA-particle, DNA-information and spin direction. They are the species-specific gametes by size and phenotype 2-cell biomorphs with generative cells. The siphonogenous cell and its remains of two prothallial cells exist. Pollen grains of pine reach 48-61 μm in length (Trenin, 1988). The pollen outside is surrounded by a rigid strong envelope – the sporoderm. The sporoderm is a complex formation and consists of intine and exine. Intine, or the inner layer of the envelope of pollen grains, is formed soon after the microspore emergence and consists of two main layers: The continuous internal and external layers. During maturation, its cytoplasm begins to accumulate nutrients – starch and lipid droplets (Tretyakova, 1990). The mature pollen has protein bodies. The pollen grain has a distally located germination furrow, through which the pollen tube comes out. During its transition to a state of rest, the connection of pollen grains with the mother plant gradually weakens, and then disappears. After disclosure of microsporangium, they leave the mother tree. Each pollen grain has its own individual way of ontogenesis, lifetime, destiny and specificity of reproductive development. Father formation of male gametophyte occurs inside each microspore. Therefore, the pollen grain is the last of the male reproductive structure of a mother tree, and the last of the generative cell of the male generative line – the nucleus of functional sperm (Figure 35-b, c).

After microsporogenesis, a reproductive process is divided on the many independent lines by a number of microspores and their reproductive development proceeds separately from each other up to pollen germination. Duration of microgametophytogenesis from the time of microspore formation and up to fertilization is ≈ 13 months, i.e., over a year. In the Voronezh region, a flowering of pine tree is annual, is sufficiently abundant and lasts for 7-11 days. According to Efimov (2010) the limits of trees on the average length of the microstrobiles is 18-40 mm and the width is 7-13 mm. The amount of pollen from the trees in a spikelet varies from 11 to 106 mg. In the optimal years, pollen quality is high and equal to the averages 80-90%. By the dates of pollination commencement, all populations are differentiated in the early-, late- and mid-flowering trees. Approximately in the middle of the pollination period, the majority of trees come into the flowering stage simultaneously. The air bags provide a high flyway potential for pollen and it transfers for considerable distances by means of wind. Within 2-4 days, 75-82% of pollen was deposited on the plantation (Efimov, 2010). Half of the pollen settles within a radius of 5-15 m from the source; its portion is reduced by 4 times at a distance of 20-30 m; and only 7% is carried over more than 200 m (Sannikov, Grishina 1979; Robledo-Amuncio, Gi, 2005; Efimov, 2010).

As was mentioned above, during microgametophytogenesis a microspore undergoes 5 activating mitosis, or the 5 successive steps of DNA-wave passage to higher energy levels (Figure 34; Table 3). Before pollination, a mature pollen grain is in a state of metabolic rest and has a species-specific genotype and phenotype (shape and size) (Figure 35b). The pollen outside is covered with a special firm envelope – the sporoderm with two air bags. A sporoderm separates the pollen grains as even stronger reproductive structures, from tissues of the mother tree, and protects them from the mechanical damages during pollination. During pollination, pollen grain contains the remains of two prothallial degenerated cells, large siphonogenous (by the terminology of Owens & Bruns (2000) what is known as a "tube cell") and a small generative cell (antheridial cell). The following two divisions (4-5 activating mitosis) occur after pollen germination. The last 5 divisions – the formation of sperms – takes place in the next vegetative season, before fertilization. Both sperms penetrate the female gametophyte. However, in syngamy one of the most energetically strong sperm participates.

MEGASPOROGENESIS

Development of male and female gametes and generative organs occurs asynchronously and independently from each other. Therefore, another situation on the female generative line takes place in the moment of pollination. The scheme of the female generative line, the number and disposition of its reproductive stages is shown in Figure 36.

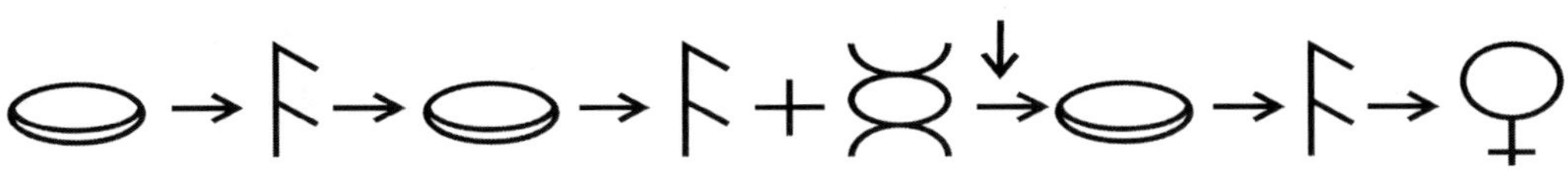

Figure 36. Scheme of female generative line of the *Pinus sylvestris* reproductive cycle.

The growth of branches with megastrobili, the formation of ovules and diploid generative cells occurs in the spring and proceeds rapidly enough. By this time, the ovule is completely formed and consists of a nucleus and an integument, at the top of which is a micropyle. The first generative cell – a primary sporogenous cell, which received a portion of biological energy during two activating mitosis of megasporangiogenesis, passes to the mitotic divisions (2nd stage ). In the center of the nucellus, 4-8 diploid sporogenous cells with the same corpuscular-wave characteristics are formed, from which only one generative cell becomes the megaspore mother cell.

Change of Space Orientation of Female Cones

During pollination, the female cones are directed upward, occupy a strictly orthotropic position and are characterized by a clearly pronounced negative geotropism. It is shown that the change in their spatial position at this time reduces the number of fertilized ovicellsby6-39% (Mc William, 1958). The period of receptivity cones is divided into three phases (Jonsson et al, 1976): 1) The initial, in which fully differentiated ovules are at the top of the megastrobili and begin to open; 2) the maximum receptivity, when the seed scales are fully open; and 3) the final, in which seed scales thicken and begin to close. The duration of the period of pollen receptivity can be different (from 8-9 days to 4 days) and it depends on the tree genotype and weather conditions (Efimov, 2010).

Once again, we will remind you that the inception period and beginning of the gametogenesis period, as compared with other periods, differs by the algorithm of development of the male and female generative sphere, the independence of two reproductive processes up to pollination and the particularity of formation of the genotype-environmental connections with respect to the climatic and substrate conditions. For example, in spring the gametogenesis by the male generative line begins from male meiosis and pollination, and by the female generative line the continuation of megasporangiogenesis process takes place. Female meiosis occurs only after pollination.

a b c d e f

Figure 37. Development of vegetative and generative sphere during geotropic motion of pine cones.

It is interesting that in relation to the weather conditions, the spatial organization of megasporogenesis (the spatial arrangement of megastrobili in the tree crown) is the same as during microsporogenesis. The development of the vegetative and generative spheres is also associated with each other (Figure 37a). Before pollination,the young shoots begin to elongate rapidly and carry the megastrobili on the outer part of the crown(Figure 31a, 37b). In this position of megastrobili, the effectiveness of cross-pollination is the maximum.

During pollination, the pollen is carried by the wind from the anther to the ovule. The megastrobili are in the phase of the "open cone" (Kotelova, 1956). The duration of this phase ranges from 4 to 7 days. In this time, female meiosis begins and the genetic material has close contact with the external environment. During leptotene, the chromosomes change their spatial position and the eukaryotic antenna array captures the new informational cosmic signal, which is fixed on the DNA-wave. After pollination, the cones are closed (the phase of a "closed cone") and begin to change their own spatial position (Figure 37c). Initially, the bending occurs in the basal part of the axis and a small angle between the shoot axis and the cone axis arises. In 3-5 days it reaches 90º, and the cone takes the horizontal position (Figure 37c). Then, during 2-3 weeks of the geotropic motion the spatial position of the megastrobili with regards to the initial changes to 180º (Figure 37d). This spatial position of megastrobili and ovules in them remains until embryogenesis ends, and the reproductive process is organized relative to their polar axis. In this upright position, a megasporocytes polar axis coincides with the gravity direction, the DNA-energy transition to higher energy levels during two meiotic divisions occurs. In the next 2-3 weeks a bend increases and cones "are pressed" to the shoot, turning its tip inside the crown (Figure 37e, f). Their polar axis again deviates from the gravity direction. The cones will remain in such a position until the spring of the next year. Therefore, the pollination occurs at a negative geotropism of the megastrobili, and meiosis takes place at a positive position (Minina & Larionova, 1979).

Simultaneously, with the geotropic motion of the megastrobili the needles of the current year begin to grow. Figure 37a shows the dynamics of their development in relation to the growth rate of needles and cone motion within a month after pollination. Figure 37d is the position of megastrobili on the 20th day after the pollination. Although the cones are turned to 180º, they are not covered with the needles. Then, the bending takes the form of arcs, and the cones are covered by a bunch of needles, which continue to growover2-3weeks(Figure 37e, f). At the end of the vegetative season, when the bumps pass into a resting state, the needles take a horizontal position. After turning, vegetative and generative buds are opened and their further development will be in direct contact with the environment. Cones, as before, separate from the environment by means of needles.

The process of megasporogenesis as the microsporogenesis is highly dependent on a weather conditions. Regarding Stupinsky's test subject, it usually occurs in late May (10-

15 days after pollination). At this time, average temperature is higher than + 10°C. Frosts are rare, and it is usually abnormally warm, sunny weather. Therefore, the main stressful factors include the solar radiation and increased temperature regime. The result of the impact of different stressors on meiosis is the expansion of the modification variability of female gametes.

Origin of Megaspore Mother Cell

In previous work, we have suggested that the choice of the megaspore mother cell from 4-8 sporogenous cells is associated with the process of pollen germination. We proceeded from the date that, firstly, the female experiences meiosis without the pollen grains in the pollen chambers, or does not occur or stops after the first meiotic division, which leads to cones falling off during the 1styearof development (Khromova, 1985). Secondly, a few pollen grains germinate in the nucellus, and participate in the fertilization of only one sperm. If this is true, then after pollination the functions of the male and female generative sphere are two interdependent and closely interrelated processes.

As was already mentioned, during germination a pollen grain consists of two cells – the energetically strong generative cell and energetically weak siphonogenous cell. These two oscillatory systems are in a rest state. In physics, the equilibrium at the critical point is described. In this state, the oscillating function of a DNA molecule stops, but the biological harmonic oscillator, as a physical system, is not destroyed. All metabolic processes are slowed down. A cell becomes a closed system. Pollen can leave the resting state only by means of an external force.

Pollen enters into the pollen chamber together with the secretory liquid, which stands out by the nucellus cells every day during all periods of pollination. Apparently, this fluid is the trigger, which awakens the pollen grain from its dormancy. Germination of pollen is realized through a special furrow which directly contacts with the cytoplasm of siphonogenous cell. The outer layer of intine in the furrow location is thinner, and consists of cellulose and pectin (Martens & Waterksyn, 1962; Trenin, 1988). Secretory liquid through intine begins to absorb by the siphonogenous cell. It gradually swells, is activated and goes out of dormancy. Investigations *in vitro* have shown that the formation of the pollen tube precedes the processes of pollen activation: Increased respiration intensity, activation of mitochondria, the use of pollen storage substances, development of membrane system of the dictyosomes, and participating in the formation of the pollen tube wall (van Went & William, 1984). Then, a siphonogenous cell begins to germinate, but the generative cell remains in its resting state until the next spring.

As it has been shown, the male gametophyte before pollination has passed the most part of the reproductive development by the male generative line (Figure 33). The female generative cells are found only in the beginning of the female generative line. We think that the entering of secretory liquid into the siphonogenous cell may be the external force that breaks the equilibrium state of its biological harmonic oscillator (at the critical point). The wave function on the DNA molecule is resumed. DNA-wave, as the self-oscillatory system, becomes a biofield source. A higher level of the biological energy of siphonogenous cells guarantees the entering of nutrients from a nucellar tissue. Biofield forces provide the attaching of pollen grains to the nucellus tissue, activation of all intracellular processes and the beginning of pollen tube formation.

The oscillated DNA-wave radiates the biofield. The radiation pulse falls on all cells of the ovule (nucellar and sporogenous cells), which are also self-oscillating systems. The DNA-wave of the siphonogenous cell in this case acts as the external force. If we proceed from the wave nature of the process, the coincidence of its frequency of oscillation with the frequency of one of the sporogenous cells leads to a resonant response and the maximum values of the amplitudes of DNA oscillations are achieved. The energy potential of this sporogenous cell is increased. It begins to grow and becomes the megaspore mother cell.

During a geotropic motion of cones, a polar axis of activated sporogenous cells ceases to coincide with the gravity direction that does not affect the process of megasporocyte formation. However, the initiation of female meiosis in these generative cells is excluded. Therefore, the cones for a period of growth and maturation of the megaspore mother cell should change their spatial position – with negative geotropism (orthotropic orientation of cones) on the positive ones.

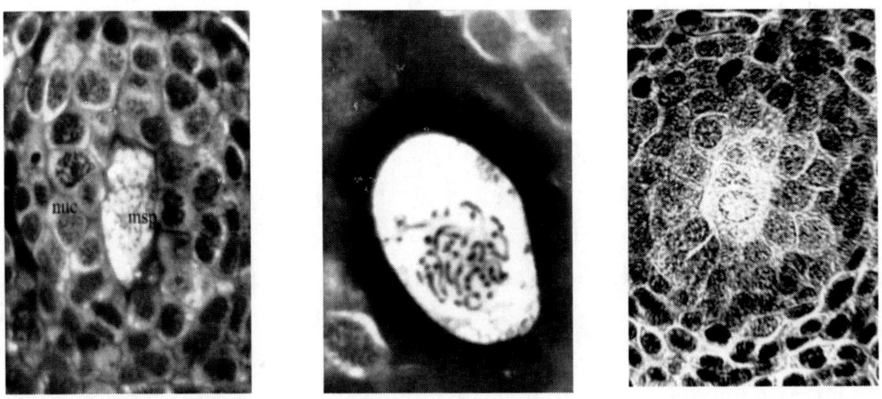

a b c

Figure 38. Megaspore mother cell before meiotic division (a); prophase of meiosis (b) and tetrad of microspores (figure c, courtesy of Dr. V.S. Svintsova).

After increasing the energy potential, the activated sporogenous cell is growing rapidly and elongates along the long axis of ovicell (Stewart & Gifford, 1967; Kozubov et al., 1982) (Figure 38a). Its cytoplasm is also polarized. The distribution of plastids and mitochondria is uneven. They are mainly concentrated in the chalazal part of the generative cell. After turning the cones and achieving a specific size, a megasporocyte undergoes two meiotic divisions (Figure 38b).

FEMALE MEIOSIS

Juel (1900) was the first who studied in detail meiosis in megasporocytes of larch. The scheme and order of its divisions are common for all conifers. In the premeiotic interphase of megasporocyte, a DNA replication takes place. After 1st meiotic division, a chalazal and a micropilar cell are formed and a reduction of chromosome number takes place. Then, chalazal and micropilar cells undergo second meiotic division and a tetrad of microspores is formed. Only one cell (lower and larger) of a tetrad becomes the functional megaspore and continues the reproductive process, and the other 3 megaspores soon die.

Female meiosis or activating meiosis represents a combination of two types of divisions – an ordinary (male) meiosis and activating mitosis (Kuznetsova, 2013). The Cell cycle of megasporocyte begins from the DNA-wave radiation. This energy initiates a process of DNA-particle doubling. The transition from mitosis to meiosis occurs in the leptotene, as a result of a change in the spatial position of chromosomes within the microsporocyte nucleus. Reception of a new information signal begins also from the leptotene. Zygotene and pachytena are the stages of realization of two important processes: Homologous chromosomes conjugation and crossing-over mechanisms. During crossing-over, a genetic material of the DNA molecule is changed and each female gamete receives the individual species-specific DNA-particle. Simultaneously, a megaspore, due to the filtration of the chromosomes in the zone of a non-uniform magnetic field becomes the sexual generative cell with a certain spin direction (with the spin↑ or with the spin ↓).

It is necessary to take into account that female meiosis, unlike male meiosis, presents the reproductive stage, which definitely organizes in the space. During activating meiosis, a movement of energetically weak daughter cells of the basal pole (chalazal cell, nonfunctional megaspore) corresponds to the direction of gravity, and a movement of energetically strong daughter cells of the apical pole (micropilar cell, functional megaspore) occurs in the opposite direction (Table 3). Meiosis ends by the formation of the functional megaspore, which are characterized by the individual species-specific DNA-particle, with a higher level of DNA-energy, renewed DNA-information and the certain spin direction.

Disposition of macrospores into a tetrad has, as a rule, the linear form (Figure 38c). The deviations of linear form occur in the second meiotic division and relate with divisions of the cells of the basal pole (Kozubov et al., 1982). IFor example, this violation of meiosis does not affect the chief direction of biological energy movement, namely, on the genetically determined process of DNA-energy increase from a megaspore mother cell to the micropilar cell, and then from a micropilar cell to the functional megaspore.

To summarize the above, it should be emphasized that female meiosis is a unique stage of the reproductive cycle. Within a single cell division,4 new basic states of the genome are simultaneously formed. All changes are connected with the DNA molecule (particle and wave). The end of this stage in the scheme of the female generative line is marked by a vertical line (Figure 36). During activating meiosis, the transition of generative cells from the diploid to the haploid state takes place. The differentiation of gametes by the spin orientation arises. DNA-wave receives the new information signal and on its basis the individual species-specific DNA-information of a female gamete is formed. DNA-energy goes down on the higher energy levels and rejuvenation of the genome of functional megaspore occurs (Table 3). Crossing-over provides a random, independent recombination of genes, the result of which is the expansion of genotypic variability of female gametes on the basis of a DNA-particle. In that a megasporocyte and a functional megaspore are two different genetic structures after meiosis, the connection between the male and female generative sphere disappears.

Megasporogenesis proceeds on a stable substrate and in changeable weather conditions (Table 4). The impact of unfavorable weather factors on the genetic material of female gametes increases the modification variability of their population gene pool (Table 4).

As was mentioned above, meiosis is a special type of division, at which the replication of chromosomes takes place only once, and the coiling of chromosomes and nucleus division occurs twice – during the 1st and the 2nd meiotic divisions. According to our understanding, at the time the coiling of chromosomes takes place the process of the accumulation of elastic energy inside the super-compressed spring of DNA molecule occurs as well. Metaphase chromosomes are the second independent source of internal energy of the eukaryotic cell. During telophase, each daughter cell receives an equal amount of free energy in the moment of the decondensation of metaphase chromosomes. Its energy is distributed through several channels, foremost among which are: The energy supply of the processes of transcription, cell growth, restoration of ATP from ADP, creation of the electromagnetic field and excitation of biomolecules. This source together with the energy of a biological harmonic oscillator, photosynthesis and respiration provide the energetic possibility of proliferous eukaryotic cells.

Metaphase chromosomes (especially bivalent ones formed during the 1st meiotic division) are always shorter and thicker than mitotic metaphase chromosomes (Darlington, 1965; Swanson & Webster, 1977; John, 1990). This fact testifies that in them is accumulated a larger amount of elastic energy. All this energy, which is liberated during telophase I and II in their uncoiling, enters into the dyad and tetrad of megaspores, and, accordingly, into the "rejuvenated" functional megaspore.

MEGAGAMETOPHYTOGENESIS

A functional megaspore is an energetically strong generative cell and the first cell of a future female gametophyte. Moreover, this gamete receives free energy during two divisions of meiosis. Therefore, it begins to grow rapidly before the division. The ultrastructure of its nucleus and cytoplasm also are changed significantly. The quantity of heterochromatin in a nucleus has decreased. The nucleus of a megaspore is large, and its size remains unchanged during cell growth. The cytoplasm in the process of its volume increase is gradually vacuolated. The nucleus is situated at its micropilar end. The nucleocytoplasmic ratio of a functional megaspore is altered, whereby before the 1st cell division the conditions are created for the beginning of the free-nuclear mitosis.

By this time, megastrobili end the geotropic motion and female cones are pressed to the shoot and separate from the environment by means of needles (Figure 37 e, f). A functional megaspore is the polarized cell. However, after cone turning its polar axis doesn‘t coincide with the direction of gravity, which is a prerequisite for the beginning of the activating mitosis. During free-nuclear mitosis the mitotic spindle of chromosomes does not coincide with the direction of gravity, the nuclei are divided without cytotomy and a free-nuclear female gametophyte is formed. Development of the female gametophyte in this period is carried out at the relatively normalized conditions. On the female generative line, this is the 3rd stage by the emergence of a new reproductive initial -  (Figure 36).

Initially, the female gametophyte is a haploid. In the ovule, it occupies a basal position. The most rapid growth of the gametophyte takes place during the first free-nuclear divisions. The first divisions of nuclei occur synchronously, but then, this synchronicity is disrupted. A gametophyte has an oval shape. In its center, a large vacuole is disposed and the nuclei with the cytoplasm occupy a parietal position. By this time, the starch disappears in the cytoplasm and the number of lipid droplets decreases. On a free-nuclear stage, a particular envelope around the female gametophyte is formed – the sporoderm. The sporoderm is found in all species of conifers. Initially, this is a thin homogeneous layer, but with the development of the female gametophyte it is thickened and differentiates into two layers – the endosporium and exosporium. The exosporium consists mainly of suberin and the endosporium – from pectin and cellulose (Trenin, 1988). According to electron microscope studies of the sporoderm, there are three layers: The inner, homogeneous; the medium, consisting of small globules; and the outer, forming protuberances (Kozubov et al., 1982; Kuznetsova, 1994). The metabolism is changed in the direction of accumulation of proteins and lipids.

The female gametophyte of *Pinus sylvestris* hibernates at the stage of 32 free nuclei (Fergusson, 1904). According to the laws of physics, the functioning of cells and plants, as the quantum-mechanical systems, is a temperature function and only in a certain temperature is diapason possible (as a rule from 0-5 to 40-45°C) (Kuznetsova, 2013).

Beyond these boundaries, the DNA oscillations are stopped. Therefore, at the end of every vegetative season the trees and their cells pass to the resting state. Transition to the resting stage is a genetically determined and environmentally dependent process. With a decrease in the duration of daylight hours and a night temperature that falls dramatically, these plants begin to prepare for winter. The radiation of their oscillators occurs less frequently, and, accordingly, the mitotic activity of cells is reduced. Plant growth is slowed down and then stops. The intensity of the metabolic processes also decreases (Henkel, 1980). The result of temperature fall below 5°C is a cessation of biological harmonic oscillator action and a biofield disappearance. An oscillator goes into a state of unstable equilibrium, but it is not destroyed. The speed of metabolic processes decreases sharply. The respiration intensity becomes less than 200-400 times. The cytoplasm moves away from the envelope. Cells and plants pass to a state of winter dormancy.

GENOTYPE-ENVIRONMENTAL CONNECTIONS IN THE YEAR OF POLLINATION

A pollen grain is the last reproductive structure on the male generative line. Each separate gamete represents the intact isolated genetic system. Therefore, pollen germination and pollen growth can be considered as the development of independent organisms–a genotype of the male gametophyte, which has an individual corpuscular-wave genome, characterizing the certain life duration and specifics of development. Competition and selection of the most viable and qualitative genotypes of male gametes occurs on the basis of their genetic variation. After pollination, the development of male and female gametes represents two closely interrelated and interdependent processes.

Formation of a population gene pool of male gametes begins with a departure of pollen from the microsporangiums. Megastrobili at this time are characterized by a maximum receptivity: The seed scales of cones are fully disclosed (Figure 31a, b.), and the micropylar channel has a maximum width. Rapidly elongating shoots of the current year carry out the megastrobili on the outer part of the tree crown. This promotes better access to the pollen during pollination. In the morning, an ovule in the micropylar channel secretes the drop of rich sugar secretory fluid, which helps to capture and transport the pollen grains to a nucellus (Mc William, 1958; Minina & Larionova, 1979). Pollen grains are wetted by a liquid, and the air bags determine their orientation with respect to the nucellus with the distal end downwards.

As a rule, 1-5pollen grains fall into the pollen chamber. At this time, the male gamete is at rest. Its germination begins in 2-3 days after contact with the nucellar tissue. In *Pinus sylvestris*, unlike a species of flowering plants, the morphological devices which prevented self-pollination are absent. Therefore, the pollen, caught in the pollen chambers during open-pollination, consists of its own (inbred) and heterogenous (outbred) pollen. Withrespect to each mother tree, its genotypic composition will be considered as the initial spectrum of genotypes of male gametes.

A gene flow, which is transferred by pollen during panmixis, is the bearer of genetic information of the species (Koski, 1970; Knox, 1990). During pollination, the pollen is carried by the wind into pollen chambers of their own and other trees of the population, within which they are able to experience panmixia. It was found that in forests with close panmictic connections,200-1000 trees are united in the area of 1-2 hectares (Wright, 1953). Each plant within a population has a certain ecological niche, and, accordingly, microstrobiles in the tree crown also occupy different "micro-niches." Hence, it follows that the initial spectrum of genotypes of male gametes in addition to the genetic information bears other information about the climate situation on the territory of the panmictic population in the year of the pollination.

After pollination, the cones are closed. The edges of integument gradually shift, a passage in the micropyle is overgrown, and secret production ceases. Further development of pollen tubes, ovicells and embryos will take place within closed systems of female cones until the departure of seeds. As it has been shown, after the geotropic motion of megastrobili all reproductive processes are moved into the crown of the tree. Running time lasts about a month. During this time, the size of cones increases more than twice (Figure 37a). A bunch of needles are separated from the environment, where the temperature diapason is smoothed by4.0-6.0ºC for the maximum and by3.0-4.0ºC for the minimum temperatures (Protopopov, 1977). Therefore, their reproductive development occurs in relatively (stable) micro-climatic conditions (Table 4).

As was mentioned above, during microsporogenesis and megasporogenesis the extension of the genotypic and modification variability of male and female gametes on the basis of a DNA-particle occurs. The internal mechanisms (crossing-over) and external factors (weather conditions, different spatial disposition of trees within a population and reproductive organs within their crown) are taking part in this process. Scots pine is a very labile species in terms of vegetative and reproductive characteristics, sexualization type, number and range of genetic abnormalities. Therefore, the curves of genotypic variation of pollen have a symmetrical distribution regarding the genotype of their own mother trees. A population gene pool of a new generation is created on the basis of the genetic material of each mother tree and carries on itself the imprints of the genetic state of pollen grains and the specificity of their genotype-environmental norm reaction.

During reproductive development, a bisexual mother tree produces two types of gene pools – pollen grains and ovicells. Unlike pollen, the process of female gamete formation on the female generative line continues untilfertilization and the ovicells do not leave their mother plant. In the optimal years, the initial spectrum of ovicell genotypes depends on various factors: The vitality state of a tree (plus, minus trees), the cone abundance of a tree, the weather conditions, the potential seed production megastrobili, the violations of genetic material, etc.

In the life cycle of this plant, the panmixia is one of the key mechanisms, which integrates the individual trees of any population into a unified whole. The ability of cross-pollination gives any population gene pool such properties as stability, uniformity, etc. (see Part II, Chapter 4). As is known, genotypes of male gametes in the process of pollination are involved only. By this time, a pool of pollen represents a plurality of independent genotypes, each of which is able to independently participate in the further reproductive process of the male generative line (formation of functional sperm and fertilization). During panmixia between genetically heterogeneous trees occupying different ecological niches within a panmictic population, the complex mutual exchange of genetic information occurs. Therefore, the initial pollen spectrum of each tree before and after pollination is significantly different by its genotypic composition. In comparison with their own pollen, the genotypes of different trees are enriched. The curve of variability of this new gene pool is a result of the amalgamation of a great number of the ecological and genetic information and norm reactions of different trees with a mode in the zone of norm-forming genotypes.

Panmixia continues within a population and is the main mechanism of genetic material mixing and formation outside of intraspecific diversity of the population gene pool (Table 4). During pollination, each mother tree creates the initial spectrum of male gamete genotypes of a higher level. In fact, this is the top of its genotypic diversity, because a pollen gene pool is formed on the basis of three types of variability – the genotypic, the modification and with the allogamy participation. Theoretically, its composition at this moment tends to be the genetic diversity of the species gene pool. The main stages of formation outside of the intraspecific diversity is, firstly, the complete removal of their own pollen from anthers, secondly, the inclusion in the reproductive process of each tree in different proportions of inbred and outbred male gametes and, thirdly, the selection among them of the most competitive, adapted, outbred and without genetic anomaly genotypes of pollen.

During pro-embryonic selection, the maximally extended initial spectrum of pollen genotypes is corrected and narrows. Therefore, the population gene pool is led to compliance with the habitat by the principle of "genotype-environment" interaction. In this plan,2-4 periods are united under a number of common features. As the analysis showed, during the formation of population gene pools a genetic material of pollen, embryos and plants undergoes 4 types of the same transformations: 1) Expansion of variability by means of internal (fertilization) and external (cross-pollination) mechanisms, and under the influence of external factors (phenotypic variability of plants); 2) autotrophic period of development (pollen tube embryo, seedling); 3) selection (direct, indirect, inbreeding depression) in the course stages; and 4) selection (direct, indirect), which is realized in the transition from one stage to another or during the next stage of the life cycle. In the end of each period, some questions of biology concerning male and female gametes, embryos and plants, and proceeding from their corpuscular-wave nature will be discussed.

AUTOTROPHIC STAGE OF POLLEN DEVELOPMENT

The germination of pollen grains and pollen tube growth presents the existence of an independent organism – the male gametophyte. In the first vegetative season, its development is usually divided into 5 stages: 1) Germination and intensive promoting (autotrophic period); 2) linear or vertical growth of pollen tubes; 3) branching tip tube or horizontal growth of pollen tubes; 4) termination of growth; and 5) rest period (Kotelova & Khromova, 1974; Kuznetsova, 1991). It is necessary to take into account the fact that the pollen germination, as a biological phenomenon, does not cause the change of the genome of the generative and siphonogenous cells. Therefore, pollen grains before and after they exit from the dormancy have approximately the same genotypic composition of the population gene pool by the state and variability of their DNA-particle, DNA-energy and DNA-information.

Pollen grains after some days after entering into the pollen chamber and attach to the nucellus surface begin to germinate (Figure 39a). Germination needs, firstly, the water and nutrients, mainly, carbohydrates and mineral salts, which are contained in the secretory fluids (Poddubnaya-Arnoldi, 1976). Secondly, the optimal temperature regime is required, because the higher (more than 30ºC) or the low (below 10ºC) temperature reduces its viability. Therefore, the temperature and the pollination fluid, secreted by the nucellus every day during pollination, are the triggers, which take out the pollen grain from its dormancy.

Nucellus has the shape of a shallow funnel (Figure 39a). Therefore, when a pollen grain gets into the camera, it is located in the center of the nucellus. Thanks to the air bags, its distal pole comes in contact with its cells, which increases the efficiency of pollen germination. When 2-5 pollen grains get into the pollen chamber, as a rule, the first of them occupies the central position on the nucellus (Figure 39a). This pollen grain plays a leading role in the choice of one of the female 4-8 sporogenous cells, its activation and its transformation into a megasporocyte. The remaining pollen is arranged around. The pollen participates in the selection of male gametes and fertilization of different archegonia, or is eliminated from the reproductive process. Therefore, it is important that the first pollen grain was a viable, outbred and of best quality sample.

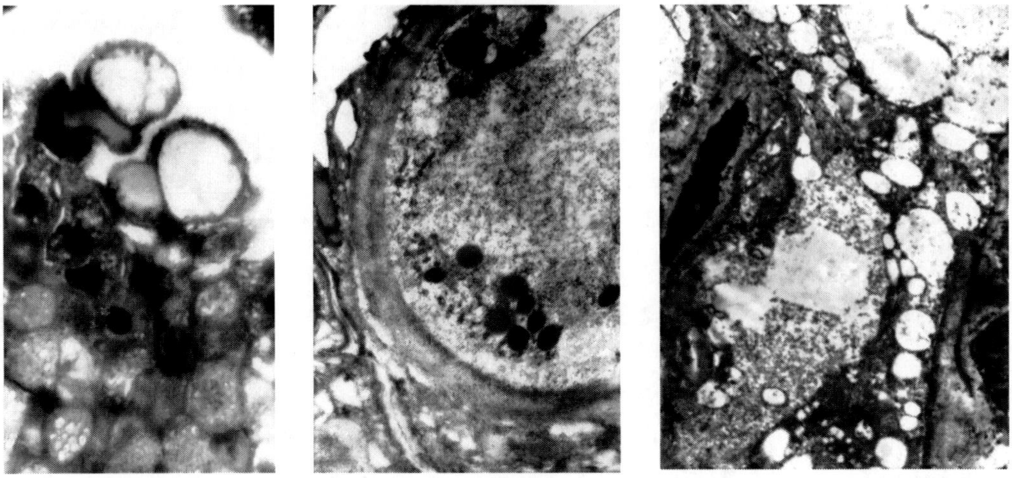

A B C

Figure 39. Pollen grain germination (a) of *Pinus sylvestris* after open pollination: ultrastructure of the cytoplasm of probing pollen tube (b) and pollen tube on the stage of unbranched vertical growth (c).

The quality of the first pollen in chambers is greatly facilitated by a feature of pollination biology. Depending on the weather conditions, the process of Scots pine pollination continues from 4-5 up to 7-10 days. Its peak is in the middle date of this period. As is known, the mature anthers of any tree have a certain number of abnormal pollen grains (Kupriyanov, 1989; Tsapenko, Sinelnikov, 2012). Most quality is the first pollen party, and lower quality is the last one. Therefore, the probability of pollination by the pollen of the first party is high enough in the early and middle flowering trees. Among the later flowering trees, a significant portion is of partially self-fertile or self-fertile forms. Secondly, the female and male spheres of the majority of trees develop asynchronously. The start of their pollination and the beginning of the pollen mass departure occurs before the "open cone" phase, when the megastrobili are active during the initial period of cone receptivity or are still closed. They are opened on the following day, or sometimes 2 days later. Therefore, self-pollination takes place. Accordingly, the inbred pollen in the pollen chambers gets from the first and the last party, which is removed from reproductive process during selection.

Germination of the pollen takes place through a special groove, which is a more thin area of exine (Kozubov et al., 1982; Trenin, 1988; Knox, 1990; Williams, 2009, etc.). Groove germination inside a pollen grain contacts the siphonogenous cell. Therefore, during germination a generative cell continues to lie in dormancy, and a siphonogenous cell is activated. At the moment of awakening from the rest state, its DNA-wave resumes the oscillating function and the biological harmonic oscillator again begins to work. A siphonogenous cell, as an oscillatory system, becomes the source of the biofield. Respiration, osmoregulation and metabolism are restored. Investigations *in vitro* have shown that the formation of the pollen tube precedes the processes of pollen activation:

Increased respiration intensity, activation of mitochondria, the use of pollen storage substances, and development of a membrane system of the dictyosomes participate in the formation of the pollen tube wall (van Went & William, 1984).

Initially, a probing pollen tube is formed – the autotrophic stage of the male gametophyte development. At this time, the growth of the tube occurs at the expense of pollen grain nutrients. Probing a pollen tube passes between the nucellus cells in this tissue. At this time, the cytoplasm of a tube is weakly vacuolated and contains a large number of organelles: Mitochondria, plastids with the starch inclusions, lipid droplets and ribosomes (Figure 39b). Its tip turns to the nucellus, where its cells are filled by a lipid and starch. During germination, a pollen tube loses the sexine. Then, the elongation of the siphonogenous cell in the pollen tube begins. Its development is done by the principal of differentiated cells. The cytoplasm is gradually vacuolated (Figure 39c). A generative cell remains in the pollen grains until the next of the vegetative season. Electron microscopic studies have shown that the differences in the ultrastructural organization of inbred and outbred tubes on the autotrophic stage are not observed (Kuznetsova, 1991, 1996).

Direct and Indirect Selection

In conifers, between the male and female gametes is a layer of diploid tissue – the nucellus. Since the nutrients, which are contained in the pollen grain, is insufficient to the formation of necessary length pollen tube, it comes into contact with the cells of nucellar tissue. It will provide its trophic function up to fertilization. As a consequence, the male gametes go over to the heterotrophic way of feeding.

The nucellus is formed during megasporangiogenesis (Table 3). In the spring, a nucellar tissue of ovules presents the undifferentiated, poorly vacuolated, meristematic cells with a nucleus, located in the center. At the time of pollination, a starch in the micropylar (apical) part of the nucellus gradually accumulates, lipid droplets appear and a cell vacuolization begins.

The pollen tube after germination and transition from the autotrophic to heterotrophic way of nourishment rapidly is elongated (a stage of unbranched vertical growth). When several pollen grains on a nucellus germinate, a leading pollen tube among them stands out, which is characterized by more rapid growth. It moves into the nucellus tissue at the greatest depth. Development of a male gametophyte is accompanied by the formation of the growth zone, vacuolization and active movement of cytoplasm (De Win et al., 1996). The area growth is small and filled with numerous vesicles, the formation of which is related with the dictyosomes activities. In this zone, a new plasma membrane and cell wall of the primary wall is formed. The cytoplasm occupies a thin parietal layer. During unbranched growth, pollen tubes are in close contact with nucellar cells. Nearby cells are destroyed, and their content is used for the metabolism. The callose layer is formed over the tip of the tube as the uneven outer layer surrounding its envelope. Callose formation connects with the incompatibility processes and incomplete destruction of nutrient substances, which in conditions of degradation are deposited on the surface of the tube (Tupy, 1959; Heslop-Harrison, 1978; Kuznetsova, 1996).

During the stage of unbranched vertical growth the differentiation of pollen tubes by the length occurs. In previous works, we suggested and showed that this phenomenon is genetically determined as a corpuscular-wave nature and laid down in the genome of the pollen. Differentiation of male gametes by the DNA-energy and DNA-particle create preconditions for the subsequent genotypic selection.

In this respect, it is necessary to distinguish the energetic differences of higher and lower levels that exist between the generative and vegetative cells, the lower and higher plants. The higher level is the differences that arise between the daughter cells after the activating mitosis and activating meiosis (generative cell – siphonogenous cell of pollen; functional magaspore – nonfunctional megaspore). Likewise, the differences between a species, which are on the different levels of evolutionary development are as follows: Lower plants experience1-3 activation stages and higher plants experience4-6 stages. A lower level of genotypic differences are the differences within a single species-specific energy level. We will describe how the differences between genotype-leaders and genotype-outsiders occur. Under the genotype-leaders and genotype-outsiders, we understand those genotypes (phenotypes) of pollen, ovicells, embryos and plants, in the genome of which a more high or more low biological energy within the range of its species-specific variability was laid, are realized in the course of their development in the form of increased or decreased morphological, functional and metabolic activity.

Trees in the struggle for existence compete with each other both directly and indirectly. The different speed of their growth depends on the nature of the genotype, and the character of genotype-environment relationships with the substrate and climatic conditions. In the first case, the stability and survival of plants provide the genotypes with a minimal number of genetic violations and with the highest level of DNA-energy. Tree differentiation by the DNA-energy is the basis for their direct competition during ontogenesis (Solbrig & Solbrig, 1982). This phenomenon has a wave nature. Genotypes-leaders form powerful trees, which provide maximized accessibility of nutrients and ensure the survival of species in the conditions of a fluctuating environment.

Direct competition between pollen tubes is absent, because the siphonogenous cells have an insignificant number of energetic levels. The indirect competition is considered as a mediated process in which the most compatible genotypes with the nucellus receive a main nutrient resource. Genotypic diversity of gametes, which is greatly expanded in the cross-pollination, is regulated by the diploid tissue of the nucellus. The growth of pollen tubes provides its nutrients and the compatibility degree of the DNA-particle of the mother tree and male gametophyte. This leads to a different speed of growth tubes and their death on the different stages of formation. Dynamics and morphology of the pine male gametophyte shows that the processes of incompatibility occur between outbred pollen tube and the nucellus (Kuznetsova, 1991, 1996, 2003, 2012). A place of their action is the tube envelope. These processes after open pollination lead to the callose accumulation on the surface of pollen tubes, to the termination of their vertical and horizontal growth, the tube transition to a rest state, and the different growth rate of gametes into the nucellus. A mother tree, a genotype of which is compatible with the environment, as a sieve passes the entire spectrum of genotype pollen through itself. Reproductive success is ensured by the male gametes, which are most compatible with them, and, as a consequence, with the environment.

Studies have shown that a cone in the stage of the tube experiences unbranched vertical growth in direct contact with the environment and their development is dependent on the climatic conditions of the year. Young needles are short and do not cover the megastrobili on the outside. A weather stress may accelerate, decelerate or cease the tube growth depending on the adaptive potential of the genotype. Pollen germination and the beginning of pollen tube growth are sensitive stages to environmental stress, which is confirmed by the date of reduction of the cone number over a period of their development. The experiment was carried out for40 of the same trees over3 years (1992, 1993, 1994). Figure 40 shows the dynamics of cone preservation during two optimal years,1992 and 1993.

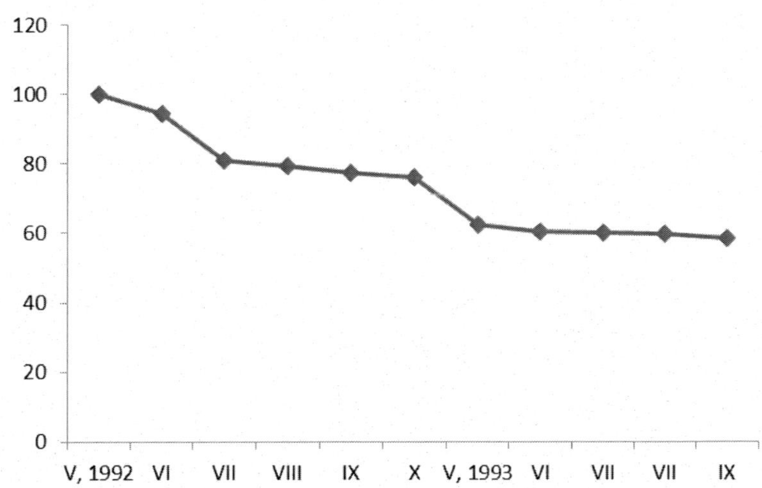

Figure 40. Dynamics of the preservation of pine cones after open pollination.

The preservation of cones depending on weather conditions is ranged from 50 to 75%. Two peaks of their diminishing are revealed – the second month after pollination, and in May of the next year. The first peak is connected with the death during pollen germination and the beginning of pollen tube growth; the second is connected to the output of pollen tubes from winter dormancy. Increased mortality of male gametes at this time leads to the death of ovules and cone fall.

The third stage of development of the male gametophyte is connected with the formation of branched pollen tubes. A tube envelope after the stopping of apical growth in the thinnest parts is broken and forms a series of cytoplasmic outgrowths that disjoin the cells of the nucellus and are wedged between them (Figure 41a-c). Vegetative nucleus of the siphonogenous cell passes in the pollen tube and remains before branching. The number of outgrowths is not more than 4-6, i.e., equal to the number of future archegonia. They grow slowly and horizontally down in the four different directions (Figure 41a, b). Probably, the tube outgrowths are oriented in the direction of the possible location of future archegonia, in order to be as close to them. Horizontal growth of pollen tubes occur in the relatively normalized conditions, because cones end geotropic turn and the long young needles are in the shade. If the male gametophytes during vertical growth tend to penetrate deeper into the nucellus, during horizontal growth they try to cover the area. Such a growth algorithm, firstly, gives them a competitive advantage during fertilization, and, secondly, the basis for a change of leadership of male gametes. At the time of horizontal growth, cytoplasmic outgrowths are gradually covered with a callose layer (with 4 stages of tube development). Then, tube growth begins slowing down, and it goes into a resting state up until spring (the last 5 stages).

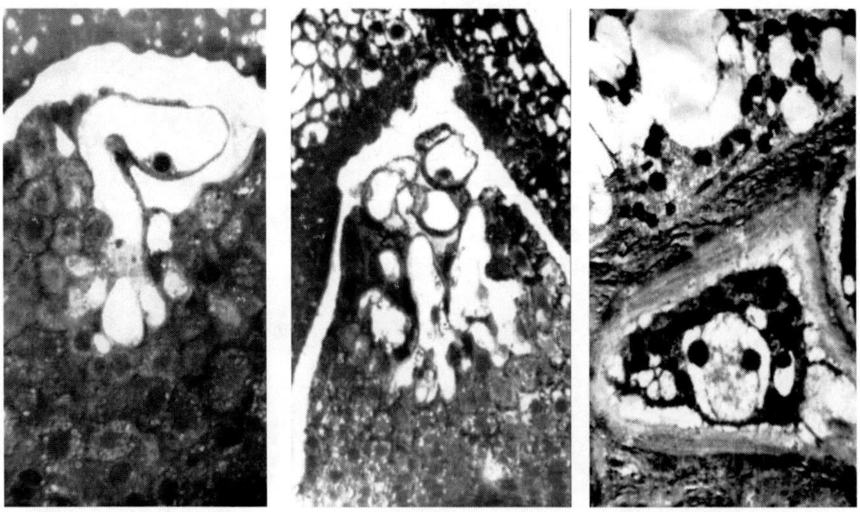

a b c

Figure 41. Formation of branched pollen tubes (a, b) and ultrastructure of cytoplasmic outgrowth (c) on the stage of horizontal growth, open-pollination.

In July, pollen tube development or a channel for syngamy of gametes is fully completed. At this time, male gametophytes form branched pollen tubes, germinated in the nucellus tissue at different depths. Male gametophytes, which are outstripped in growth rates, have a greater chance to become the genotype-leaders, with the lagging experienced by the genotype-outsiders. Their energetically strong generative cells reside inside a pollen grain during the dormancy until the following spring.

The population gene pool of pollen, depending on their reaction norms, is differentiated on the stable, modal and sensitive groups of genotypes that have different resistance to drought, frost and other regional stressors. During the 1st vegetative season, megastrobili have small sizes and are in direct contact with the external environment. Therefore, the influence of weather factors on the processes of survival, selection and development of male gametes is sufficiently large. The selective advantages have mostly adapted (by the principle of the "genotype-environment" interaction) to the regional weather condition genotypes.

During the vegetation season, a selection of the most viable ovules takes place. In this regard, an important parameter of the generative sphere is the ratio of seed scales in the fertile area and their total number in the cone, which is the relative value of the fertile zone. In the Voronezh region, these parameters are in the range of 18.3 – 38.0%, and the average value of the sample is 25.0%. Its formation and size depends on a number of external and internal factors. There are data of genotypic conditionality of the trait concerning the "relative value of the fertile zone" (Abaturova et al., 2002).

Table 5. Influence of mother trees on the structural and dimensional traits of the megastrobiles of *Pinus sylvestris* in the Voronezh region

№	Trait	Influence on the factor, %
1	Potential seed production	56.7
2	Relative value of the fertile zone	44.4
3	Total number of seed scales	28.6
4	Megastrobili length	22.7
5	Length of seed scales in the fertile zone	66.3
6	Length of ovule	72.7

Investigations of seed reproduction in the pollination year (year n+1), carried out in the forest zone have shown that the potential seed production of Scots pine is in the range of 20-50 ovules in the strobile (Romanovsky, 1997). In the forest-steppe zone (Stupinsky test-object), this trait has a wider border and is in the range 14-76 ovules. The potential seed production is characterized by higher variability than parameters of the relative value of the fertile zone (Svintsova, 2002). Other traits of megastrobili have a relatively low level of variability in the 1st vegetative season. The variability of the trait "megastrobil length" for the whole sample is 10.0%; the "megastrobil width" is 5.9%. It should be noted that the influence of individual peculiarities of the mother trees on megastrobile length in this period is only 22.7%. For the other structural traits of cones, the significant influence of individual peculiarities of the mother plants has been established (Table 5).

Table 6. Character of the connection between the morphometric parameters of *Pinus sylvestris* megastrobili in the year of pollination and fertilization

Traits	Values of coefficient linear correlation, R, %
August end	
Length of megastrobil – the total number of seed scales	0,45*
Length of megastrobil – the relative value of fertile zone	0,58*
Length of megastrobil – the potential seed production	0,67*
Width of megastrobile – the total amount of seed scales	0,38*
Width of megastrobile – the relative value of fertile zone	0,61*
Width of megastrobile – the potential seed production	0,69*
May end of next year	
Length of cone – the total number of seed scales	0,71**
Length of cone – the relative value of fertile zone	0,56*
Length of cone – the potential seed production	0,69*
Width of cone – the total amount of seed scales	0,38*
Width of cone – the relative value of fertile zone	0,56*
Width of megastrobile – the potential seed production	0,70*

Note: * - $P \leq 0{,}05$; ** - $P \leq 0{,}1$.

In the fertile zone, a low level of the variability of size seed scales is revealed, and coefficients of variation did not exceed 8%. The greatest influence on the variability of megastrobili – the length of fertile seed scales and ovules – have the individual peculiarities of mother trees. The strength of the influence of this factor is, accordingly, 66.3% and 72.7%. Influence of the differences at the cone position in the crown on the variability of the sizes of fertile seed scales and ovules is considered. Endogenous variability is very low,no more than 3%. A positive correlation between the dimension and structural traits of megastrobili is shown (Table 6).

In optimal weather conditions, the reduction of potential seed production (in the period from pollination to fertilization) is relatively low – 6-20% (Svintsova, 2002). Apparently, the high viability of ovules has caused a sufficient concentration of pollen on the plantation. At the same time, the differences in the degree of ovule elimination between individual trees within one year, as well as between generations of ovules in various years are established.

Analysis of the morphology and size of ovules has shown that the cessation of ovule development in the fertile zone of megastrobile occurs at the beginning of gametogenesis and, apparently, due to their insufficient pollination. Trees of any population in with ahigh concentration of pollen in the air can be in different aerodynamic conditions. Therefore, parts of ovules is un-pollinated areas and their development is stopped. These ovules in the mature cones represent the "point" structures, as a lionfish is not separated from the seed scale. The pest damage of 1-year cones is the additional factor (with un-pollination) for the decline of potential seed production of pines.

To summarize the above, it should be noted that after pollination a new population gene pool of male gametes into pollen chambers of mother trees has formed. This genotypic composition is characterized by the higher level of genetic diversity, because it represents the spectrum of the pollen of outbred and inbred genotypes, i.e., the outbred male gametophytes from different mother trees of panmictic population and the inbred gametophytes from various ecological micro-niches of their own trees. A nucellus of the mother plant after pollen germination is the habitat of this new gene pool. During gametogenesis, the male gametes appear with each other in the indirect competitive relationships. The nucellus towards their genotypes displays the adaptive substrate, because mother trees passing through all stages of natural selection are well acclimatized to the regional conditions.

Megastrobiles make the geotropic turn in pollen germination and vertical growth of the pollen tube and are in close contact with the environment. In May, the weather is very unstable. Therefore, a development of male gametophytes during this period is highly dependent on weather conditions as well as the phenocritical stages of the reproductive cycle. The selective advantages have more adapted genotypes.

On the male generative lines, open pollination produces the high dispersion of reactions by the compatibility with the nucellar tissue – from the inbreeding compatibility to the outbreeding incompatibility of pollen. As a rule, one nucellus germinates from 1-4 pollen tubes, which are beginning to compete with each other directly and indirectly – through the genotype –via environmental contact. The normalized environment by the substrate allows us to better identify the genetic differences of gametes and pick out among them the genotype-leaders. Probably, during competitive development of pollen tubes the change of the leading position among male gametophytes can occur. The pollen tube of the first inbred pollen grain is replaced on the leading outbred tube.

INBREEDING DEPRESSION

Now we will move on to the analysis of inbreeding depression in the Scots pine, which takes place during pollen tube growth during annual pollination. This phenomenon in coniferous trees has its own specifics. Therefore, in the literature the reaction to inbreeding is usually described as a phenomenon of self-fertility, or the ability of a tree to produce viable seeds after self-pollination. The nature of inbreeding depression is still discussed. Currently, there are two widely used concepts – an embryonic lethal system (Koski, 1971) and a genetic incompatibility system (De Nettancourt, 1977, 1999). For the interpretation of our experimental data, we adhere to the theory of incompatibility.

A genetic incompatibility system of angiosperms is realized on the interspecies and intraspecific levels. On the interspecies level, all species of flowering plants by the reaction to self-pollination can be divided into three groups: Self-incompatibility (SI), self-fertility (SF) and self-compatibility (SC). The self-incompatibility group includes over half of all families of flowering plants (de Nettancourt, 1977; Vishnyakova, 1989; Surikov, 1991; Franklin-Tong, 2008, etc.). The SI system include complex genetic and physiological mechanisms, which prevent self-fertilization and enforce outbreeding in alternation of generations. SF species are the cross-pollinated plants with lower intensity of inbreeding depression. The level of self-pollination in them directly depends on the annual weather conditions. SC species include the typical self-pollinators. They prefer self-pollination and are more adapted to different environmental conditions (Agadzhanyan, 1992). It is considered that cross-pollination was the initial breeding system (de Nettancourt, 1977; Vishnyakova, 1996). In angiosperms, the general direction of evolution development of this system is the transition of species from self-incompatibility to self-compatibility. A genetic incompatibility system is the system whose main task is the regulating of close relationships.

Self-incompatibility is a genetically determined trait. Its genetic factor in the SI species is termed by the S symbol based on protein-protein interactions and has many different alleles. According to the general concept of its nature, the growth of inbred pollen tubes, having a certain allele of the S locus, is inhibited in the pistil tissues (stigma, column) of those plants, in which the cells contain the same allele. The reproductive process after self-pollination is broken down between pollination and fertilization and, consequently, the gametic syngamy does not occur. Self-incompatibility, depending on the time of the S genes action and localization of their products, can be of two types: 1) A gametophyte (GSI), which is controlled by its own haploid genotype, and 2) a sporophyte (SSI), which is caused by the diploid genotype of an anther, i.e., a sporophyte.

All coniferous plants are cross-pollinated species. Their genetic system of incompatibility is in the earlier stage of evolutionary development and is realized only on the intraspecific level. Numerous experiments of artificial self-pollination have shown that the populations of coniferous plants are characterized by a wide range of polymorphism concerning the self-fertility level (Barnes et al., 1962; Park & Fowler, 1984; Williams & Savolainen, 1996; Kuznetsova, 2009, 2012; et al.). The populations of Scots pine according to the inbreeding reaction disintegrate into three types of genotypes – self-sterile (*SS*), partly self-fertile (*PSF*) and self-fertile (*SF*) forms (Kuznetsova & Isakov, 1996). The self-fertility level (*Rsf*), or the degree of self-compatibility of a pine, is determined by the formula:

$$Rsf = \frac{Psp}{Pop}$$

where *Psp* is a percentage of plump seeds after self-pollination and *Pop* is a percentage of plump seeds after artificial open-pollination.

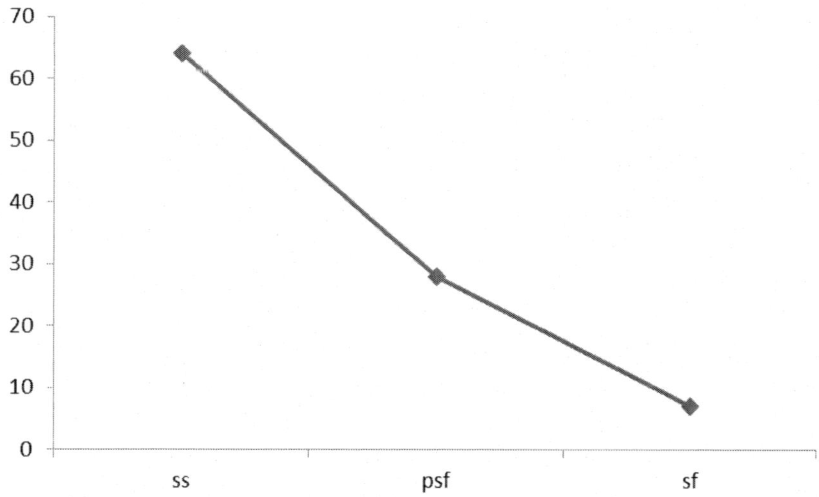

Figure 42. Distribution of pine trees by self-fertility level on Stupinsky test-object.

Quantitative evaluation of a self-fertility trait was carried out within the following limits: Self-sterile – $0 \leq Rsf < 0.1$; partly self-fertile – $0.1 \leq Rsf < 0.5$; self-fertility forms – $0.5 \leq Rsf < 0.8$ and higher (Kuznetsova, 2010, 2014). In the optimal years, self-fertility levels of pine vary from 0 to 0.75. The average *Rsf* of a sample is low(0.13-0.14). The frequency distribution of genotypes has an asymmetric character, in which more than ¾ of trees belongs to the self-sterile group. The relationship between forms in the optimal years is 8*ss*: *4psf*: 1*sf* (Figure 42).

In coniferous plants, an incompatibility is the genetically determined trait according to our 6-year experimental data (Kuznetsova, 2009, 2012, 2015). However, its genetic nature and mechanism action are not known. In optimal years the self-fertility level include sufficiently stable characteristics of a tree genotype. Each tree inside the sample has its own rank, and its ranking position does not change by years. Spearman's rank correlation coefficient varies from 0.89 to 0.96.

Another principal difference between species of flowering and coniferous plants, that the genetic system of incompatibility of SI and SF flowering species destroys a reproductive cycle between pollination and fertilization. During their evolution, the different mechanisms and devices have been developed to improve the efficiency of cross-pollination. Two principally different types of such mechanisms and devices exist. This is: 1) A set of morphological devices of the flower and the temporal characteristics of its development; and 2) a genetic incompatibility system, which prevents self-fertilization (De Nettancourt, 1977; Agadzhanyan, 1992; Vishnyakova, 1996; etc.).

In conifers, there are only two devices, which are poorly developed and inefficient. This is, firstly, the differentiation of generative organs within the tree crown: The female generative organs are located mainly in the upper part of the crown and the anthers in the lower part, which overlap in the center of the crown. Secondly, the pollination begins ½-1 day before the flowering (stage "open cone") of the tree. However, since the pollination process continues during several days, the probability of getting its own pollen into the pollen chamber of a tree is very high. As a consequence, the effective morphological devices, which prevent self-pollination and self-fertilization in conifers are absent. Therefore, during open pollination pollen falls into the pollen chamber both via their own and other trees within a panmictic population. The proportion of inbreed pollen in the pollen pool of the tree and plantation as a whole is the variable trait. It largely depends on the weather conditions (rain, wind direction, etc.), sexualization the type of tree, and its surroundings.

As mentioned, in flowering plants a genetic incompatibility system on the intraspecific level destroys the reproductive cycle between pollination and fertilization. The main task of this system is the regulating of close relationships and providing of heterogeneity and heterozygosis optimal levels in populations. Numerous cytological studies have shown that the lack of fertilization is due to the death of pollen tubes during their growth in the column tissues. The absence of self-fertilization is connected with the high intensity of incompatibility processes. Therefore, in SI and SF species fruits and seeds after self-pollination are absent or are formed in small amounts.

In conifers, self-pollination has little influence on the cone yield of trees, and the seed production in them is decreased by 10 or more times. The data of cone preservation in the year of pollination and seed formation after self-pollination are different. Belostotskaya (1979) showed a decrease in cone yield in self-pollination compared to the cross-pollination more than 25%; in the experiments of Kotelova (1956) a cone shedding was 65% and 39%, accordingly, at self-pollination and open pollination. According to the works of other researchers, an inbreeding depression occurs only during embryogenesis (Hagman & Mikkola, 1963; Koski, 1971; Romanovsky & Khromova, 1992). We have traced the dynamics of the number of cones reduced during the 1st and 2nd year of their development on the same 26 trees of Scots pine after self-pollination and open-pollination. The summarized results of this analysis are shown in Figure 43. It is established that, depending on weather conditions, a preservation of cones varies in the open-pollination from 50 to 75%, in the self-pollination from 25 to 55%, with a wide polymorphism of trees in the sample (0-94%). The cones after self-pollination are smaller and variable in size. Dynamics of cone shedding demonstrate the uneven passage of this process during two vegetative seasons. The graph shows the two peaks of their shedding in both variants of pollination – during the 2nd month after pollination, and in May of the following year after the beginning of pollen tubes growth. In the second month after the pollination of pine occurs, the most intense period of their shedding occurs. At this time, the difference between the experiment variants is detected, which persists until cone maturation. These data suggest that in that year the inbreeding depression in the pine takes place. However, the intensity of the incompatibility processes is not enough to fully halt the development of inbred pollen tubes into the nucellus.

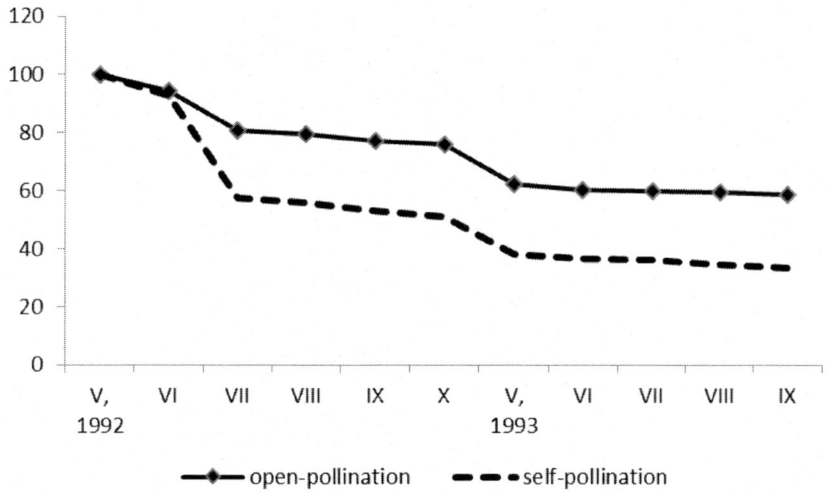

Figure 43. Dynamics of cone preservation after self- and open pollination.

Still, Lewis (1944, 1949) considered a diploid tissue between the male and female gamete as a system, which controls the processes of sexual reproduction of higher plants. Taking into account that during meiosis, the genetic material of microspores is changed as a result of gene recombination and they become unidentical by the DNA-particle, both with each other and with the genotype of the mother tree. Therefore, the interaction between an inbred pollen tube and diploid tissue of a column or nucellus has the character of contact between two heterogeneous but closely relationships systems.

The nutrients into the pollen grains are not enough to form a pollen tube of the required length. Therefore,the nutrients of nearby cells of the column tissue use and pass on the heterotrophic way of nutrition. It is established that after pollen tubes reach the embryo sac, the sugar content in the column is reduced by two or more times (Juhri & Vasil, 1961). The place of an inbred pollen tube stop depends directly on the intensity of incompatibility reactions. These processes are accompanied by callose deposits in a zone of their contact (Poddubnaya-Arnoldi, 1976). Pollen tubes due to tissue incompatibility are characterized by the limited ability to utilize the column nutrients, which are deposited on their common border in the form of the callose layer. Cork-like structures appear inside and on the surface of the tube, which divide it into segments. With the formation of callose plugs, the cytoplasm and nucleus are moved to the tip of the tube, while the others parts gradually die. The intensity of this process directly depends on the nature of the genotype. Most parts of outbred pollen tubes have a callose layer of optimal thickness, form the pollen tube of necessary length and participate in fertilization. Inbred pollen tubes contain two times more callose per unit length than in outbreed tubes (Tupy, 1959). Massive deposits of callose slow down and stop their growth and cause their death.

Our light- and electron-microscopic studies have shown that in Scots pine the incompatibility processes between inbred pollen tubes and nucellus also take place (Kuznetsova, 1991, 1996, 2012). Proceeding from the morphology of the apical zone of pollen tubes, their growth is accompanied by the overgrowth of outer envelopes, and in self-pollination this process is more intensive as compared with open-pollination (Figure 39-b; 44-a). In the 1st vegetative season, three groups of inbred pollen tubes by the reaction ofself-pollination is allocated.

1) Pollen tubes with a pronounced reaction to the inbreeding. They either do not germinate into nucellus tissue (tube horizontal growth, Figure 44-b) or the death occurs shortly after transition from the autotrophic to heterotrophic nutrition.
2) A pollen tube whose death comes in the end of 1st vegetative season (Figure 44-c). The inhibition occurs in the course of tube growth, as the result of its metabolism. After transition to the heterotrophic nutrition, they are able to use only part of the components of the nucellar tissue. The other part of biomolecules, attracted to the gametophyte, has been deposited on the border of the two tissues as the callose layer. The width of this layer is gradually increased. During tube vertical growth, its thickness in the apical zone is approximately two times greater than in open-pollination. During horizontal growth, the internal cavity of branches are filled by a callose layer (Figure 44-d). The increasing layer of callose becomes a mechanical barrier to the movement of nutrient substances. We think that violation of trophic function, due to the incompatibility of the gametophyte and sporophyte is the main cause of inbred pollen tube death.
3) A large part of pollen tubes is characterized by the growth close to normal. The incompatibility between them and nucellus also take place. However, the intensity of inbreeding depression is not so great, and a male gametophyte has time to complete the cycle of its development. These pollen tubes in spite of callose deposition remain a viable source. The loss of tubes of the 1st and 2nd group leads to the death of ovules and reduce the seed number in the cones, which usually takes place in self-pollination. The 3rd group of inbred tubes will probably continue its growth in the following vegetation season and will participate in fertilization.

The characteristic peculiarity of the incompatibility system of flowering plants is both a genetically determined and environmentally dependent trait. Even in the SI species, you can sometimes get a small number of seeds from self-pollination. Such shifts of seed reproduction systems happen when specific external influences are imposed on them. East (1929) showed that the intensity of incompatibility reactions varies definitely depending on environmental conditions (first of all, a temperature factor): At the optimal temperature, it is at the maximum intensity, and at a deviation toward the lower and higher values it becomes lower. At the present time, the environmental dependence of genetic incompatibility systems has been widely used in breeding practices. For example, the cultivation of different sorts of the sugar beets, which is a typical cross-pollinated plant, in the conditions of highlands (1800 and 2200 m above sea level) leads depending on the sort to a weak, partial or complete removal of their barriers of self-incompatibility (Maletsky S.I. & Denisova, 1974).

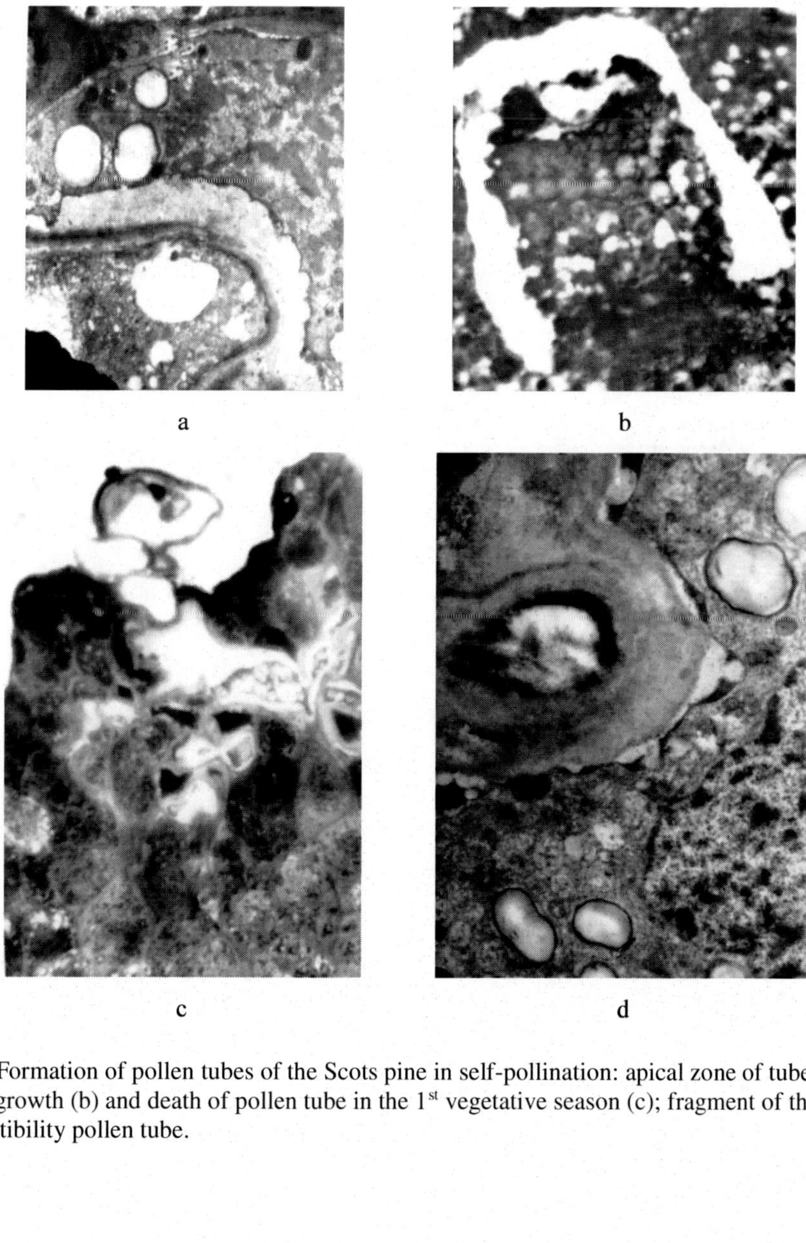

Figure 44. Formation of pollen tubes of the Scots pine in self-pollination: apical zone of tube (a); horizontal growth (b) and death of pollen tube in the 1st vegetative season (c); fragment of the branches of incompatibility pollen tube.

In pine, we found a similar connection between the inbreeding depression and regional temperature (Kuznetsova, 2009, 2012, 2015). It is established that in the optimum years, the self-fertility is a stable enough trait. Every tree in the population sample has its own rank, and its rank position by years is not changed. The rank correlation coefficient of Spearman is 0.89-0.96. During such years, a sample is characterized by the lowest level of self-fertility (K_{sf} = 0.13-0.14) and the highest intensity of incompatibility reactions. A relationship between genotypes is not fixed and varies around the average norm of 8 *cc*: 4 *psf*: 1 *sf*. Self-fertile plants in the abnormal weather conditions (warm, cold, droughty years) become a labile ecologically dependent trait. It reacts flexibly on the weather conditions and is characterized by a high threshold of sensitivity to changes in temperature. No matter how the temperature is deflected (a decrease or increase concerning the regional norm), a self-fertility trait definitely reacts to its change. In abnormal years, its parameters always change in the same way – in the direction of greater self-fertility. The more vegetation conditions deviate from the regional norm, the more the level of self-fertility ($K_{sf}$of population sample) is higher.Accordingly, the intensity of incompatibility reactions is decreased.

Thus, according to the experimental datafertilization in coniferous plants after self-pollination as a rule takes place and the inbreeding depression is realized at a later stage of the life cycle – during embryogenesis. In conifers, the diploid nucellar tissue between pollen and the ovicell is situated also. Nevertheless, their inbred pollen is able to germinate and form the normal pollen tubes. Let us consider the reasons for the lowered intensity of inbreeding depression.

A number of factors influence the intensity of inbreeding depression (Lewis, 1944, 1949): 1) Temperature (a speed of callose deposition increases in proportion to temperature growth); 2) time (the shorter interval between pollination and fertilization, the higher intensity of incompatibility reactions); and 3) distance (the thicker diploid tissue between pollen and ovicell, the greater probability of inhibition of pollen tube growth). The incompatibility is controlled by special genes whose action mechanisms are described in a number of works (De Nettancourt, 1977, 1999; Surikov, 1991; Franklin-Tong, 2008).If you do not take into account the last two factors, all coniferous plants compared to flowering plants are characterized by a more prolonged interval between pollination and fertilization, and, as a consequence, a low intensity of self-incompatibility processes. One of the reasons for such a long time, apparently, is the presence in the matrix reproductive cycle of all gymnosperms the stage of megasporangiogenesis (2nd stage  on the female generative line, see Figure 36). It is known that a formation of female gametophytes is preceded by the beginning of oogenesis and ovicell maturation (2nd stage, Figure 36; Table 4) and continues 20-30 days in *Abies sibirica* (Nekrasova & Ryabinkov, 1978), and until the year in *Pinus sylvestris, Pinus sibirica* (Trenin, 1988, Kuznetsova, 1994). Therefore, the speed of incompatibility processes is not enough for stopping the growth of inbred pollen tubes in the nucellus tissue. Therefore, most of them are formed as viable male gametophytes, successfully reaching the ovicell and participate in syngamy gametes.

To summarize the above, it should be noted that the origin of the genetic incompatibility system of angiosperms from gymnosperms is undoubtedly present. Analysis of seed reproduction systems in flowering and coniferous plants gives an understanding about their evolution. This transition was accompanied by the differentiation of plants on the SI, SF and SC species, the appearance of special mechanisms and devices that deliver maximum efficiency, correspondingly, the cross-pollination for SI species and the self-pollination for SF species.Since the evolution of SI species went in the direction of an intensification of incompatibility processes, the time for the realization of inbreeding depression from the embryogenesis on the gametogenesis stage occurred.

The intensification of incompatibility reactions could be carried out in two ways. Firstly, by decreasing the time between pollination and fertilization, which led to the appearance of incompatibility without heterostyly (gametophytic type of incompatibility). Gametophyte incompatibility is now the most widespread type among flowering plants. Secondly, by increasing or decreasing the length of the column, through the diploid tissue of which the pollen tube passes after getting the stigma. This way has led to the appearance of incompatibility with the heterostyly or sporophytic type of incompatibility. Flowers differentiated on the long- and short-columnar, or due to unequal length of columns in the flowers on different plants of the same species.

Conclusion

Up until pollination, the development of male and female gametes are two independent multi-stage processes. "Bio-build" genotypes are carried out by the species-specific male and female generative lines, which differ by the number and sequence of reproductive stages, by the initiation date of the reproductive process and its rates, by the localization of generative organs in a tree crown, etc. Female meiosis, unlike male meiosis, is a strictly oriented division relative to the gravity. Activating meiosis can be initiated only in case of positive geotropism of megastrobiles, but a reception of a species' cosmic signal takes place only at the negative geotropism. Therefore, the cones change their spatial position after pollination. The 1st pollen grain germination is, apparently, the trigger mechanism of activation of sporogenous cells and their transition megasporocyte formation.

A beginning of formation in the population gene pool of a new generation is associated with a significant expansion of genotypic and modification variability of male and female gametes (on the basis of DNA-particle, DNA-energy and DNA-information). Up to pollination, their development takes place in the conditions of stable substrate and varying climate. During microsporogenesis and megasporogenesis, the high positive and low temperatures in spring lead to the death of non-adapted genotypes, and extend the modification variability of surviving male and female gametes within the norm reaction of species. These gametes, included in the reproductive processes by the male and female generative lines, carry the information about regional conditions in the year of pollination. This confirms the involvement of environmental factors in the processes of formation of the population gene pools of pollen and future ovicells.

Biology Gametogenesis and Formation of Genotype-Environmental Connections in the Year of Fertilization

The biological harmonic oscillator is a quantum object and its operation is possible only in a certain temperature regime (0-40°C). In winter, the plant is in dormancy, in which its potential energy is in the state of a local maximum, the DNA-wave function, intracellular metabolism and exchange between environment is suspended, and the structural and functional components of cells, and its viability is retained. This type of equilibrium is characterized by the instability, so that any external disturbance can break it. In the spring, the positive temperature acts as such a trigger. In April, when the average daily temperature exceeds 5°C, a biological harmonic oscillator of vegetative and generative cells goes out of balance, and the growth and development of the tree continues.

Any tree as the quantum-mechanical system possesses a certain amount of energy. Its cell systems according to the gradient of the decrease of biological energy can be divided into 4 categories: Generative, meristematic, differentiated and dead cells. We think that the biological age of a tree, which reflects the degree of its morphological development and life state should be determined by the total level of biological energy of all its cells, including the dead cells, the number of which increases with age.

All living cells at the time when the tree awakens from dormancy have a certain quantity of internal energy (DNA-energy). Structural and functional components of the cytoplasm and nucleus of resting cells are not disrupted. Their volume outside is not fixed rigidly. Therefore, in every spring the vegetative and generative cells after recovery of DNA-wave and biofield are able to move to active metabolism. According to the 2nd law of thermodynamics, a biological harmonic oscillator, which has a certain energy, does not save it. Therefore, an oscillator soon after its awakening from the dormancy passes to the lower energy level and a portion of free energy is radiated. This energy enters into the cell and initiates 1st mitotic division and the beginning of tree development in the next vegetative season.

a b c

Figure 45. Change of the form and size cones during fertilization year.

In the spring, the growing processes in the cones are most active. In late May, the sizes of normally developing cones and their elements (seed scales and ovules) are significantly increased in comparison with the size of corresponding cone structures of the autumn period: A megastrobile's length is an average of 320%, a seed scale's length increases by 470%, and ovules increase by 330%. Simultaneously, with the cone growth its spatial position is also changed. Before the spring development, they are pressed to the escape, have an oval shape and are located inside the beam of long needles (Figure 45a). At the initial stage of growth, their morphology does not change significantly (Figure 45b). They are increased evenly in length and width and are still separated from the environment by means of needles. By the end of May, the cone is lengthened and the typical conical form is increasing, the vertical axis of which coincides with the direction of gravity (Figure 45c). Before fertilization, cones have the typical axis for each tree shape and size, and their growth is stopped.

Significant differences by the intensity of the cone development are observed both within the crown, and between the trees. The levels of individual variability of analyzed traits are higher. Variety of "cone length" trait for the sample is 22.0%. The values of endogenous variability are fluctuated from very low in the individual trees (not above 5%) to a relatively high level (≈ 20%). This causes an increase of the strength effect of the mother plant factor on the variability cone sizes from 22.7 in late August to 74.5%.

In May, together with normally developing ovules and a cease in the development during the previous year, there is one more type – the ovules, which have lagged behind in growth from the normal time span. Disturbances in their development are not associated with pollination; they are defined by lethality of female gametophytes (Romanovsky, 1997). For example, the size of fertile, "normal ovules" are as follows: Length – 2.25 and width 1.50 mm, the size of which has lagged behind in the growth ovules is 1.50 and 0.75 mm, respectively. Female gametophytes completely degenerate and other structures (integument and apical part of nucellus) continue to grow, although much slower compared to the norm. Such ovules in the mature cones form a category of underdeveloped seeds – the sterile structures, having a sclerogenic seed coat and small sizes compared to the normal seeds. The share of such ovules in the megastrobiles varies greatly from year to year. In May 2011 it was small – ≈ 2% for the sample. This year, the degree of realization of the potential seed production on this stage is shown in Figure 46. These data of the megastrobile structure of *Pinus sylvestris* is in large part similar with the results of other researches, carried out in a different way by yearly weather conditions (Svintsova, 2002). This testifies to the stability of the gametogenesis stage to the regional weather stress.

As was mentioned above, the reproductive process by the male generative line has stopped at the stage of microsporangiogenesis, after a branched pollen tube of certain length was fully formed. All this time, the generative cells were inside the pollen grain. With onset of spring warmth, their position has not changed – the generative cell remained inside the pollen grain at a rest state. With respect to the reasons, we can say the following.

Plants and their reproductive structures (seeds, pollen grains, etc.) in adverse conditions of the environment are able to survive only in the case of transitioning to dormancy. This is a complex genetically determined and relatively long process. We think that it is based on the change of cells' physical state under the influence of 4 main factors – light, pressure, temperature and volume. This is caused by the work of a biological harmonic oscillator, as the physical system depends on the state and interconnections of these factors.

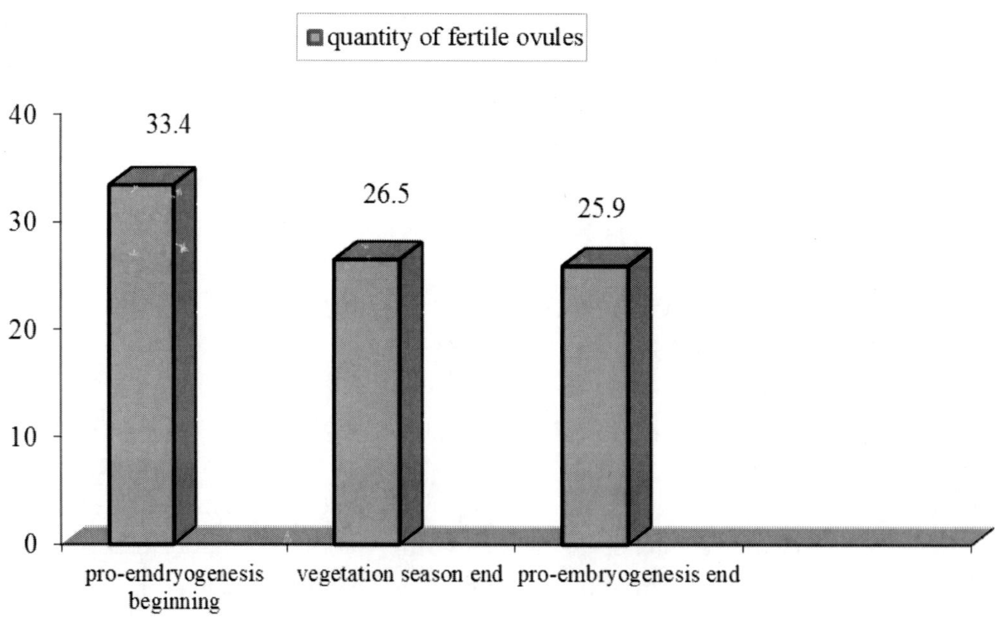

Ordinate axis: quantity of fertile ovules, pcs.

Figure 46. The character of realization of the potential seed production of *Pinus sylvestris* during proembryogenesis.

The transition into a state of rest is a collective process, when the living cells of a plant (all or a specific part) act organized and interconnected. It is not possible without the preparation period during which the physical conditions into all cells changes a certain way under the influence of internal and/or external reasons. A biological harmonic oscillator in species-specific points of the upper and lower limits of temperature and intracellular pressure goes out of balance and exists in the regime of unstable equilibrium. Its work is stopped, but the oscillator is not destroyed. The plant's vital activity (or separate organs, cells) without a DNA-wave and biofield is almost completely stopped. Living systems can be in this state for a long time, and capable of carrying enough great stress loads. The oscillator itself is unable to find equilibrium. Its return to the normal operation is possible, but this requires the external or internal impact on the system.

At the beginning of the next vegetative season, the pollen grain contains only a generative cell, the volume of which is much more its own size. A nucleus of a siphonogenous cell by this time has been destroyed. A pollen tube is a hollow or partially filled callose channel. Generative cells could not awaken from dormancy with the coming spring's warmth. It biological harmonic oscillator on all sides "is compressed" by the chemically immobile storage substances of its cytoplasm (lipids, proteins, etc.). Moreover, the pollen maturation occurred in May at a positive temperature, and during the summer generative cells remained in the resting state into the pollen grain. Internal forces, which could destabilize its system from the inside, are also absent: In the pollen grain, a generative cell is separated from the pollen chamber by a sporoderm and from the nucellar tissue by the tube cavity.

Another situation takes place via the female generative line. Development of a female gametophyte continues until the 1st vegetative season ends. During this time, a functional megaspore undergoes 4 free-nuclear divisions and a gametophyte overwinters on the stages of 32-nuclear pro-embryo (Ferguson, 1904). From summer's end, a gametophyte begins to pass on into a resting state, which is characterized by the following deep intracellular changes: The radiation of a biological harmonic oscillator is slowed until it stops; a starch in the cells is accumulated; the intensity of metabolism decreases sharply; and the divisions of nuclei are absent. The main requirement for a cell passing into a resting state is that the processes should change gradually. In this case, the transition will occur without any injury to the genetic material. Since a sporoderm has not yet formed, the light and temperature are the main environmental factors for female gametophyte passage to the resting state. Accordingly, they are capable of leaving a dormant state under the influence of light and a positive temperature regime. Therefore, with the coming of spring temperatures nuclear division and the development of female gametophytes continues.

In spring, a nuclei undergoes the 6 free-nuclear divisions and a female gametophyte is rapidly increased in size. The trophycal function is provided by means of the nearby energetically weak cells of nucellar tissue. In the zone of contact, they have the ultrastructure of the cells of a secretory type, which are characterized by the high synthetic activity. All nutrients are moved by the vector of a stronger biofield of quickly growing female gametophytes and are used for its nutrition. The gametophyte consists of the haploid nuclei, surrounded by the cytoplasm islets. The nuclei are located in a wall boundary layer. Chromatin is uniformly dispersed throughout the volume of the nucleus and it is weakly stained by the nuclear dyes. The growth of the female gametophyte occurs from the basal to the apical pole of the ovule. In May, there is a typical growth rate for all coniferous and they have an elongated oval shape. So, the top of female gametophytes are gradually approaching towards a male gamete, which is still inside the pollen grain.

In a spruce, the ratio of a megaspore volume to the volume of cellular gametophytes is equal to 1:35000 (Kozubov et al., 1982). According to Ferguson (1904), the number of free nuclei in the female gametophyte of a pine before the formation of the cell walls is about 2000. For example, the megagametophytogenesis (⬭stage) includes ≈ 11-12 free-nucleus divisions. Each division of the nuclei is preceded by a radiation of free energy quanta and a transition of their biological harmonic oscillators down by one energetic level. Hence, it follows that during the process of female gametophyte formation the DNA-energy of its nuclei and, accordingly, archegonial initial become lower than the potential of functional megaspores at the 11-12 levels. Evolution of the reproductive process is directed at the increasing of the energetic potential of species and at the reduction of energy losses, which is confirmed by the evolutionary changes of female gametophytes. So, the number of free-nuclear divisions in the species of the Zamia and Ginkgo families to the 8 (or 256) free nuclei is decreased (Konar & Oberoi, 1969; Singh, 1978).

Now, we will move on to the analysis of some peculiarities of female gametophyte formation. The female gametophyte in the course of its development has 4 stages: 1) An exponential growth; 2) a stationary stage; 3) an internal differentiation; and 4) a physiological rest. In the 1st vegetative season, when the thickness of an envelope was small and is capable of exercising its influence on gametophyte growth was insignificant. It is the stage of exponential growth. Also, take into account that all nuclei of the one female gametophyte are the identical Fermi particles with the same spin direction. According to the Pauli exclusionary principle, a nuclei with an identical direction of spin cannot approach each other and are in the same place within the gametophyte. Therefore, afterward each free-nuclear division is evenly distributed around the perimeter of the gametophyte.

In the 2nd vegetative season of nuclei division, the growth of female gametophytes continue. The decrease of their biological energy promotes the formation of highly durable envelopes known assporoderm. We think that its formation, as well as a sporoderm of pollen grain, is connected with the action of a genetic incompatibility system. The processes of incompatibilities take place on the border of two heterogeneous cells by the DNA-particle tissues and two different energy potential systems – an energetically strong female gametophyte and an energetically weak nucellar tissue. The outer side of the gametophyte envelope is the zone where the biomolecules moving from the nucellar tissue are detained and settle. The sedimentary biomolecules, densely pressed into each other by means of biofield forces, begin to polymerize and form a sporoderm around gametophyte.

Sporoderm in the process of thickening more and more limits the growth of female gametophytes. Energetic gradients between a female gametophyte and a nucellus gradually decreases. The rate of gametophyte growth slows down. The volume of nuclei begins to increase more rapidly than the volume of its cytoplasm. This leads to the normalization of the nucleocytoplasmic ratio and to a transition of a female gametophyte from a stage of logarithmic growth to the stationary phase. All nuclei, formed during free-nuclear divisions, are distributed along the envelope of the gametophyte. When their number is 2,000, the formation of envelopes between nuclei begins. The envelope is formed initially between the nuclei of a parietal layer. The inner walls of the first cells of an endosperm are at first remain open, and then the periclinal envelopes are produced (Kozubov et al., 1982).

The development of female gametophytes at the stationary stage is carried out in a partially fixed volume. In all species of *Pinaceae,* a gametophyte in this period has anoval form. A large vacuole occupies the most part of its volume (Allen & Ovens, 1972; Singh, 1978). Cells of an endosperm are more and more filled with the inner cavity of a gametophyte. The volume of the cytoplasm increases gradually, the central vacuole decreases and the degree of vocalization cells is reduced. Via low gas, the number of molecules in a volume of liquid is directly connected with the pressure. Hence, it follows that a continuous cell division leads to the increase of osmotic pressure inside the female gametophyte. Its growth initiates its transition from the stationary stage to the internal differentiation. The gametophyte growth slows down even more, and then it stops.

In the next vegetative season, the cones grow rapidly. Intensive growth processes occur in the seed scales, cone axes and ovules with all of them developing in structure. At the end of May, cones take the vertical position, in which their longitudinal axes coincide with the gravity direction. The longitudinal axes of gametophytes, ovules and cones have the same direction. Female gametophytes in the normally developing ovules of largest megastrobiles has sizes of1200 mkm in length and 900 mkm in width. The first endosperm cells of cellular gametophytes have the same size and ultrastructure. The cell envelopes between the nuclei are formed centripetally; therefore, the cells in a superficial layer represent a radial row (Figure 47a). These are highly vacuolated cells with thin cell envelopes and one nucleus (Figure 47b). This time, the gametophytes are differentiated by the archegonial initials. The archegonial initials are the 1-4 endosperm cells located in a superficial layer of the micropilar zone of gametophytes. Usually, these are the pyramidal, early closed generative cells, the longitudinal axis of symmetry of which coincides with the direction of gravity.

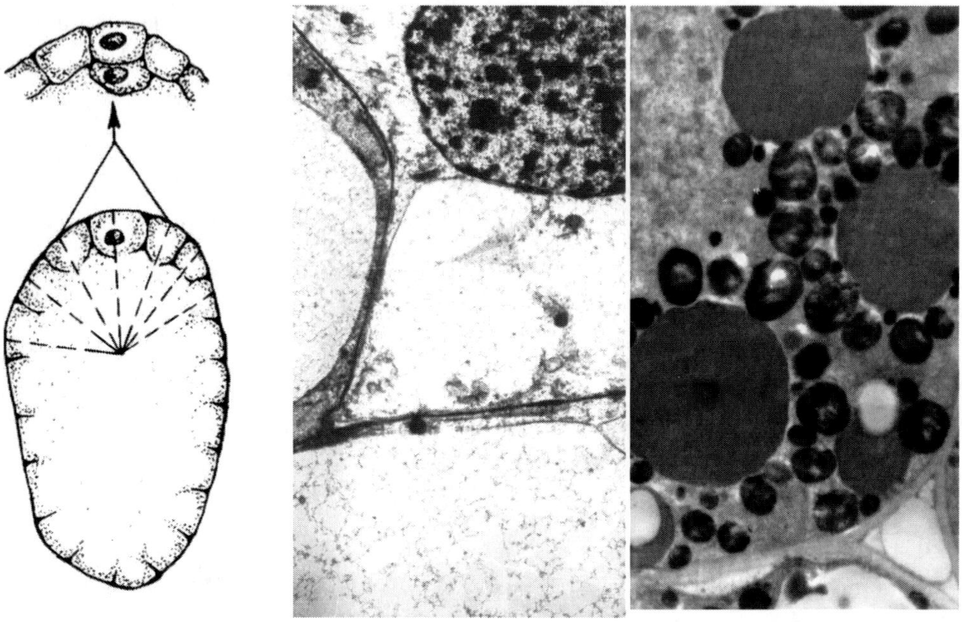

a b c

Figure 47. Female gametophyte in the cellular stage: a – emergence of archegonial initial; b – ultrastructure of the first and c – mature endosperm cells.

The leading event in the period of internal differentiation of female gametophytes is oogenesis. This is the 3rd activation stage (ᚨ, Figure 36, Table 3) by the female generative line. Oogenesis begins from the division of archegonial initials. The initial undergoes six divisions of two types: Two activating mitosis and four endomitosis. The last generative cell is the ovicell, which participates in the fertilization. During first activating mitosis, the initial cell divides by a transverse partition and gives rise to an energetically weak primary neck cell and an energetically strong central cell. A primary neck cell, in turn, after the passing of two divisions (transverse and longitudinal) comes down to the zero energy level and forms the four neck cells. These cells remain on the top of the ovicell and degrade soon after fertilization. The central cell, which has received the portion of biological energy, begins to increase rapidly in size. Moreover, initially its growth occurs mainly due to the increase of cytoplasm volume. An algorithm of the nucleus and cell growth at the different stages of female gamete formation in the Siberian larch is shown in Table 7 (Trenin, 1986). The average cell volume of one of the superficial cells has been used as the size of the archegonial initial and the unit of counting. These data give a vision about the rate of growth of generative cells during oogenesis and the change of nucleocytoplasmic ratio in the conifers.

Table 7. The area of cells and nuclei of the archegonium of Siberian larch (Trenin, 1986)

Area	Arhegonialinitial	Central cell		Matureovicell
		Early stage	Late stage	
Cell	1.0	7.7	75.0	145.0
Nucleus	0.15	0.2	0.4	18.1

Sporoderm are characterized by high stability. Sporoderm has two main layers – the exosporium, consisting of a difference in the size of osmiophilic globules and the homogenous endosporium. Therefore, the cell divisions are continued inside the fixed volume, gradually filling all of the volume of female gametophytes. Endospermogenesis are proceeding under the conditions of progressively rising intracellular pressure and an increased number of molecules, density of the nucleoplasm and cytoplasm. The volume of cytoplasm increases and the vacuoles divide into parts. The speed of molecules is directly connected with their ability to enter into a chemical reaction. Therefore, the generative cells begin a biosynthesis of chemically inert substances such as starch, storage protein and lipids. Within ≈2 weeks the endosperm cells fill up with proteins and lipid drops (Figure 47c). The cells of the outer layer are smaller and are completely filled by the proteolipides. The internal cells are formed later. These are the larger cells by size, having a vacuole and considerable amounts of starch.

According to experimental data, the initially haploid cells of functional megaspores during endospermogenesis are replaced by 3 types of cell divisions – free-nucleus mitosis, typical mitosis and endomitosis (Fujita, 1961; Pichot & Maätaoui, 1997). The genetic material of mature endosperm represents the system that multiply ploidy levels, nuclear fusion, incomplete divisions, etc. (Nagl, 1978). It should be noted that the two main tendencies in the development of relations of male and female gametes with the mother tree – an increasing isolation of gametes from the plant internal environment – and on the other hand, more and more of their dependence from the nutrient tissue of microsporangium and ovules.

The result of the further growth of the intracellular pressure is the cessation of cell divisions and the transfer of gametophytes to the rest state. This is the 4th and last stage of the development of female gametophytes. In dormancy, the biological harmonic oscillators of all its cells are in equilibrium at the critical point. They can come out from this state only under external or internal forces. From the aforementioned point, it becomes obvious that the endospermogenesis and maturation of pollen by their nature and mechanisms of transition to the resting state are very similar. Only the male gametes come out from dormancy, and the endosperm cells do not. The nutrients of endosperm provide a trophic function of eggs, embryos and seedlings in the autotrophic period of their development.

When a central cell during activating mitosis received a portion of biological energy, it becomes the energetically strong cell and gets an energy advantage over other cells of endosperm. As a consequence, the balance inside an ovule, which has been reached between the tissues of the nucellus and the gametophyte during sporoderm formation is lost. A rejuvenating central cell has the great potential for growth and development, and is characterized by a more powerful biofield. Therefore, the nearby energetically weak cells of an endosperm react to its origination in a certain manner. In a zone of contact, they are transformed into the secretory cells. Under the influence of a powerful biofield their metabolism, the longitudinal axes of symmetry and the planes of cell divisions change. They turn and are drawn to a central cell, surrounding it by the concentric continuous layer of jacket cells. The jacket cells are characterized by a large number of organelles with the specific ultrastructure and increased secretory activity. They operate in a very intensive regime. All synthesized substances pass by the vector of a stronger biofield and are used for the nutrition of a quickly growing central cell. This regime is harmful for the jacket cells. Therefore, they are rapidly destroyed, and their intracellular contents are absorbed. Then, the next layer of secretory cells is formed, which occupies their place.

The central cell grows very rapidly. Its volume with respect to the archegonial initial is increased dozens of times, mainly due to the cytoplasm (Table 7).As the development of a central cell as well as endosperm cells occurs in conditions of gradually increasing intracellular pressure, a cytoplasm is filled by vacuoles and a generative cell takes the "foamy" structure (Figure 48a). A nucleus is increased not so significantly and remains in its apical part under the neck cells. Then, a central cell undergoes the 2nd activating mitosis. An energetically weak ventral canal cell and an energetically strong ovicell in the result of this division are formed before fertilization. A ventral canal cell is the ephemeral structure and quickly dies. The capacity of biological harmonic oscillators of ovicells in comparison with the central cell becomes higher. The intensity and extent of biofield is significantly increased. The energy gradient between the endosperm cells and ovicell becomes even higher. This leads to an increase of nutrient streams towards the ovicell. An ovicell begins to grow, its size before fertilization is doubled and it becomes the largest cell of a female gametophyte.

According to a low gas volume, the high proliferative and synthetic activity of endosperm cells inside the partially fixed, and then in the constant volume of female gametophytes leads to the increase of intracellular pressure and cytoplasm viscosity. Since the ovicell formation occurs in the conditions of constantly rising intracellular pressure, the process of its maturation, as well as the endospermogenesis, we will consider as the path of gametophyte transition to the resting state. Table 7 shows that in comparison with the central cell, the rate of ovicell growth is significantly lower. Morphology and metabolism of the ovicell undergo the same changes as the endosperm cells. Volume of vacuoles is decreased rapidly. The cytoplasm is compacted and filled with the nutrients. Radiation of a DNA-wave occurs less frequently. The biofield is getting smaller. The result of a change in physical conditions is the transition of the ovicell from the activating mitosis to endomitosis.

An ovicell passes through four divisions of endomitosis. During this time, its maturation and transition to the dormancy takes place. The biofield is gradually compressed to the ovicell envelope. It is differentiated into two zones: An inner or perinuclear zone and an outer zone (Figure 48b). The distinctive feature of this type of division is as follows: The radiation of a biological harmonic oscillator and doubling DNA-particles happens, and the nucleus and cell divisions do not occur. In the course of endomitosis, the polytene chromosomes with 32C DNA are formed (Ermakov et al., 1981; Barantseva, 1992). During ovicell maturation, a nucleus is growing rapidly. In comparison to the nucleus of a central cell its size increases to more than 20 times (Table 7). In Figure 48a and48b,the changes in the size of the nucleus during this period can be seen. The formation of polytene chromosomes leads to the weight gain of a nucleus, and it migrates to the center of the ovicell. The volume of the nucleus per one genome is 2.5 times higher in a mature ovicell than in the central cell (Ermakov et al., 1980).

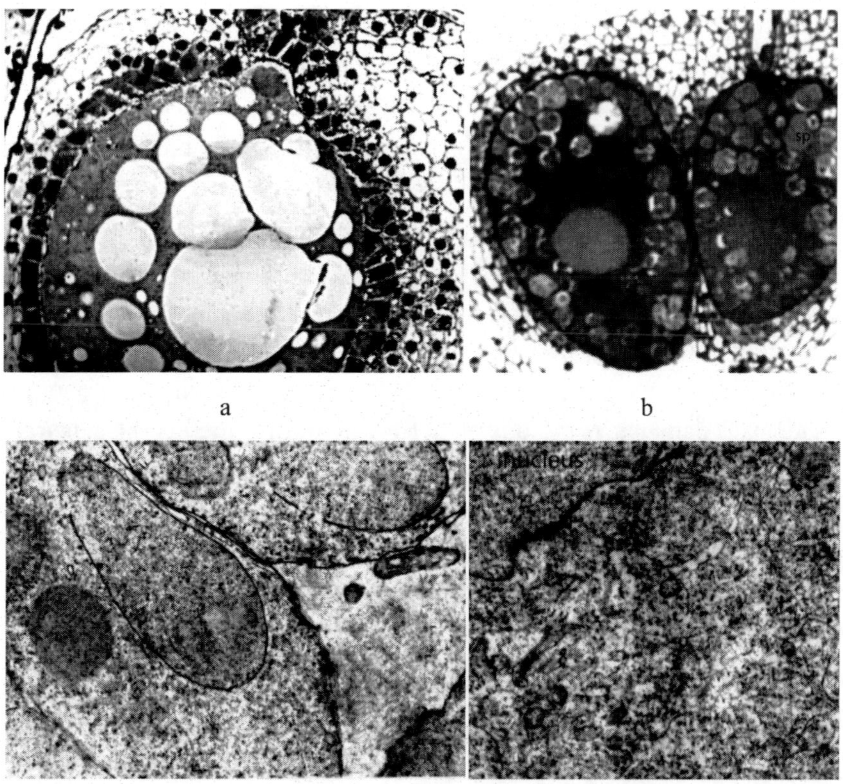

Figure 48. Centeral cell (a) and ovicell during fertilization: overall appearance (a) and fragment of cytoplasm from the outer (c) and perinuclear (d) layer of ovicell.

The vicell of a pine is characterized by the bilateral symmetry, polarity and certainly oriented in the direction of gravity. Because of this, its biofield has two zones of action: The close-range, changing as $1/r^2$, $1/r^3$ and the far-range, changing as 1/r, its cytoplasm is differentiated on the inside (perinuclear) and an outer zone with a different ultrastructure. The differentiation of the cytoplasm of animal oocytes can be changed through centrifugation, but after its termination they go back to the original position (Raven, 1961). The outer zone is organized in a number of large and small inclusions, which are the specialized reserve substances of a protein nature. The great inclusions, or the Hofmeister bodies, have a complex structure (Figure 48c). Apparently, they are formed by the partial filling vacuoles by the protein granules. The perinuclear layer has a width of 8-10mkm. There are a large number of ribosomes, elements of endoplasmic reticulum, mitochondria and microvesicles (Figure 48d). Hofmeister bodies are absent. The centripetal direction of the biofield gives it the appearance of protuberances. At this time, the inner layer has an intense basophilic and Feulgen-positive coloring (Takao, 1960). A nucleus is located in the center of the ovicell. Chromatin is presented only in the form of euchromatin. In the nucleoplasm, a large number of nucleus-like formations are founded. The ovicell is surrounded by an envelope on the outside, which is common for the ovicell and the jacket cells. The envelope consists of alternating thickened areas and pore fields. Pore fields are the places of active transport for the metabolites.

A further growth of pressure leads to the transition of gametophytes into dormancy. The biofield is absent. Endosperm cells and ovicells are completely filled with the nutrients. The speeds of assimilation and dissimilation processes are minimized (Tokin, 1987). Jacket cells are vacuolated, which indicates the absence of exchange between the gametophyte and the external environment. The nucleus of the ovicell from the Feulgen-positive becomes the Feulgen-negative (Shimamura, 1956). The change in the entropy, pressure and volume of the gametophyte and all its components tends to be zero. In this period, the ovicell is in a rest state and at the same time in a state of maximum receptivity. Female gametes come out of dormancy during the sperm penetration in its nucleus. Without fertilization it, as well as all endosperm cells, is doomed to expire.

The fixation is always founded by a certain number of ovules, in which a part of the archegonium remains without fertilization. Their number increases considerably in the years with insufficient pollination. Unfertilized ovicells are destroyed within two weeks after fertilization. By this time, fertilized ovicells have transparent cytoplasm, because their embryos are already utilized by most parts of the nutrients. Unfertilized ovicells on their background stand out by the intensely stained cytoplasm. The irreversible changes appear in 7-10 days after pollination (Kuznetsova, 1994). The perinuclear layer is absent, and the nucleus is directly contacted with the degraded cytoplasm (Figure 49). In the cytoplasm, the vesicular fragmentation of membranes and the intensification of the lyses process take place. Then they, under the pressure of developing embryos, are destroyed and disappear.

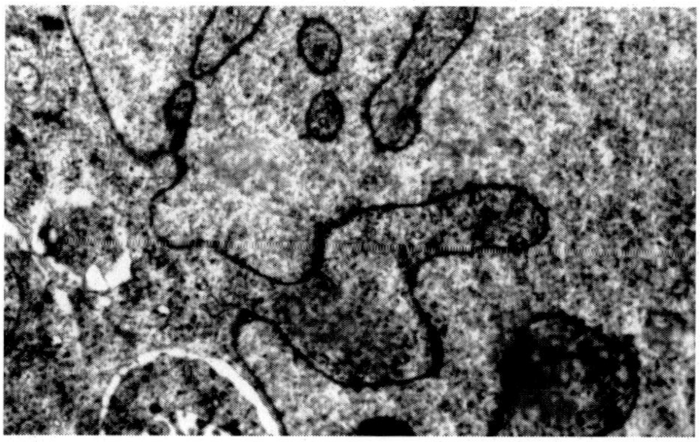

Figure 49. Unfertilized ovicell in *Pinus sylvestris.*

Now we mill return to the analysis of interrelations between male and female gametes during oogenesis. You remember that after pollen germination a resting generative cell remains in the pollen grain. The following spring, it continues to be in the rest state. Siphonogenous cells have already died. A small generative cell occupies an empty pollen grain via the same spatial position as before the pollen germination. It is attached to its envelope from the side opposite the furrow germination.

In the spring, a female gametophytes pass from a free-nuclear to a cellular stage. They grow rapidly in the direction from the chalaza to the micropyle before the ovicell reaches maximum maturation size, i.e., the male and female gametes are approaching each other. The nucleus of the central cell is located in its apical part on the top of the gametophyte. If we are to proceed from the assumption that the male and female gametes are the Fermi particles, there is a situation when the nucleus of a resting generative cell and the nucleus of the central cell draw together to such a distance, and between them the spin-spin interactions arise.

We think that the forces of mutual attraction between male and female gametes with the opposite spins are enough for connecting them together. However, the generative cell by a size is much smaller than the central cells. In addition, the central cell is closely pressed to the sporoderm. Therefore, the spin-spin interactions, which act through a separating medium (nucellus, sporodem, etc.) will be able to set in motion only a male gamete. It is detached from the sporoderm of a pollen grain and hits on the opposite side. The hit takes out a generative cell from dormancy. As a result, the wave function on the DNA molecule is restored. The biological harmonic oscillator is returned to the ordinary active state and becomes a source of biofield. The radiation of the DNA-wave after a certain time, and the entering into the generative cell of quantum free energy initiates the 4th division of microgametophyitogenesis. As a result of the activating mitosis, the energetically strong spermatogenous cell and the energetically weak sterile cell are formed, in which the sterile cell soon dies. During this time, the central cell undergoes the 2nd division of oogenesis (activating mitosis) and gives rise to the energetically strong ovicell. As a consequence, the mutual spin-spin interactions between the spermatogenous cell and the ovicell become stronger so that it provides the passage out of a spermatogenous cell from a pollen grain inside the pollen tube. The male gamete is moved along the tube toward the female gamete, and stops before its branching. The tip of the tube is filled with callose. During this time, an ovicell is undergoing a period of maturation –the complex transformations of cytoplasm and nucleus: An endomitosis, an accumulation of reserve nutrients, and a transition to the resting state. Before fertilization, a spermatogenous cell is divided again. Two nuclei without cytokinesis – energetically weak and strong sperms – in the result of the 5th activating mitosis of microgametophytogenesis are formed. During fertilization, only one energetically strong sperm participates. The gamete syngamy after formation of the energetically strong sperm and the transition of the ovicell in the dormancy state (the state of its maximum receptivity) is an inevitable process.

It should be said that there is a consequence, which follows from the quantum nature of gametes. The microspore, pollen, and sperm are characterized by a certain spin direction: 50% of them with spin-up (↑), and 50% with spin-down (↓). With regards to female gametophytes, the spin direction of all endosperm cells is identical (spin ↑, or spin ↓). For example, the initial archegonium has the same spin direction as the functional megaspore, from which this gametophyte originates. It should seem that only half of the resting generative cells can continue the reproductive process, because the male and female gametes are attracted to each other with the opposite spins. The rest will die. However, it is not. If we proceed from the Pauli exclusion principle, when the male and female gametes come together, trying to get to the same place, there is interaction which aspires to place their spins towards each other (Feynman et al., 1965). This interaction causes the spins of the male gametes to take the opposite direction. Therefore, the majority of the generative cells coming out of dormancy are included in the reproductive process, which is confirmed by experimental data.

One more factor, which determines reproductive success, is the number of archegonium into the gametophyte. It is established that the potential possibility for multiple fertilization in Scots pine of the Voronezh population is 95% (Svintsova, 2002). The largest gametophytes have a greater number of archegonia. In some plants, the number of archegonium on the gametophyte is in the range of 1.7 - 3.3. There is a positive relationship between the size of the female gametophyte and the number of archegonium: $r_s = 0.62$ ($P < 0.05$).

Due to the availability of the polyarhegonial system, one female gametophyte develops several ovicells and germinates, as rule, from one to four pollen tubes, which are not identical in genetic constitution. Genotypes of gametes are differentiated on the genotype-leaders and genotype-outsiders, depending on their level of DNA-energy. The distribution of their genotypic composition by the level of DNA-energy has the form of a standard normal curve (Figure 50a). Differences by the energy potential are manifested in the form of increased or decreased morphological, functional and metabolic activity of male and female gametes. Genotype-leaders of pollen form a more powerful pollen tube and stably occupy a leading position between their competitors. Genotype-leaders of functional megaspores give rise to the energetically strong female gametophytes, which grow more rapidly; early on, they begin a period of internal differentiation and fertilization.

A few archegonium of one gametophyte, as the generative cells with the same direction of spin, do not interact with each other. Accordingly, the ovicells also do not interact among themselves. Generative cells are dormantly inside the pollen grains and their spatial position is fixed. They have the pollen tubes, the length and width of which corresponds to the compatibility degree of male gametes with a nucellus tissue. At the time of the exit of spermatogenous cells from the pollen grains, a direct and indirect selection of sexual gametes take place, but these processes are manifested in another form.

A female gametophyte during its growth reaches the zone, where the forces of spin-spin interactions between the male and female gametes begin to act. The differences in their energy potential creates the situation of polyvariant (non-random) selection of partners for syngamy. When given a choice between a few central cells and a few generative cells, the most energetically strong female gametes are likely to "attract" the most energetically strong male gametes. The remaining pairs will be created in accordance with their energy gradient. The fertilization in the variant of one generative cell and a few central cells occurs for the most energetically strong ovicell, and vice versa.

A pollen tube is the channel through which the movement of male gametes to the ovicells occurs. The indirect type of competition is defined by the length and width of tubes, which depend on two factors – weather conditions and a degree of compatibility/incompatibility of pollen genotypes with the nucellar tissue of mother trees. Therefore, a wider tube with the thin callose layer will ensure the rapid motion of gametes, and narrow tubes with the powerful callose sediments, on the contrary, will slow down. The competitive advantage will also receive the sperm of leading pollen tubes, as they are located closer to the ovicells. Hence, it follows that the male and female gametes, which were realized in the fertilization are for the most part outbred genotypes, which are compatible with mother trees and the environment by the principle of "genotype-environment" interaction. Owing to the removal of un-adapted and inbred gametes, the stability and adaptability of the population gene pool is increased.

It is experimentally proved that an insufficient level of ovule pollination has a negative effect on the growth and viability of megastrobili of Scots pine (Svintsova, 2002). In the year of pollination, this effect has a trend: The rank correlation coefficient of Spearman between the length of megasrtobiles and the level ovules (%) stopped in their development is 0.43. At the end of May next year, the correlation increases significantly and amounts to 0.90 (significant at the $P < 0.01$).

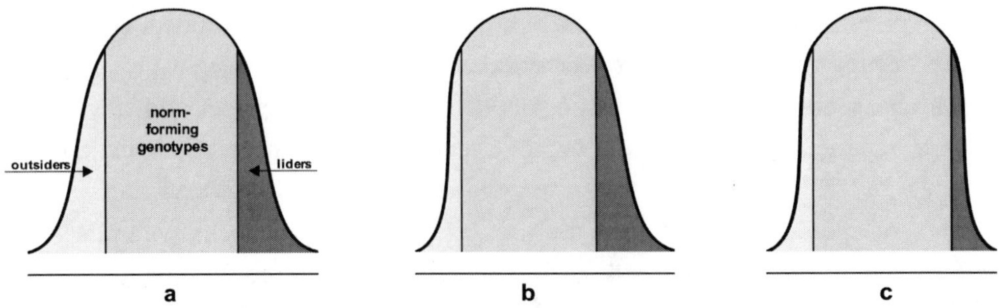

Figure 50. The distribution of the genotypic composition of male and female gametes, embryos before and after the direct spin-spin and indirect selection.

If we proceed from this scheme, the unfertilized ovicells and the unnecessary pollen are mainly experienced via genetic damage, energetically weak, inbred, non-adapted male and female gametes. During direct and indirect selection, the initial spectrum of the genotypes of pollen and ovicells is narrowed a certain way. Accordingly, the connection of the gene pool with the environment and nucellus of the mother tree are increased. In this way, a genotype-environment interaction arises.

During direct spin-spin selection, the initial spectrum of male and female gametes is not only narrowed, but a one-sided shifts (Figure 50b). This takes out their genetic system from the balance. Its stability in the non-equilibrium state is provided, on the one hand, by the normalized conditions of the microclimate, which are created inside the cones. On the other hand, the close connection of genotypic composition of gametes with their habitat, due to which in the course of gametogenesis is created a set of points of sustainable local balance between a microsporangium and a archesporium, a nucellus and a pollen tubes, a female gametophyte and a nucellus, etc.

In order for the development of gametes to occur by the certain scheme, they must have the necessary information since inception. Korogodin & Korogodina (2000) consider the information as a “guide to action.” This property of information belongs only to living matter. During the analysis of the generative sphere, we found many examples confirming this property. This is a list of some of them: 1. Inception of generative organs and initiation of a reproductive process happens at the dates and the location that provide the stability of genetic material. 2. During male and female meiosis, the genetic material has direct contact with the external environment, due to what the variability (genotypic, modify) of a popular gene pool is maximally expanded. 3. In the 1st vegetative season, megastrobiles carry out the geotropic motion, the result of which all activation stages have the same orientation with regard to the gravity direction.

It seems obvious that a reproductive cycle can be characterized as a system, in which generative organs and reproductive processes are ordered in a space and in a time. Property information “guide to action” is in its essence a biotechnology *ex situ,* which was created by nature itself. At all levels of the organization of genetic material, it operates strictly in accordance with the fundamental laws, on the basis of the close relationship with a generative and vegetative sphere of plants through the formation of certain genotype-environment relations. At present, our knowledge in this area of biology is not yet quite complete. Many phenomena and processes have the corpuscular-wave nature. All this must take into account the scientists working in biotechnology *in vitro*.

FERTILIZATION AND PROEMBRYOGENESIS

It is generally agreed today that the beginning of fertilization is the moment of fusion between the male sex cell (spermatozoid, sperm) and the female sex cells (eggs, ovicell), which leads to the diploid zygote formation. It is preceded by the processes of insemination in animals and pollination in plants. As a rule, the fertilization of pine occurs between late May and the first half of June when the sum of effective temperatures reaches 510-550°C and the growth of cones has already ended.

Scientists have long shown that between eggs and spermatozoids, there is a relationship in distance and the contact (Tokin, 1987; Belousov, 2005). The hypothesis of the chemical nature of "attraction" of gametes is currently the most widespread conception in embryology. In pines, the main way for fertilization to occur is monospermy, when only one sperm participates in singamy. For this purpose, a central cell (female gamete) outputs from a resting state the only certain generative cell (male gamete) inside one pollen grain. Its derivative – a spermatogenous cell –overcomes the first barrier towards fertilization. It comes out from a pollen grain into the pollen tube through the narrow germination furrow of a sporoderm. This phenomenon cannot be explained as the effect of special chemical substances, released by the central cell and/or spermatogenous cell. We think that the more powerful forces in the gamete attraction should act.

A pollen tube is only the first part of the distance towards the female gamete. The male gamete needs to overcome a few more barriers, because between an apical end of the tube and a female gamete are located in the callose layer, nucellar tissue, sporoderm and ovicell cytoplasm, which is densely packed with nutrients. A spermatogenous cell is not able to overcome these barriers and stops in the apical part of the tube. It undergoes the 5th activating mitosis and gives rise to a functional sperm, which is the last and most energetically strong generative cell by the male generative line. The sperm motion towards the ovicell resumes and accelerates. If you look at the sperm trajectory, beginning from a pollen grain up to an ovicell, the trace represents a direct channel like that of a cannonball (Figure 48b).

Special devices for sperm penetration inside a female gametophyte and ovicell are absent. Nevertheless, the interaction between the gametes really occurs; for this reason, fertilization always takes place. The forces acting between them drive the sperm towards a nucleus of the ovicell. If we proceed from the basic concepts of quantum mechanics, they give the understanding about the nature of fertilization mechanisms. The answer is composed from the following effects: 1) The ovicell and sperm are Fermi particles; 2) the Pauli exclusion principle; and 3) gametes only with opposite spin directions are attracted between themselves.

The spin-spin interaction between the nucleus of the generative cell and the nucleus of the central cell is a straight line, coinciding with the direction of gravity. It should be noted that in this direction all the final events before fertilization take place. This is the disposition of a mitotic spindle during 1-5th division of the microspore, two activating mitosis during oogenesis and the movement of male gametes into the pollen tube. Sperm enters into the archegonium with the micropilar end between the neck cells. Then it is moved down to the nucleus of the ovicell, where the copulation of the nuclei takes place (Figure 48b, 51).

During this time, an ovicell begins to enter into a resting state, or in the equilibrium state at the critical point. The ovicell in this moment possesses a raised receptivity. Apparently, it is caused by the specifity of this state, at which the biofield is concentrated mainly close to the nucleus. Therefore, the ovicell biofield cannot prevent a rapprochement of the gametes and a penetration of the sperm into cytoplasm, and then into a nucleus. The result is two most energetically strong gametes by the male and female generative lines that participate in fertilization and give rise to a zygote (Figure 51b).

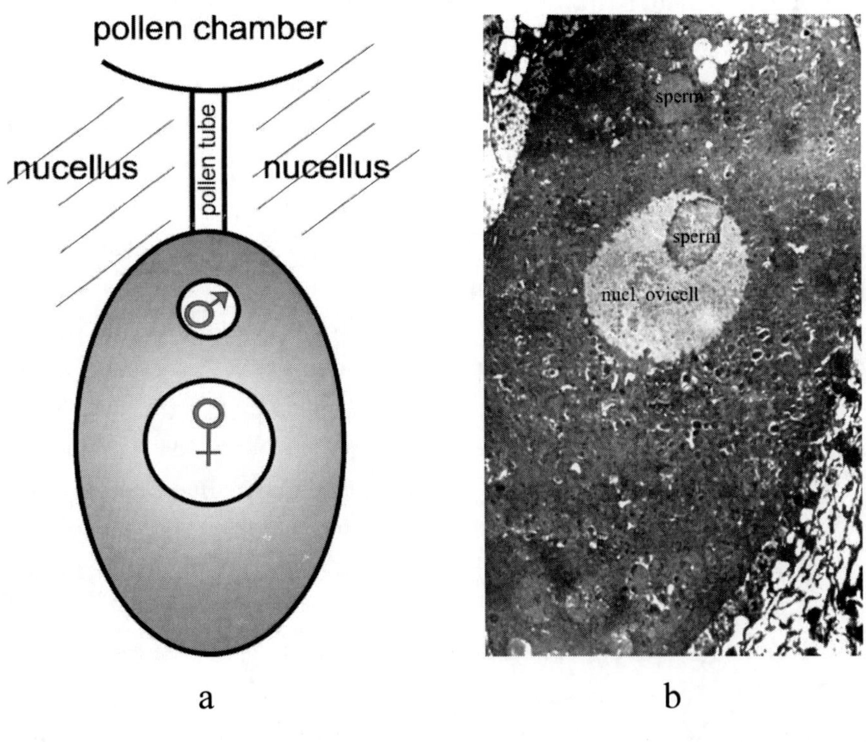

Figure 51. Fertilization and zygote formation in *Pinus sylvestris.*

Summary

The main events of gametogenesis are the following:

1) The four basic states of the genome are changing for the creation of energetic, corpuscular and informational programs of the future plant. There is an increase of DNA-energy gametes at three stages of the reproductive cycle (microgametophytogenesis, female meiosis, and oogenesis) (Table 1). During meiosis, the genotypes with non-identical DNA-particles (crossing-over) and spin direction (differentiated by sex) are formed. At the same time, the eukaryotic antenna array turns to receiving the regime, captures a species informational signal from space, and fixes and strengthens it on the DNA-wave of male and female gametes.
2) The interaction of male and female generative spheres begins and ensures the choice of one megasporocyte from 4-8 sporogenous cells, the output of generative cell from the pollen grain into the pollen tube, and gamete syngamy.
3) The enlargement of genotypic diversity on the basis of a multitude of microstrobiles, megastrobiles, and sexual gametes differing by DNA-energy, DNA-particle, DMA-information and direction, as well as with the participation of cross-pollination occurs.

Chapter 5

Creation of Corpuscular, Energetic and Informational Matrix during Embryogenesis

In the description of the embryogenesis of *Pinus sylvestris* L., we sought to use commonly accepted classification of reproductive stages and terminology, developed by Buchholtz (1920, 1946) and supplemented by Singh (1978). According to this classification, embryogenesis is divided into 3 stages: Proembryogenesis, early embryogenesis and late embryogenesis.

The process of fertilization occurs in a similar way. The penetration of male gametes into the ovicell changes the physical situation inside the female gamete. Firstly, a sperm penetration causes the short-lived contact of a gametophyte with the pollen chamber via the pollen tube and an intracellular pressure inside the ovicell is decreases. Secondly, the sperm is a high-frequency harmonic oscillator, and its action is the main external factor for the exit of the ovicell from dormancy. This process is very precise for the intended purpose. It activates only one cell of the female gametophyte – the ovicell. The endosperm cells do not experience its impact and continue to remain at the rest state. Therefore, the penetration of the sperm into the ovicell should be considered as a process of female gamete outputs from dormancy and its activation, but not as fertilization.

Cytological, morphological and biochemical changes due to fertilization processes are studied mainly in animals and less in plants. It has been established that shortly after spermatozoid penetration a series of biochemical changes occur in the egg: Rising of respiration intensity, permeability of egg membranes, increasing of carbon, phosphate and etc. metabolism, stimulation of Ca^{2+}-dependent processes (Tokin, 1987; Tokmanov et al., 2002, Belousov, 2005; etc.). Then, the biosynthesis of DNA, proteins, etc. begins. During the first division, the DNA content increases insignificantly, and then very quickly (Brachet, 1960).

Ovicells and sperm are the highly specialized reproductive structures, which are substantially different by size, morphology and metabolism. In addition, their genomes differ by the state of DNA-particle, DNA-energy and DNA-information. Therefore, the zygote of pine cannot be the first cell of a plant. A diploid genome cannot arise for the following reasons. Firstly, an ovicell is a very large immovable cell, which is in its resting state. Sperm represents a working biological harmonic oscillator. It is much smaller than the ovicell oscillator. In the first two hours, the nuclei of *Drosophyla* are divided every 9.6 minutes (Zhimulev, 2003), and they testify about a high frequency of oscillation for its DNA-wave. Apparently, an energetic potential of sperm is higher too. Secondly, the genomes of male and female gametes have different DNA-particles. The amount of DNA in a haploid nucleus of sperm is equal to C, and in the polytene chromosomes of the ovicell after endomitosis divisions – 32C (Ermakov et al., 1981: Barantseva, 1992). Thirdly, the first nuclear divisions of zygote quickly follow each other. According to Ermakov et al. (1981, 1982), the proembryo development of *Pinus sibirica* is not accompanied by the DNA replication and is not controlled by the zygote nucleus. They hypothesized that the first divisions of the zygote are determined in the process of ovicell maturation. This implies that the zygote is not the last cell of the reproductive cycle. The "bio-build" process of the genotype continues during proembryogenesis (Figure 28).

PROEMBRYOGENESIS

Changing of the physical situation after sperm penetration inside the ovicell destabilizes its genome, and it exits a state of balance. The wave function of its many-threaded polytene chromosomes is restored. In pine, polytene chromosomes contain 16 times more DNA than a diploid cell. Therefore, energy potential of sperm will be higher than in the ovicell in 5 levels. Suppose that the activated ovicell continues a cell cycle with the G_2 period of interphase. The pine reproductive cycle for alignment of the genetic material of sperm and ovicells has a special type of cell division known as cleavage. In detail, the mechanism of cleavage division in animals was described in our previous work (Kuznetsova, 2013). A simplified scheme of DNA-particle changes during 4th divisions of proembryogenesis is shown in Figure 52.

The end result of proembryogenesis is the fertilization of syngamy of the genetic material of male and female gametes and the creation of the diploid genome of future plants. Cleavage is a complex corpuscular-wave process which continues during 4 divisions of zygotes within the ovicell. In the course of cleavage, the male and female gametes participate with different programs (Figure 52), and the combinations of the three types of divisions are used – free-nuclear mitosis, activating mitosis and cleavage in the next combination of divisions: Free-nuclear mitosis + cleavage and activating mitosis + cleavage.

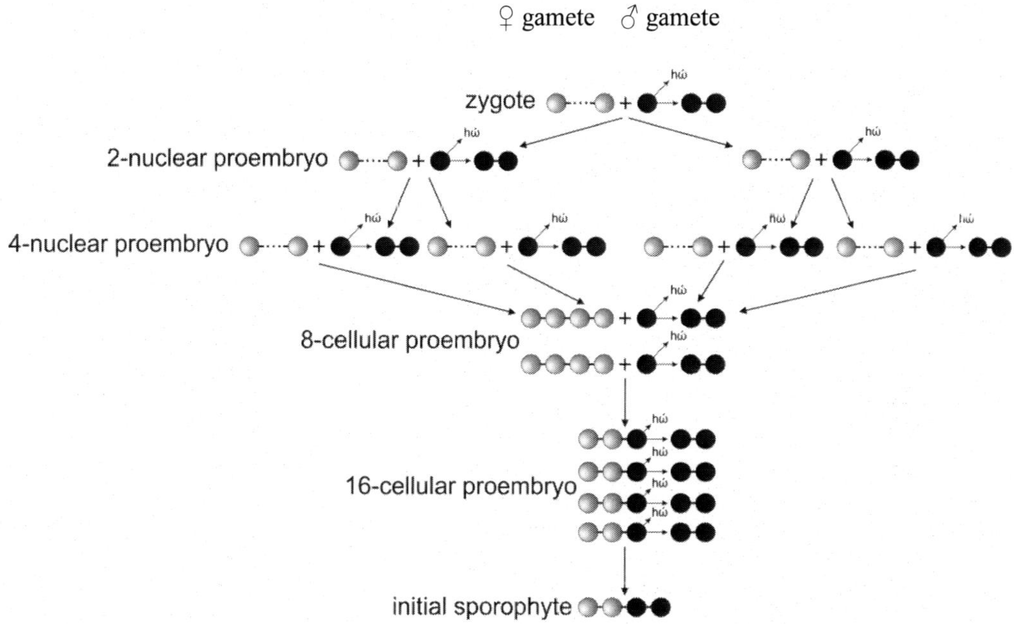

Figure 52. Scheme of the change of sperm and ovicell DNA-particle during proembryogenesis.

Proembryogenesis is the stage of zygote division inside the ovicell: 2- and 4-nucleous proembryo, primary and secondary proembryo. The starting position of a genome at the moment of nuclei copulation is as follows: The nucleus of ovicell contains 32C DNA, the nucleus of sperm – 1C DNA. A sperm after the penetration into nucleus of an ovicell goes down in an energy level and radiates a quantum of free energy. This energy initiates DNA sperm doubling and its amount becomes 2C. Since the frequency oscillation of a sperm is high, the duration of S-and G_2 periods is shorter. Then, the zygote undergoes the 1st division by means of free-nuclear mitosis + cleavage. As a result of this division,2-nuclear proembryos are formed, with each nucleus containing two haploid sets of chromosomes: The polytene 16C DNA-particles of ovicells (division without the doubling) and 1C DNA DNA-particles of sperm. After division, the two nuclei pass in the G_1 period of the interphase. The co-oscillatory motion of cumbersome polytene chromosomes of an ovicell with the high frequency chromosomes of sperm influence and destabilize each other. The result of mutual destabilization of their state is the 2nd transition of male chromosomes to the lower energetic level and radiation of a free energy quantum. This energy causes a doubling of sperm DNA-particle and the passage of a 2-nuclear proembryo to the next division. At this time, the set of female chromosomes also undergoes the cleavage without doubling. During this free-nuclear mitosis + cleavage, a splitting in half of polytene DNA-particles and division of doubled DNA-particles of sperm take place. Accordingly, 4-nuclear proembryo contains 8C DNA from an ovicell and the 1C DNA from a sperm. An 8-nuclear proembryo is formed after the 3rd DNA-wave radiation of sperm and causes the same sequence of events, which occur in a similar manner. At this time, the cell envelopes are formed, giving rise to the primary 8-cellular proembryo. Its nuclei include the 4C DNA of ovicell and the 1C DNA of sperm. 4th transition of sperm to the lower energetic level initiates the last division of proembryogenesis, during which the secondary 16-cellular proembryo is formed. The nucleus of each of its cell has 2C DNA from an ovicell and 1C DNA from a sperm. DNA-particle of the sperm is doubled again to the 2C DNA after the 5th radiation of a DNA-wave, and the genetic material of both sets of chromosomes is the same by the DNA-particle and DNA-energy. From this time, the process of the formation of a diploid genotype of the initial sporophyte and diploid matrix of the future embryo and plant by a DNA-particle begins, which continues up to the 1st division of cells of an embryonic tier of 16-cellular embryos.

A nuclei fusion of functional sperm and an ovicell occurs in the center of the female gamete and lasts for 10 to 24 hours (Ermakov et al., 1981). The proembryogenesis duration is 7-8 days. Half of this time falls on a stage of a 16-cellular proembryo. The first two divisions of the zygote follow each other during ≈ 24 hours, which is approximately two times shorter than the mitotic cycle of a pine.

Let us define more precisely the details of the proembryogenesis, taking into account the corpuscular-wave changes of genetic material during this stage.A well-known cytological scheme of a proembryogenesis of *Pinus sylvestris* is shown in Figure 53.

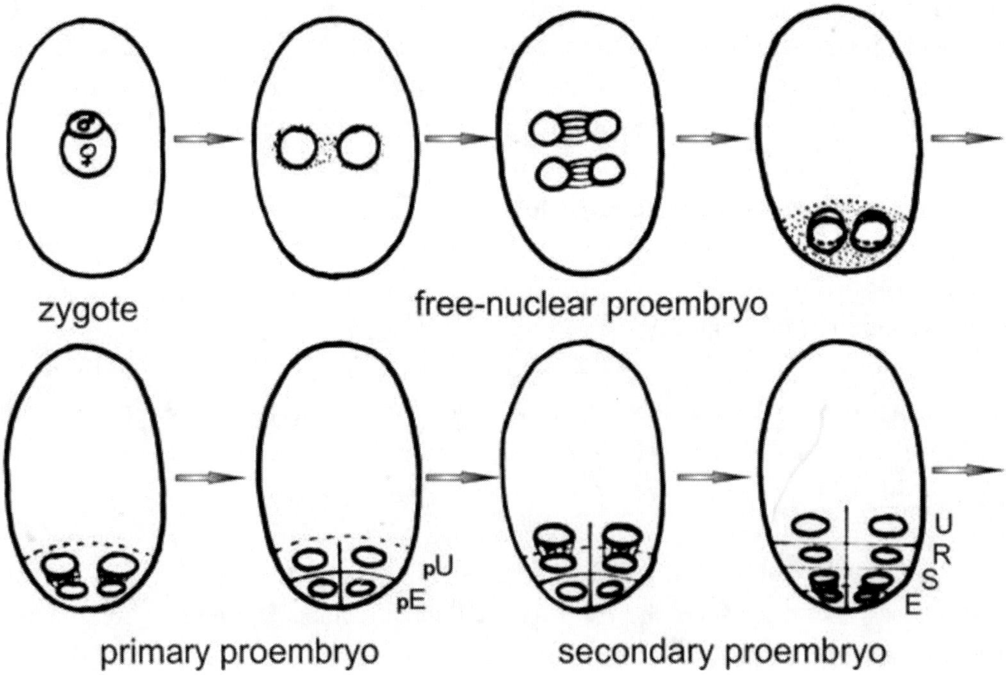

Figure 53. The scheme of the proembryogenesis of *Pinus sylvestris*.

All newly formed zygotes without a rest period immediately undergo the 1st free-nuclear division in the center of the ovicell. The envelope of the sperm nucleus is destroyed in the moment of union of male and female nuclei or shortly thereafter (Chesnoy, 1969; Trenin, 1988) (Figure 51a). After their syngamy, the two groups of chromosomes are visible (Singh, 1978; Trenin, 1988).During the 1st division,the multipolar spindle is formed, which soon is united into a common spindle (McWilliam & Mergen, 1958). The pairs of chromosomes are attracted to each other, but the true diploid set of chromosomes at this time is not yet formed. Then, the phases of the first mitosis quickly follow one another, and, as a rule, without any violations.

The nucleocytoplasmic ratio of an ovicell has changed. Therefore, the first two divisions of a zygote take place in the center of the ovicell, and are not accompanied by the formation of a cell envelope. Accordingly, a free-nuclear mitosis and cleavage occur simultaneously and give rise to 2-nuclear proembryo. Nuclei of proembryo are largeand identical by the ultrastructure, with the significant invaginations of the envelope (Figure 54a). Their inner content presents only euchromatin. In fact, a nuclei of a free nuclear proembryo are the two powerful biological harmonic oscillators. Every oscillator has its own biofield, which attracts cytoplasm from the perinuclear zone of the ovicell and creates around each nucleus a concentric cytoplasmic layer. This layer provides the trophic function of an embryo during pro-embryonic development. The perinuclear layer of cytoplasm disappears if the embryo death occurs during this stage.

The 2nd division of a zygote is ongoing and identical to the 1st, with the exception of metaphase chromosomes becoming smaller. At the same time, there is a decrease of nuclei size (Shimamura, 1956). Four free nuclei arise in the result of this combined (free-nuclear mitosis + cleavage) division, which forms a horizontal plane in the center of the ovicell. A characteristic feature is the absence of nucleus-like formations, nucleoli and heterochromatin. On micrographs, a layer of a perinuclear cytoplasmic layer around each nucleus is clearly visible and is intensely painted with dyes (Figure 54b). The width of this layer is 10-15 microns. Perinuclear cytoplasm is characterized by a large number of chromosomes, mitochondria and elements of endoplasmic reticulum. Mitochondria in shape and size do not differ from the ovicell mitochondria (differences are not statistically significant).

A biofield, which is created by a zygote, has 2- and 4-nuclear proembryosthat extend beyond the perinuclear zone of a mature ovicell. Its cytoplasm undergoes the ultrastructural changes, associated with the processes of lysis and moving of nutrients to the proembryo. At first, these changes appear on the border of a perinuclear zone. Then, the metabolic activity and destructive processes are extended to the entire cytoplasm. A density of cytoplasm decreases. Inclusions are fragmented into smaller segments. More heavy 4-nuclear proembryos are lowered in the basal part of the ovicell. The death of 2- and 4-nuclear proembryos rarely happens (≈ 5%) and meets with equal frequency in self-pollination and open pollination. If a proembryo death occurs, the perinuclear layer of the cytoplasm disappears.

The period of free-nuclear divisions during proembryogenesis is the 6thand final stage by the emergence of a new reproductive initial (⬭, Figure 27). In *Pinus sylvestris,* a number of divisions at this stage are reduced to two. A 4-nuclear proembryo is the new reproductive initial for the 6th activation stage. Four of its nuclei are equally hereditary, i.e., the morphogenetic possibilities of cell lines originating from them are the same. This defines the vertical symmetry in the proembryo organization, which is typical for all species of fam. *Pinaceae.*

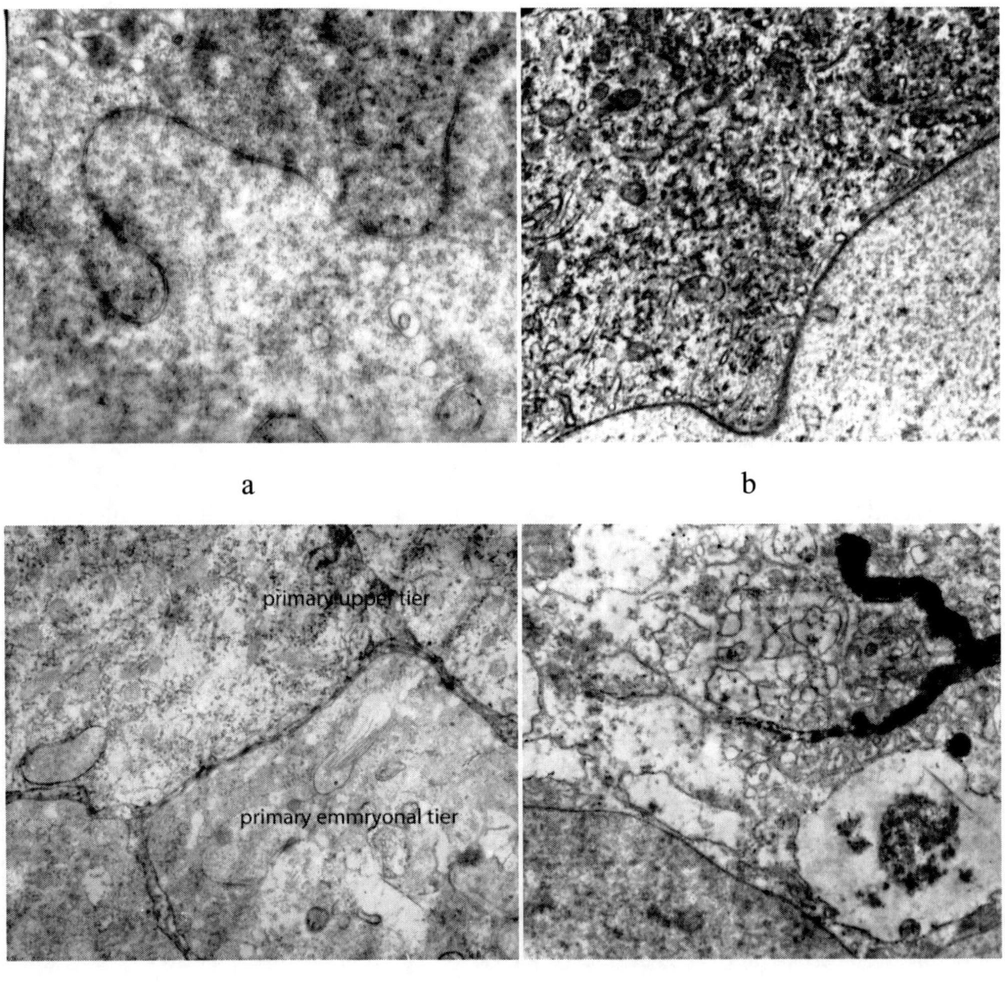

Figure 54. Ultrastructure of 2-, 4-nuclear and 8-cellular proembryo in the norm (a-c) and 8-cellular proembryo in the degradation (d).

The scheme of the last two divisions of proembryogenesis is shown in Table 3. This is the last 6th activation stage (ᚨ) in the pine reproductive cycle (Figure 27; Table 3). The 3rd division of proembryogenesis is a combination of two types of mitosis – activating mitosis and cleavage. 4-nuclear proembryos undergo the activating mitosis, giving rise to an 8-nuclear or a primary proembryo, and passes from the free-nuclear to the cellular stage of development. The perinuclear cytoplasmic layer is included inside the cells. Primarily, the horizontal wall between the nuclei is formed, and afterwards the vertical. Primary proembryo is organized spatially in two tiers with four cells in each. Tiers differ significantly by the cell size, ultrastructure, functional load and subsequent fate, and the cells within the tier are identical by the structure and function.

The cells by the vertical axis lost the property of totipotency. The embryo is formed only from the embryonic tier. Its cells are smaller by size. Cytoplasm is characterized by a dense homogeneous matrix and a small number of organelles (Figure 54c). Nuclei are larger. Chromatin is in the decondensed state and uniformly dispersed throughout the nucleus volume. After activating mitosis, the cells of the primary upper tier lose the biological energy, going down to the 1st energetic level. They have the weakly vacuolated and more electronic-light cytoplasm. Its ultrastructure is similar to the perinuclear layer and is characterized by a large number of ribosomes, mitochondria, elements of endoplasmic reticulum and small number by the size plastids. It is interesting to note that after embryo death, cells of both tiers are vacuolated and their ultrastructure becomes identical with the cytoplasm of a degraded ovicell (Figure 54d).

A 16-cellular or secondary proembryo is formed after 2nd activating mitosis. The secondary proembryo consists of the four tiers by 4 cells in each. These include an embryonic, suspensor, rosette and upper "open" tiers. The generative cells, which are located in a horizontal plane, possess the identical ultrastructure and function. In comparison with the primary proembryo a differentiation between the apical and basal poles of an embryo is increased. The embryonic tier of a 16-cellular embryo is the future sporophyte initial. The sporophyte initial is both the latest generative and the first vegetative cell, the genome of which contains the individual species-specific energetic, corpuscular and informational program of the development of the future plant. They are the most energetically strong cells in the reproductive cycle. The nuclei are characterized by the deep invaginations, which greatly increase the surface of contact with a cytoplasm and testify to the high metabolic activity (Figure 55a). The cells of three tiers of the basal pole are functionally inert (except a suspensor cells), nonviable and die. All potency of a new organism to the growth and morphogenesis are concentrated in the apical pole.

The functional activity of pine embryonic cells is confirmed by the study of their nuclei during proembryogenesis and early embryogenesis (Svintsova & Butorina, 2002). It is known that the morphological characteristics of nucleolar are related with the important molecular-genetic processes of a cell and reflect the peculiarities of its metabolism (Alberts et al., 2002; Chentsov, 2004). Scots pine has 6 pairs of chromosomes with the secondary constrictions and the maximum number of nucleolar organizer regions corresponding to this number (Muratova, 1983). The nucleoli of the nucleonemny morphofunctional type for the intensively proliferating embryonic cells are characteristic (Chelidze & Zatsepina, 1988). In the cells of the embryonic tier, the average number of nucleoli on the nucleus is 5.1 ± 0.2, and their number during early embryogenesis does not change (Svintsova & Butorina, 2002). Distribution of cells by this parameter corresponds to the norm.

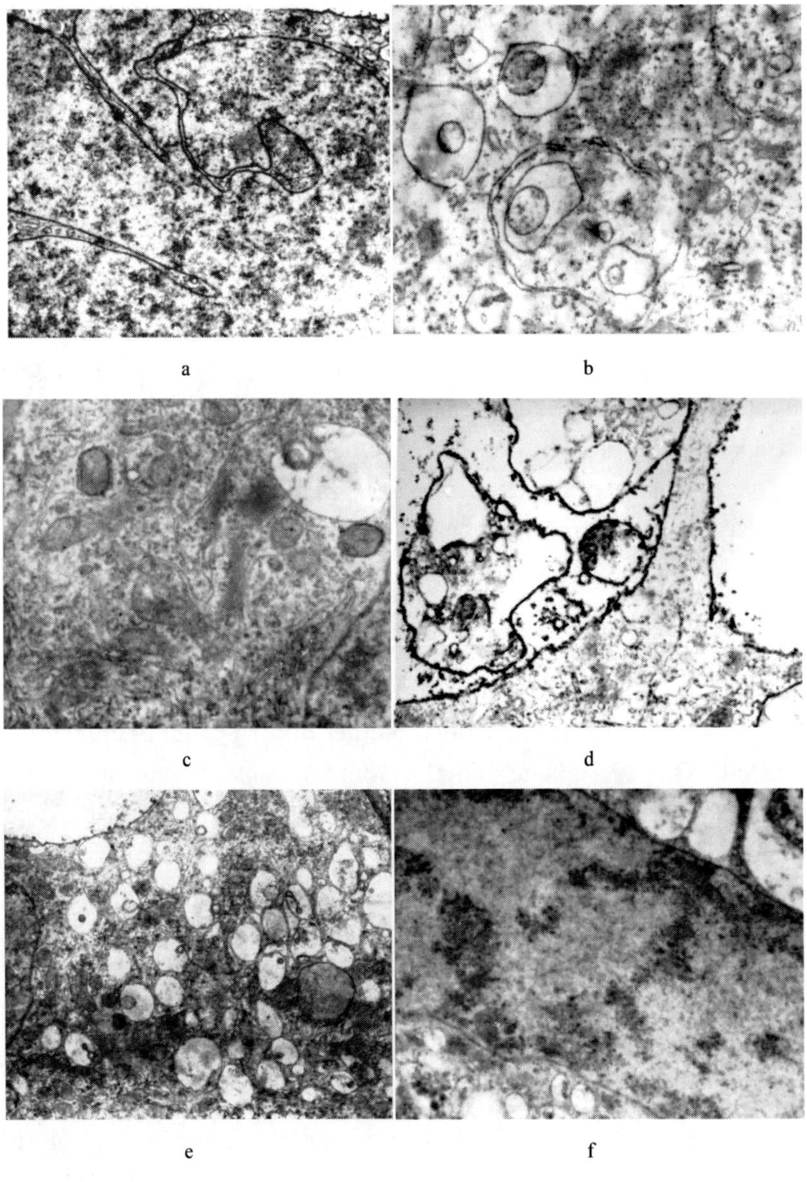

Figure 55. Cell ultrastructure of the embryonic (a-c), suspensor (d), rosette (e) and upper open tiers of 16-cellular proembryo.

As already mentioned, the cytoplasm of the perinuclear layer of the ovicell in the cells of primary proembryos entered. The formation of cytoplasm in the embryo takes place on the 16-cellular stage (Kuznetsova, 1994). The intensity of these processes is decreased from the embryonic to the upper "open" tiers and shows all its stages. Restructuring begins with the appearance in the cytoplasm of a large number of lysosomes. Initially, they are the small (0.6 micron) vesicles of spherical or oval forms. Then, in the center of lysosomes the thin fibrillar inclusions, enclosed in a single membrane (Figure 55b) appear. Gradually, lysosomes fill the entire volume of embryonic cells and the autolysis of cytoplasm begins.Sometimes, one or more layers of endoplasmic reticulum isolate the larger sectors of cytoplasm and the content of formed autophagic vacuole undergoes lysis. The cells are quickly vacuolated. Newly formed vacuoles often contain cytolisosomes in which the lysis of its contents continues. Simultaneously, the reproduction of new cytoplasm and organelles under the nucleus control occurs. At the end of proembryogenesis, the cells of the embryonic tier have the ultrastructure of the meristematic type, the main features of which are preserved after proembryo death (Figure 55c). A newly formed cytoplasm is the poorly vacuolated and does not contain lysosomes.

Suspensor cells, lengthening, and transfer of the embryonic initials from an ovicell into the female gametophyte tissue occurs. A suspensor tier consists of weak energy cells (≈4-5 energy levels) and their development occurs by the type of differentiated cells. There is a decrease of the nucleoli number from 8.7 ± 0.2 to 4.8 ± 0.3 (Svintsova, Butorina, 2002). During the cell cycle, the morphology of the nucleoli is changed and the annual nucleoli with a central vacuole appears, which are characteristic for the cells with a delayed type of development. In the suspensor cells, a restructuring of cytoplasm also occurs. However, the intensity of these processes and the speed of new cytoplasm formation are much lower. Therefore, the volume of vacuome is constantly increasing. Vacuoles often contain the autophagic vacuoles (Figure 55d). Cytoplasm only occupies the parietal zone.

The cells of the rosette tier are passively carried out from an ovicell and together with the suspensor cells are linked between the ovicell and the embryo and then die. This is the energy of weaker cells, which are characterized by electron-dense cytoplasm and a large number of organelles (Figure 55e). The vacuoles are located by the periphery of cells and have a tendency to unite. The cytoplasm forms a rather wide parietal layer for connecting. The lysosomes are concentrated mainly around the nucleus or along the vacuoles. Lipid droplets are not numerous. Plastids are encountered very rarely, which was mentioned earlier by Thomas (1973). The cytoplasm of the rosetter cells of a degrading proembryo and ovicell have the same ultrastructure.

The upper "open" tier is an example of a cell system of the last zero energy level. It does not bear a functional load. Its cells remain in the ovicell and often do not form the cell envelopes. Therefore, the nuclei are in direct contact with the cytoplasm of an ovicell

(Figure 55f). They have large sizes and an oval shape without invaginations. Chromatin is characterized by the high level of compaction and usually located along the nuclear envelope, which confirms the low metabolic activity of the nuclei. At the beginning of early embryogenesis, a nuclei of this tier gradually disintegrates and disappears.

As was mentioned above, at the end of proembryogenesis the chromosome sets of both gametes have the same corpuscular-wave state (Figure 52). The DNA amount in the ovicell chromosomes after the 4th division of the zygote (activating mitosis + cleavage) comes back to the norm (2C), in the sperm chromosomes as before(1C). After the next (5th) DNA-wave radiation and DNA-particle, the doubled DNA quantity in the sperm chromosomes becomes equal to 2C. In this state, the cells of the embryonic tier enter in the G_2 period, which lasts 4-5 days. Therefore, the energetically strong embryonic cells enter into the final step of the reproductive cycle, which is the formation of a diploid genome of the sporophyte initial.

In the previous work (Kuznetsova, 2013), we suggest that during meiosis the sexless diploid maternal cell gives rise to four sexual haploid cells of two types –sperm and ovicells with the spin upward (↑) and with the spin downwards (↓). During fertilization, an asexual diploid state of a genome is restored, because the spins of the homologous chromosomes of the sporophyte initial look towards each other (↑↓). If it is true that the corpuscular-wave differences between the homologous chromosomes (interspecific incompatibility, such violations as translocations, etc.), then they are barriers for the gamete syngamy and formation of stable diploid genomes. The results of interspecific crossing confirm the existence of mechanisms, which secure the functional autonomy of species parental genomes (Bennett, 1988; Monahova, 1990; Zhao & Davidson, 2014; etc.).

As is known, the chromosomes of sexual gametes differ by the following traits: Morphology, size, centromere location, presence of secondary constrictions and satellites, and their disposition. The chromosomes of each type in the diploid genome are presented twice. Homologous chromosomes have the same morphology, size, and opposite spin direction – one of them is brought about by a female gamete, the other by a male during fertilization. A fundamental feature of the spatial organization of a diploid interphase nucleus is the approximation of homologous chromosomes (Ashley & Pocock, 1981; Bennett, 1987; Monakhova, 1990). In association, they maintain the functional and structural discreteness of each chromosome, their individual phenotype and at the same time the integrity and high level of the orderliness of a diploid genome. Experimentally, it is confirmed that two DNA sequences are present in each somatic cell, the quantitative relation of which is identical in the different cells of an organism (McCarthy & Hoyer, 1964).

If we proceed from the quantum nature of chromosomes, a syngamy of sexual gametes presents a longitudinal (two by two) association of homologous parental chromosomes by means of formation of spin-spin connections between their homogenous nucleotides. These processes satisfy the physics laws and biology experimental data. This is, firstly, the attraction of chromosomes when the spins are oppositely directed; secondly, the fixation of interphase chromosomes with the nuclear envelope; and thirdly, the paired localization of homologous chromosomes in the nucleus. It is considered that the principle of the paired localization of homologous chromosomes is laid in the nucleus of a zygote during fertilization (Ashley & Pocock, 1981). The same chromosomal regions (telomeric with telomeric, centromeric with centromeric, interstitial with interstitial) come into association (Monahova, 1990) as they are experimentally proven. Spatial organization of chromosomes, which was created in the zygote, is preserved in the somatic cells during interphase. In the embryonic cells the process of the formation of spin-spin connections between the homologous chromosomes occurs during the G_2 period, when the genetic material of gametes has the same corpuscular-wave characteristics.

We should take into account that a genetic material of male and female gametes has also the corpuscular-wave differences of a lower level. These differences are laid within the norm reaction of species in the form of intraspecific variations of DNA-energy, DNA-particle and DNA-information. However, we think, that these differences do not prevent the spin-spin attraction of homologous chromosomes. During G_2 period, the sets of male and female chromosomes undergoes the process of internal coiling. In the course of internal coiling, the paired arrangement of homologues creates a physical basis for the formation of sustainable spin-spin connections between the nucleotides, because their spins look towards each other. Probably, the attraction of complementary nucleotides occurs by simultaneously entering into the internal coiling zone of the 8 nucleotides (4 from a sperm and 4 ovicell chromosomes). As the probability of this event is smaller, this greatly increases the duration of G_2 period (4-5 days). Nucleotides are bound together by means of spin-spin connections along the entire length of homologous chromosomes before the first division of a sporophyte initial. The high synchronicity of this process indicates the following fact. Usually, the 4 cells of an embryonic tier simultaneously undergo the mitotic division, giving rise to a single embryo and every cell of embryonic tier as a result of cleavage polyembriony forms only sometimes a separate embryo.

The embryonic cells of a 16-cellular proembryo are characterized by the small size and an optimum nucleocytoplasmic ratio. Their own cytoplasm is formed. These 4 embryonic cells become the sporophyte initial. During fertilization, the two combinations of a sporophyte initial depend on the spin direction (♂ ↑ + ♀ ↓ or ♂ ↓ + ♀ ↑) are created. Both these combinations give rise to the amonoecious bisexual plants. Consequently, the genome of a sporophyte initial and all vegetative cells arising from it have instead only four of three basic states – DNA-energy, DNA-particle and DNA-information.

The sporophyte initial is passed from an ovicell to a corrosion chamber by means of lengthening suspensor cells. As a rule, all generative cells within one tier have the same DNA-energy, DNA-particle and DNA-information, and between tiers they differ only by

DNA-energy. Variability by the DNA-energy determines the metabolic activity, growth speed and further fate of the cells of embryonic and suspensor tiers. Development of cells of suspensors, as the energetically weak cells, undergoes by the type of differentiated cells, when the DNA-wave radiation is not accompanied by the cell division. As a rule, the growth speed of all suspensor cells within a tier is the same. In this case, four cells of embryonic tiers give rise to a single embryo. If a suspensor cells have the various amount of biological energy, their speed of growth is different, that creates the basis and possibility for a disconnection of generative cells embryonic tiers and their selections. During embryogenesis, this is leads to the separation of embryonic cells on the four independent lines and formation of 4 cleavage embryos. From them, only one – the most energetically strong leading embryo – survives and all others die.

Embryogenesis begins with the simultaneous asymmetric mitosis of the 4 cells of a sporophyte initial. We suppose that the spin-spin connection is responsible for the stability of the diploid state genome of cell systems in ontogenesis. The spin-spin interaction between the homologous chromosomes is much stronger than the chemical bond between sister chromatids (or DNA-particles) in the doubled set of metaphase chromosomes. This explains for some reason during anaphase why a diploid set of chromatids, but not a doubled haploid set of chromosomes moves to the opposite poles of a cell. Why only the gametes of opposite sex (♂ + ♀) may participate in the fertilization and the unisexual combinations of gametes (♂ + ♂ and ♀ + ♀) crossing does not occur. Therefore, if a homologous chromosomes and sexual gametes are the spin objects, the quantum-mechanical nature and Pauli exclusion principle together support the orderliness and continuity of the genetic processes of vegetative and generative cells in the alteration generations.

PROEMBRYONIC SELECTION

Development of male and female gametes initially occurs in the microsporangium and ovules, which are located on the same maternal tree. Pollen after maturation leaves the microsporangiums and each tree loses half of its gene pool. At the same time, during pollination into the pollen chamber the plant gets pollen of different origination –a mix of their own and other trees within a panmictic population. A genotypic diversity of male gametes as the result of cross-pollination becomes much higher than before pollination. Therefore, a huge number of combinations of genotype-environmental interactions are created – from self-incompatibility to cross-incompatibility. Each maternal tree, as a sieve selects them for compatibility with a diploid nucellar tissue during pollen germination and pollen tube growth. Maternal trees present the adaptive formation with reference to the pollen pool after open pollination. Their genotypic composition is adapted to the regional conditions by the principle of "genotype-environment" interaction.

Haploid pollen and diploid nucellus tissue differ by the DNA-particle. According to the physics laws between them as the laws between two heterogeneous tissues with different energy potential, the centripetal stream of metabolites must increase, which provide the trophic function of male gametophytes. The semi-permeable cell membrane of the tube is located on the path's movement of nutrients. Therefore, a part of the substances does not pass through it and are deposited around the tube in the form of the callose layer. The thickness of callose depends on the incompatibility degree of two tissues. Incompatible inbred and outbred tubes die during germination of pollen grains or during growth in the 1st vegetative season.

In fact, a pollen tube is the channel by which the spermatogenous cell moves after exiting from the pollen grain. Therefore, this channel should be of a certain length and thickness, in order not to delay its movement. The callose layer of optimum thickness does not affect the speed of the motion of generative cell. In this case, the inner diameter of the tube is comparable with the size of generative and spermatogenous cells. Partially compatible combinations form on the boundary of tubes of more powerful callose deposits, which do not lead to tube death. More powerful depositions decrease in thickness. As a rule, a powerful callose layer is formed around the inbred pollen tubes. The motion of generative cells occurs more slowly. The processes of sperm formation and fertilization take place at a later date, then in the outbred genotypes. As a consequence, the inbred embryos corrode later and the chances to take a leading position in the inbred embryos among outbred competitors become low.

The population gene pool of zygotes after gamete copulation has been created, the formation of which is attended not attended by all genotypes of sperms and ovicells. The energetically weak, genetic disorders and inbred genotypes from the reproductive process were mainly eliminated. The genetic system in a result of proembryonic selection is out of the balance state. Proembryogenesis occurs in a closed system of the ovicell and represents a series of genetically determined divisions. Its main purpose is the formation of the diploid genome of an initial sporophyte. Death of the proembryo is very rare (≈5%). Therefore, it can be assumed that the new population gene pool of sporophyte initials (future embryos) is also in the non-equilibrium system. Let us consider whether this is true or not.

The non-equilibrium state is unstable. It can disrupt any internal or external disturbance. If a closed system at some point is not in the equilibrium, then most likely in the subsequent time it will begin to pass into an equilibrium state. Therefore, the population gene pool, which is a self-regulating system, should return also to equilibrium. Apparently, this transition is initiated from the inside and the major destabilizing factor is fertilization (or, more precisely, the formation of spin-spin connections between homologous chromosomes). Fertilization is a very complex process and it makes high demands to the genetic material of gametes. They must comply with the specific norm for all three components of the genome – DNA-energy, DNA-particle and DNA-information.

Interspecies differences between the chromosomes (number, size, morphology, etc.) are very great. In this case, the possibility of homologous chromosomes rapprochement formation between them as a stable diploid genome is not allowed. The violations of the intraspecific level lead to the local separation of the homologues, the disturbance waves in the places of their convergence, which should be accompanied by functional and structural changes and may be the reasons for numerous cytogenetic pathologies.

The analysis showed that during embryogenesis the two different processes occur simultaneously – the formation of a new population gene pool of embryos and the returning of the genetic system in the equilibrium. In order to understand the nature of these processes, let us return to meiosis from the position of DNA-energy, DNA-particles and DNA-information.

DNA-Energy

During meiosis, a population gene pool of two types – microspores and megaspores – has been formed on each maternal tree. Male and female gametes are not identical by the DNA-energy. Their level of biological energy varies within species-specific norm reactions. The differentiation of the gametes by the DNA-energy corresponds to a normal distribution, when the majority of genotypes are the norm-forming genotypes. The boundaries of the trait are limited by certain limits. The maximum amount of biological energy has the genotype-leaders, the minimum of which are the genotype-outsiders. Norm-forming genotypes have the level of DNA-energy close to the average species norm. Sperm and ovicell genotype-outsiders before gamete copulation from a reproductive process are removed mostly. Genotype-leaders are also realized on the limit of the species norm of DNA-energy. Therefore, we have assumed that the genetic apparatus of embryos, arising from the fusion of two genotype-leaders should be vulnerable to fertilization.

The peculiarity of the development of pine ovules is that their seed capsule reaches the size of a mature seed prior to fertilization (Sarvas, 1962). In order to recognize, what category of genotypes undergoes natural selection, a seed production analysis was done for 16 model trees of pine (195 cones, about 5000 seeds, from which 729 were empty seeds) (Figure 56). It takes into account the size of the capsules of which were greater or equal to the size of the smallest seed the cone (separately for each cone of the tree). Analyses of empty seeds showed that the vector of selection in relation to the genotype-leaders of gametogenesis changes its direction at the time of transition to the embryonic stage (Kuznetsova, 2003). The empty seeds depend on the capsule size, which were distributed in a ratio of 350 large:189 medium:190 small. About half of the dead seeds belong to the "large" category. Moreover, about 1/3 of them are the "largest," because their capsules were larger than all plump seeds of the cones.

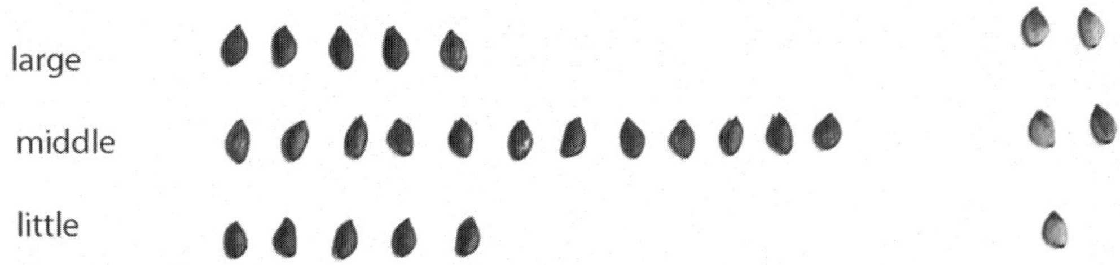

Figure 56. Seed production of the one cone in according with the sizes of its plump end empty seeds.

Although gametes with the identical DNA-waves and DNA-particles do not exist, the DNA-wave of homologous chromosomes within the nucleus oscillates approximately the same speed and wave length. Any deviations of the wave characteristics of sperm or ovicell chromosomes from the norm (in phase, frequency, amplitude) disturbed the state of the zygote. We assume that a breach in the interaction of homologues causes the resonance radiation, during which a transition of biological harmonic oscillators occurs in the lower energy levels. As a result, the norm reaction of embryos by the DNA-energy becomes somewhat narrower. The effect of biological energy reduction may be different – some embryos remain within the same category (genotype-leaders, norm-forming or genotype-outsiders); the others are moved to the lower levels. That is why the large, medium and small capsules in the fraction of dead seeds are always found (Figure 56).

The genetic material of gametes, which are realized at the upper and lower limits of a species norm reaction is the most vulnerable. If the two energetic strong gametes in fertilization participate, the effect of the superposition disorders (overlay) of waves is increased. 3 variants of the further fate of genotype-leaders are possible. Depending on the amount of lost energy, they remain genotype-leaders of the next stage, move into the other life category (norm-forming genotypes or genotype-outsiders) or die during embryogenesis. Hence, it follows that the leadership of energy strong sperms and ovicells during gametogenesis do not always go over automatically to the embryonic stage. The result of the lopsided energetic loss is the return of the population gene pool in a state of equilibrium and, as a consequence, an increase in its stability.

During fertilization, a new type of population gene pool is created –diploid genetically non-identical genotypes of embryos. Their energetic potential varies within quite wide limits – from the genotype-leaders to the genotype-outsiders. After the formation of spin-spin connections between the homologous chromosomes, the range of genotypic variation by the level of DNA-energy becomes less in comparison with the variability of male and female gametes.

DNA-Particle

Homologous chromosomes have approximately the same length, the position of a centromere and equal picture during coloring. Only the same chromosomal regions and paired genes may enter into the association (Monahova, 1990) and spin-spin connections. Perhaps, the mechanism of formation of the spin-spin connections is very sensitive to the state of the DNA-particle.

The structural chromosome mutations break the functioning of any genome (Stegniy, 1993). The ability to mutate is the same important feature of genetic material in terms of stability. A diploid set of the chromosomes of *Pinus sylvestris* L. is 24 (2n = 24). Poliploid and aneuploid plants are extremely rare. They differ significantly from the diploid ones by many morphological traits and are characterized by a lower viability, weakly developed root system and more short needles (Kozubov, Muratova, 1986).

Fertilization admits only a certain level of genetic abnormalities of gametes. The generative cells, having unbalanced chromosomal abnormalities, are eliminated during reproductive development by the male and female generative lines. The various irregularities in their cellular lines were examined by many researchers. It was shown that in the generative cells of the 2nd generation, the ratio of various types of violations is different than in the initial cells (Petrov, 1973). In subsequent cellular generations, sustainable chromosome aberrations prevail. According to Butorina et al. (2001, 2005), the frequency of PMs in Scots pine in central Russia does not exceed 5%and can be considered as the norm, i.e., it does not extend beyond a normal level of spontaneous mutation.

As is known, meiosis is one of the most environmentally-dependent stages of the life cycle. During meiotic division, microstrobiles and megastrobiles are in close contact with the external environment, which leads to the expansion of genotypic and modifiable variability of microspores and megaspores. Male meiosis occurs in the beginning of the vegetative season. Frequency and spectrum of chromosomal abnormalities in the abnormal weather conditions (sudden temperature changes, often night frosts) are increased, which leads to the reduction of pollen fertility, cone yield and seed quality. In diakinesis – metaphase I, the frequency of univalents significantly increase and the processes of synapsis and conjugation are disturbed. This causes the formation of incorrect chromosome segregation at the later stages (Borodin, 1987; Dyuzhikova, 1997). Abnormal chromosome segregation in parts of the cells is accompanied by blocking one or two divisions of meiosis. Formation of dyads, triads, pentads and hexads occurs. In some years, the violations can be observed in 30-80% of pollen grains (Noskova &,

Tretyakova, 2006). Therefore, the gametes, having the certain number of genomic, structural and complex genetic abnormalities, together with the normal conditions participate in the process of fertilization.

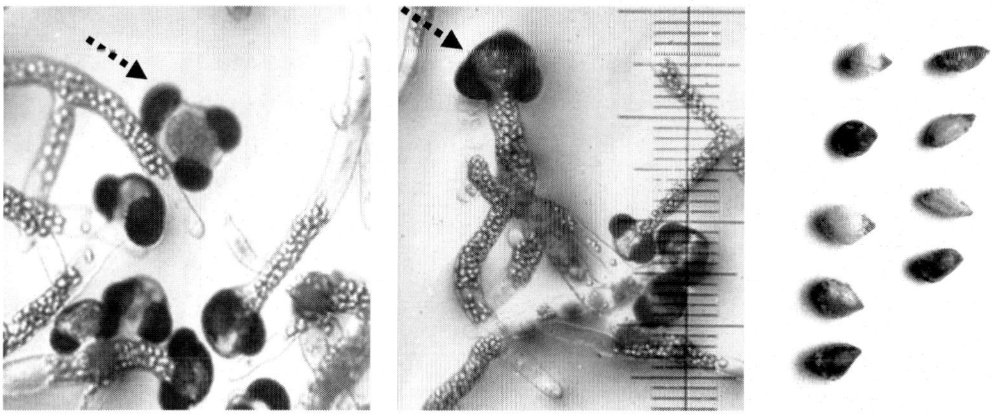

a b c

Figure 57. Germinated (a) and non-germinated (b) abnormal large pollen grains of *Pinus sylvestris*. Empty capsules of died seeds of abnormal shape and size (c).

The sensitivity of microspores to the violations of a DNA-particle is great. In the Scots pine, abnormal large pollen grains with 3-4 air bags are very rare (Figure 57a) (Noskova & Tretyakova). The morphology and sizes of this pollen can assume the increased ploidy. Usually, it does not germinate. When the abnormal grains with the unbalanced chromosome sets germinate, the majority of them degenerate and become shrunken lumps. Germination of this type of pollen and intensive growth of pollen tubes were found in pine (Figure 57b).

Megaspores are more than stable genetic systems. Their chromosomal abnormalities lead to ovule death and appear as small undeveloped capsules and empty seeds with abnormal shape and size. Figure 57c shows the two most typical forms of violations of dead seeds – they are very wide, sometimes with a raised crest (on the left) and long narrow capsules (on the right).

Female meiosis takes place in June. At this time, the main stress factor is hydrothermal stress and drought. In the forest-steppe zone, usually, the droughty years happen 1-3 times per the decade. Formation of megaspores in the other years occurs in the optimal weather conditions and the number of abnormalities of their genetic material decreases. Therefore, in the typical years the most part of violations, occurring during male meiosis, are found in the heterozygous state, when the spin-spin connections form between the horologes.

A distinctive feature of the molecular organization of chromosomes is the rotation of repeatable and non-repeatable sequences, which make up 20-80% of the genome of eukaryotes (Bostock, Sumner, 1981). Therefore, the segments located along male and female chromosomes have the possibility to participate in the spin-spin connections with the corresponding regions of non-homologous chromosomes. This leads to the exchange of the segments between the non-homologous chromosomes, or the segment of the one chromosome is transferred to another part of the same chromosome (translocation); or one segment on the same chromosome repeats twice (duplication); or the segment of chromosomes is rotated by 180° (inversion).Methods of genetic analysis were shown in Drosophila, having the inversions and translocations in the heterozygous state, and this is observed in non-homologous pairing and exchanges in both mitotic and meiotic cycles (Chadov, 2011). As a rule, these are known as short-living chromosome aberrations, which are eliminated in the process of the next divisions and cell differentiation.

Structural, genomic and complex types of violations lead to the local dissociation of homologues and different modulated waves. These processes should be accompanied by an increasing of the functional and structural changes of chromosomes within the species norm reaction and beyond its limits, which causes the more complex types of pathologies. Violations, reflecting the degree of genetic apparatus damage of male and female gametes, often prevent the association between the homologous chromosomes. Their dissociation can be the cause of the transition of the biological harmonic oscillator of embryos to the lower energetic levels. The probability of this transition is increased depending on the degree of chromosome damage as the particles. These embryos after the biological energy loss go into a lower life category – the norm-forming or energetically weak genotypes. During embryonic selection, most of their parts are removed from the reproductive process and the gene pool is released from the genetic load. It follows that the fertilization is a conservative mechanism, because it returns the population gene pool not only in the equilibrium, but a variability of its DNA-particles in the state of a species corpuscular norm.

DNA-Information

The wave nature of the genome provides an understanding about the role of DNA, as the physical bearer of individual species-specific information. The process of information system creation is realized step by step. The first step is a gametogenesis period. DNA-information represents the species of the informational signal, which was received by the formed microspores and megaspores from a space during male and female meiosis. The genetic apparatus of the meiotic prophase cell in this time has a set of special devices for its reception and fixation on the DNA-wave. The second step is DNA-information of the sporophyte initial. This type of DNA-information arises during gamete syngamy by a superimposing of two waves – a DNA-wave of the sperm and a DNA-wave of the ovicell. The third step is an informational matrix of the embryo during late embryogenesis. The distribution by the DNA-information within the embryo body is a genetically determined process. In ontogenesis, the histogenesis and morphogenesis will be held on the basis of the informational matrix and the morphogenetic field created by it. Its physical nature connects with the irreversible modulation of DNA-waves (Part I, Chapter 2).

We think that the species information in the succeeding generations is protected from degeneration by an unchanging of species signals, which generative cells capture during meiosis from the space. Then, after signal fixation on a DNA-wave, as on a physical carrier, the probability of its distortion is quite great. Moreover, we can predict that meiosis is the "bottleneck" method, where initial species information is most distorted or even partial loss. The main reason of modification for bad weather conditions at the time of signal reception. Different violations of DNA-particles, such as translocations, inversions, etc., also create the significant disturbances and noises in the moment of formation of the modulated wave (product of two wave interaction – a space signal with a DNA-wave of microspore or megaspore). A large part of gametes with these defects are eliminated during selection. The variability of the trait admits their partial incompatibility, which gives the congruity (compatibility) between the information signals of sperm and ovicells. The creation of a diploid genome in the conditions of stronger informational violations leads to the transition of DNA-energy at the lower energy levels and loss of the energetic potential of future embryos and plants. These assumptions are consistent with experimental data showing that different morphoses and deformities are generally connected with stunting and lower viability.

As shown above, during proembryogenesis or 4 divisions of zygotes the formation of the diploid matrix occurs and the reproductive development of future genotypes continues. A scheme of proembryogenesis contains 3 stages: ,  and 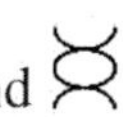 (Figure 27). This period of embryogenesis ends by the formation of the diploid genome of the sporophyte initial – 4 embryonic cells of a 16-cellular proembryo, whose homologous chromosomes are interconnected by means of spin-spin connections. If a proembryogenesis stage connects with the inception of the three genetic programs of the future plant, embryogenesis is a process of the formation of corpuscular, energetic and an informational matrix during development of the sporophyte initial. Then, an initial passes to the next stage of development –early embryogenesis, which occurs in the endosperm of the female gametophyte.

EARLY EMBRYOGENESIS

The sporophyte initial is the most energetically powerful cell in the life cycle of any plant. The initial of the pine could be 1 or all 4 embryonic cells of the 16-cellular proembryo. This is both the last generative and the first vegetative cells. As a rule, all the cells of the embryonic tier are totipotent and give rise to a single embryo. Its energetic, corpuscular, and informational programs are fully formed. Before embryogenesis, the suspensor cells are extended and carry out into the corrosion chamber the embryonic cells at their ends. An embryo possesses the great potency to cell reproduction. The biological harmonic oscillator, which during reproductive development has received the maximal species-specific level of biological energy, does not preserve and tends to reach equilibrium in accordance with the 2nd law of thermodynamics. Therefore, the DNA-energy of an embryo comes down on the lower energy level and radiates the quanta of free energy, initiating its first cell division.

During embryogenesis, the potency to the growth and development are realized only partially. Embryo development takes place on the mother tree inside female gametophyte, which carries out the transfer of an embryo in dormancy. The resting period is a specific state of cell or cell systems, when growth and division are stopped, and the intensity of the metabolism sharply decreases but the viability potentialities to development are not lost. Transition of energetically strong systems into dormancy is a genetically determined process, which we called a restriction stage (Kuznetsova, 2013). During this stage, all energetically strong systems are due to several stages: 1) Exponential growth; 2) stationary phase; 3) internal differentiation; and 4) physiological resting.

The state of a functional megaspore, increased during meiosis during its energy potential and received from the uncoiling chromosomes from a portion of free energy, is changed. Its biological harmonic oscillator creates the centripetal streams of water, biomolecules and ions from the nearby energetically weak cells of a nucellar tapetum. This leads to an increase of intracellular pressure and the megaspore begins the free-nuclear divisions and growth. A female gametophyte is rapidly increasing in size (the phase of its exponential growth). The growth of the gametophyte after the transition from a free-nuclear to a cellular stage gradually slows down and stops and it passes to the stationary phase. After sporoderm formation, divisions of endosperm cells continue in a fixed volume. By the gas laws, this should lead to an increase of pressure inside the cavity of the gametophyte. Then, a gametophyte passes to the stage of internal differentiation and gives rise to the archegonial initials and ovicells. After that, divisions of endosperm cells cease, they are filled with nutrients and they move to the final phase – physiological rest.

Similar processes take place during archegonium formation. Ovicells are formed during oogenesis. The central cell and ovicell, received during oogenesis portions of biological energy, become the energetically strong cell of female gametophyte and begin to grow rapidly (exponential growth phase). During maturation, the ovicell stops growing (stationary phase). Cytoplasm is filled with nutrients. The ovicell undergoes a certain number of endomitosis divisions (internal differentiation phase). Then, the ovicell passes into a resting state.

All processes according to the 2nd law of thermodynamics should proceed in the direction of equilibrium. The considerable increase of biological energy level in the functional megaspore and central cell lead to a reproductive system out of an equilibrium state. At first, these non-equilibrium systems consist of one or several energetically strong cells and a large number of energetically weak cells: Functional megaspore – nucellus; central cell – endosperm; and initial sporophyte – endosperm. The new energetically strong structure quickly grows. At the same time, the abscission layer on the boundary of two heterogenous tissues is formed. The envelope during thickening gradually reduces the intensity of the biofield around the new energetically strong structures to the level of the neighboring tissues of the nucellus and endosperm, stops their growth and translates them into a rest state. Therefore, the restraining stage is the process, which intends to restore the lost balance. As a consequence, the reproductive system as a whole comes back again in a condition of the average distribution of energy to the end of development on a maternal plant. In equilibrium, they are not capable of doing work and cannot leave the rest state without external or internal assistance. All of them are doomed to death, if they remain in a former state (as the cells of endosperm tissue). That is why non-fertilized ovicells and non-germinated seeds always perish.

Before fertilization, the reproductive complex – a female gametophyte and ovicells – are in dormancy. Compared to the endosperm tissue, an ovicell has a very high level of biology energy. Stopping the oscillator from working is required to create the higher intracellular pressure into an ovicell than in the cells of the endosperm. The energy balance between these systems was formed during oogenesis. For this purpose, a powerful biofield of the ovicell provides the streams of molecules from an endosperm. Synthetic processes increase the concentration of biomolecules too. Therefore, during maturation the intracellular pressure of the ovicell goes up and its cytoplasm is completely filled with chemically inert nutrients.

The return of the ovicell from dormancy is initiated from the outside. The external trigger is a high frequency oscillator of the sperm. It penetrates into the ovicell, activates its nucleus and gives rise to a zygote. The wave function on the DNA is resumed. The combined genome starts to operate and becomes a source of the biofield. Copulation of two energetically strong gametes increases the capacity of the biological harmonic oscillator of the zygote. Therefore, its development can take place inside the ovicell under the conditions of very high intracellular pressure.

At the time of fertilization, the endosperm is divided into two zones. The outer layer cells appear earlier. They have a smaller size, without vacuoles and are completely filled by the protein and lipid granules. The nutrients in this zone are spent by the embryo during seed germination. In the central part of the gametophyte below the archegonia is where the corrosion chamber is located, which has the form of a cone. The cells of this zone arise later, have larger sizes and fill amyloplasts with large starch grains (Figure 58a). Before the embryo exits into the corrosion chamber, a part of its cells is destroyed, forming a cavity. Further embryo development will be carried out in the nutrient-rich area.

The embryo after the transition into a corrosion chamber continues to feed off an ovicell through the suspensor system. However, these nutrients are not enough, as most of them were used during proembryogenesis. Then, between the energetically strong embryo and the energetically weak cells of the endosperm, the processes of active metabolism begins and the embryo passes to the heterotrophic way of nourishment. Several cell layers of the corrosion chamber are participated simultaneously in its trophic area. Primarily from the amyloplasts of neighboring cells, a dissociation of starch molecules into a simpler compounds and their absorption by a growing embryo occurs. Then, the evacuation of other plastic substances takes place. It is shown that the content of total nitrogen and protein in the endosperm is reduced after fertilization, and in the cells of embryo is increased (Sorokina et al., 1975). The processes of the active dissociation of starch into simple compounds in the endosperm cells occur (Konar, 1958).

The nearest cells of the endosperm react to the appearance of the corrosion chamber and the energetically strong embryo in the same way. They turn to them, pass in the regime of increased metabolic activity and are transformed in the secretory cells (Kozubov et al., 1982). The secretory cells are characterized by a specific ultrastructure and high synthetic activity. The intensive regime is harmful for the secretory cells. Their lysis and signs of degeneration have appeared, are accelerated and lead them to death. The cells of this layer are squeezed and nutritious substances are absorbed by the embryo. Then, cells from the next layer occupy their place. During open pollination, a part of component nutrient cells due to the incompatibility of the tissues of the embryo and endosperm are deposited around the embryo envelope as a small callose layer, which does not prevent its active growth and trophic function (Figure 58b).

The process of embryo formation inside the female gametophyte by its nature is a restriction stage. The embryo after the transition in the corrosion chamber gets in the conditions of lower pressure in comparison with the ovicell. On the one hand, the lower pressure (relative to the ovicell level) enables to form viable seeds, and on the other hand, the high pressure (relative to the mother plant level) limits the embryo growth of the frameworks of seed.

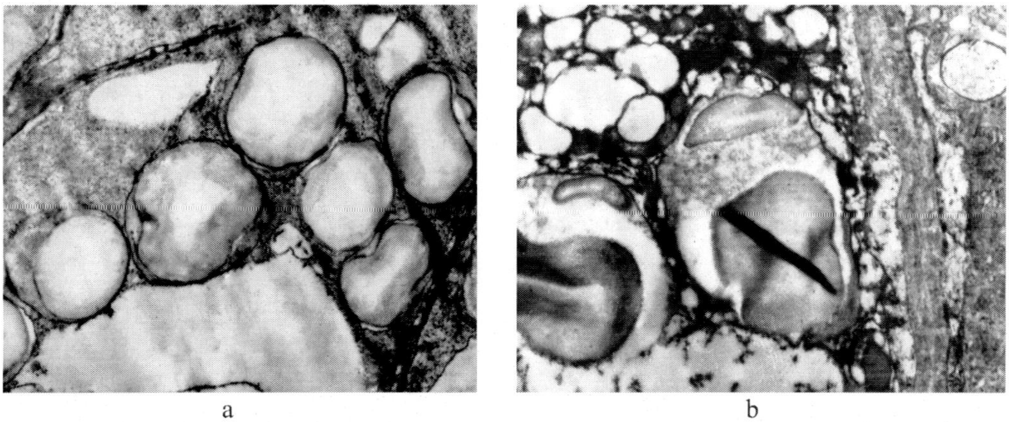

Figure 58. Endosperm cells of corrosion chamber before fertilization and in the zone contact with embryo during early embryogenesis.

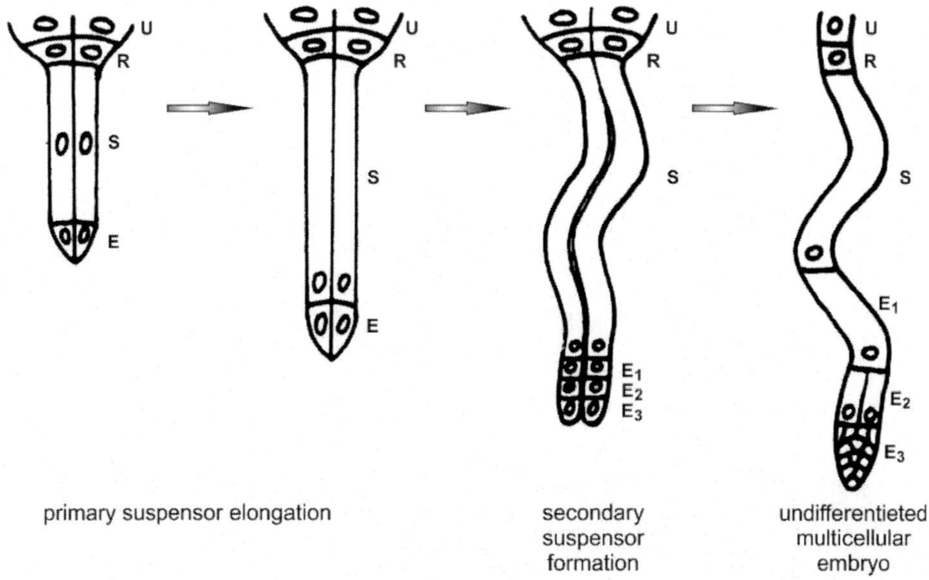

Figure 59. Scheme of embryo development during early embryogenesis.

Early embryogenesis begins from the phase of exponential growth of the embryo. It is usually divided into two stages: a) Elongation of primary and secondary suspensors; and 2) formation of undifferentiated multicellular embryo. The scheme of embryo divisions during early embryogenesis is shown in Figure 59.

The embryo development begins from 2-3 activating mitosis. It is caused by the physical situation that has turned out inside the gametophyte in this time. Firstly, a female gametophyte is characterized by increased pressure. Secondly, the elongation of suspensor cells occurs vertically downwards. Therefore, the spatial disposition of embryonic cells and spindles of their division coincide with the direction of the ovicell, cells of primary and secondary proembryos and with the direction of gravity. These are the main requirements, which are needed for the starting of activating mitosis. During activating mitosis, the distal and proximal poles of an embryo are formed, respectively, with a higher and lower level of biological energy of their cells.

Cytoembryology of first divisions of the embryonic cells and formation of the suspensors of later generations (embryonic syspensors) was studied by Buchholtz (1918). He found that the 2/3 embryos form the first inclined wall after the three divisions of the initial, and the ¼ embryos after the two. It should be noted that the sum of the length of 3 secondary suspensor cells (before their extension) almost equals to the length of the first embryonic cell (Thomas, 1973). Secondary suspensor becomes longer and pushes the embryo into the gametophyte. The length of embryonic tubes is not the same in different embryos. During this time, the separation of cleavage embryos happens. Their disconnection, as a result of the horizontal heterogeneity of 4 embryonic cells (byDNA-energy, mitotic activity, etc.), leads to the destruction of a middle plate before elongation of secondary suspensors.

During early embryogenesis, a suspensor cell grows quickly. They are the highly vacuolated cells and are characterized by the thin envelopes. This is why they begin to bend, forming loops inside the corrosion chamber (Figure 34a); the spanial position of embryo is also constantly shifted. As a result, the polar axis and division plane of embryonic cells cease to coincide with the gravity direction. This leads to changing the division type of the cells of the embryo from an activating mitosis to mitosis, as a rule, after the 2nd division of the sporophyte initial (Figure 60).

The formation of an undifferentiated multicellular embryo begins with a mitotic division of apical cells of the distal pole by the inclined wall. Then, their derivatives are divided in different directions. This is the beginning of an exponential growth embryo, when the number of meristematic cells is rapidly increasing without the signs of internal differentiation. The distal pole of an embryo becomes a semi-circular shape soon after. At the opposite proximal pole of an embryo, near the suspensor, another situation takes place. By this time, the cells of the primary suspensor are destroyed, and the embryo loses the connection with the ovicell. Even so, the spatial position of the last cells of the suspensor coincides with the direction of gravity. This suspensory cell continues to divide by means of activating mitosis, giving rise to the cells with different functional activity. Basal derivatives of these divisions are added to the cells of the embryonic suspensor. Apical derivatives are located on the border of cells in the basal part of the distal pole. It should be noted that the crisp transition in the structural organization of the meristematic cells of the proximal and distal pole of an embryo at this time is not observed.

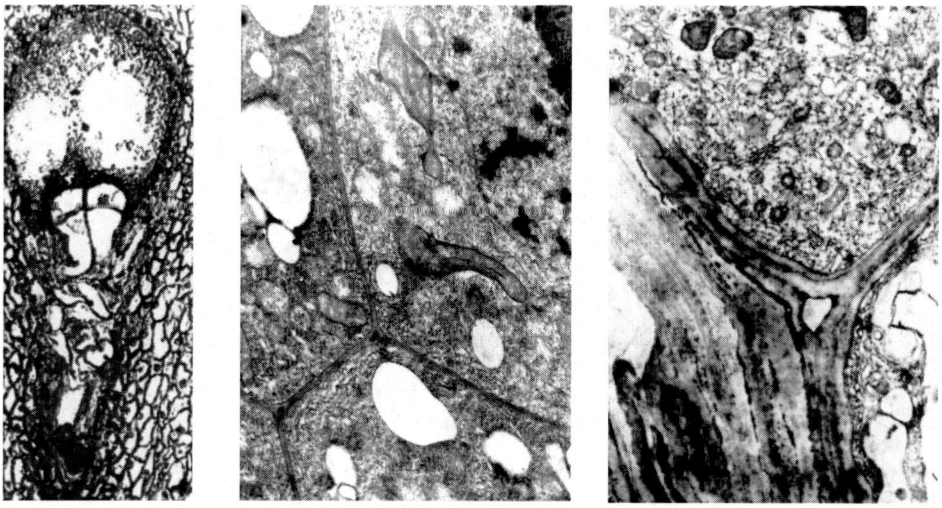

a b c

Figure 60. Early embryogenesis: overall appearance of embryo (a); ultrastructure of the cells of distal (a) and proximal (c) zone.

We have drawn attention to the fact that the embryogenesis of gymnosperms and angiosperms has many common traits – the same plan of the embryo structure, the similarity of early stage development, etc. Sporophyte initial (zygote in angiosperms and 16-cellular proembryo in gymnosperms) is characterized by the axial symmetry between the chalasal and micropylar poles. In angiosperms the chalasal upper pole gives rise to the embryo, and the lower pole –the suspensors, which perform the trophic and vegetative function (Esau, 1980). In gymnosperms, the tier of embryonic cells (chalasal pole) is also the pole for embryo formation, and the micropylar pole is a device for their movement into the endosperm tissue. In angiosperms and gymnosperms, first embryo division by the transverse wall occurs. The apical cell of this division gives rise to the undifferentiated embryo, and a basal cell form the first cell of embryonic suspensors. The embryo quickly grows and keeps an oval or circle form. When its growth comes to an end, the embryonic complex consists of the nondifferentiated meristematic cells with the same ultrastructure.

The meristematic cells of the distal pole of multicellular undifferentiated embryos by the ultrastructure resemble the sporogenous cells of microsporangium. They have the isodiametric polygonal shape, straight thin envelopes and are held tightly to each other (Figure 60b). In the center, a large nucleus with 1-2 larger nucleoli is located. The chromatin is represented by two forms: A heterochromatin and euchromatin, but the degree of its compaction is not great. Cytoplasm is poorly vacuolated, electron dense, and rich in terms of organelles.

Let us examine the algorithm of biological energy movement during early embryogenesis. The sporophyte initial prior to the first oblique wall passes 2-3 activating mitosis. During these divisions, the DNA-energy of embryonic cells greatly increases, and two energetically weak cells of the suspensor on the basal pole are formed. Then the embryonic cells pass from the activating to the usual mitosis. This type of mitosis ends with the formation of identical cells by the DNA-energy, DNA-particle and DNA-information, and their DNA-energy becomes less on the one level in comparison with the mother cell. The lowering of DNA-energy happens during every cell division. This leads, firstly, to the formation of the local complex of meristematic cells on the distal or chalasal pole of the embryo and, secondly, to the gradual decrease of the energy potential of embryonic cells. Therefore, the movement of biology energy during formation of the distal zone of an undifferentiated multicellular embryo has a descending characteristic.

The proximal or micropillar pole of the embryo originates from the cells of the embryonic suspensor of the secondary order (Figure 60c). According to Svintsova (2002), these suspensors have a higher metabolic activity than the primary suspensor. It is shown that the nucleoli number in the suspensor cells of first generation is 8.2 ± 0.2 at the nucleus. During the elongation of the 16-cellular proembryo, their number is decreased to 4.8 ± 0.3. Ring-shaped nucleoli with a central vacuole in the nuclei appear, indicating a decrease of cell metabolic activity. The suspensors of the second and third order differ slightly in the nucleoli number, respectively, by 5.6 ± 0.1 and 5.2 ± 0.2 nucleoli at the nucleus. Before elongation of the secondary suspensor, the number of nucleoli is 2.2 ± 0.2 at the nucleus. The amount of cells with the heteromorphic nucleoli increases by25%. The large (up to 18 micrometers in diameter) compactly-nucleonemic nucleoli appear, indicating an increase in their metabolic activity. Nucleolar activity of the suspensions of the third generation is even greater. The total volume of nucleoli during their differentiation is increased by more than 4 times. The detected differences of nucleolar characteristics demonstrate the higher functional activity of suspensors of the 2-3 generations in comparison with a primary suspensor of a 16-cellular proembryo.

Cells of embryonic suspensor divide by means of the activating and usual mitosis, respectively, the transverse and longitudinal division. The movement of biological energy in the proximal zone the embryo has in general an upward direction, because the activating mitosis is the main type of cell division. This results in the cells of apical pole passing to the higher energy levels, and eventually become the zone of root apical meristem. The cells of basal pole, which have lost the biological energy, are transformed into the root cap cells. Mitotic divisions in the proximal zone occur rarely. Mitosis increases the number of energetically strong cells of proximal pole, and their DNA-energy is lower on the one energy level.

The growth of embryo is limited to the volume of embryonic chamber, endosperm and seed size (Buchholtz, 1943). A female gametophyte reaches the size of mature a seed in early embryogenesis. Initially embryonic complex has the oval shape, because the embryo grows quickly in length and width. When it reaches the borders of the external zone of endosperm, the growth in the width slows down. Cells of this zone densely adjoin to each other and are completely filled with nutrients. Afterwards the embryo continues to grow, but mainly in length. Positive values of the parameters of its allometric growth (allometry index is 1.581) support our conclusion about the predominance of growth vector along the axis of chalasa – micropile (Willemse & Vent, 1990; Svintsova, 2002).

The change of growth direction is associated with the beginning of embryo differentiation. The direction of spindle division and the shape of meristematic cells change during diametrical compression of its body and they begin to coincide with gravity direction. This leads to a change of type division: From usual to asymmetric mitosis. Asymmetric mitosis ends by the formation of two daughter cells: One daughter cellis the identical with a mother cell by a DNA-energy, DNA-particle and DNA-information. The second daughter cell is identical with it by the DNA-particle, DNA-information, but DNA-energy is lower on the two energy levels. The first cell, where the DNA-energy of which does not change, becomes the cell of apical meristem of shoot. The second cell, which has lost the energy, follows the path of differentiation and gives rise to the tissue of hypocotyls. There are consequences of changing of the type division and direction of embryo growth: The first is the emergence on its distal pole the zone of high metabolic activity – the shoot apical meristem. The second is the separation of local complex on the meristematic and differentiated cells. Due to the asymmetric mitosis, the initial energy potential of meristematic tissue maintains at the same high level and the totypotency of meristematic cells is preserved.

Despite the fact that all embryonic cells remain meristematic, the motion of biological energy is characterized by the exact order and coherence. Tree type divisions (activating mitosis, mitosis and asymmetric mitosis) have a significant role in the spatial location of tissues on the distal (chalasal) pole of the embryo. They lead to the determined separation of an embryo body on a few meristematic regions – without differentiation and with various degree of differentiation.

Figure 61 shows the scheme of distribution of the energy gradients into the embryonic complex in the end of early embryogenesis. The arrow indicates a relative gravity to the spatial position of the embryo. The energetic matrix is built on the principle of bipolarity and in accordance with the law of energy conservation. The most energetically weak group of cells originates in the course of activating mitosis and gives rise to the root cap cells (the 4th lightest area on the scheme, Figure 61). Next via the energy gradient of the cell group is the differentiated meristem (the 3rd darker area). This meristem arises during asymmetric mitosis and occupies a large part of embryo body. It should be noted that this group of meristematic cells are characterized by a rather low level of differentiation. This is followed by the gradient of the two zones of energetically strong cells – an apical meristem of a shoot and an apical meristem of the root. These two meristematic zones (the 1st and 2nd dark areas) are located on the lower and upper end of the embryonic complex, on the opposite sides of the future hypocotyle. The meristematic tissues remain in an undifferentiated state and preserve the initial potency of the genome (totipotency quality) during all plant life.

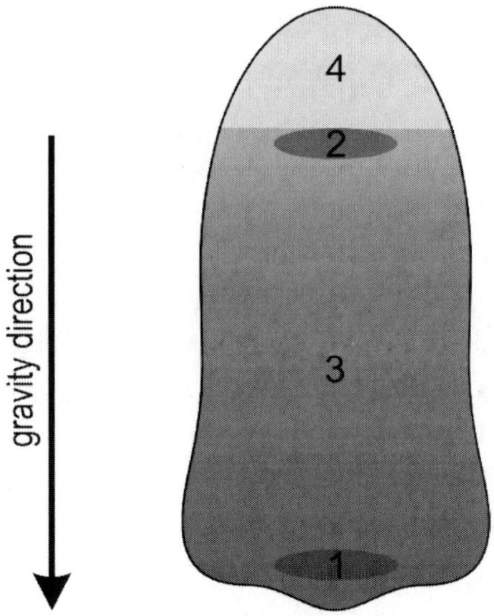

Figure 61. Schema of energetic matrix of an undifferentiated multicellular embryo: 1 – apical meristem shoot; 2 – apical meristem root; 3 – differentiation zone; 4 – root cap.

The embryo is characterized by a high orderliness of development in space and time. The supporting of orderliness of the cellular systems, metabolic activity of intracellular processes and plant vitality requires the certain energy. Undoubtedly, all these processes to a great extent are determined by the DNA-energy level of cell and plant genomes. The transition to the morphogenetic process starts only after end of energetic matrix formation has occurred. Morphogenesis proceeds only on the basis of a created system of energy gradients inside undifferentiated multicellular embryos (Figure 61). This energetic matrix is transformed into the qualitative (structural and functional) differences of cells and tissues. This is the plan of development and future inner organization of embryos and plant bodies. All subsequent species-specific processes of specialization and histogenesis on this matrix will be imposed.

In this regard, our data are not in contradiction with Child's theory of physiological gradients (Child, 1920). Its initial position is the statement that before the qualitative differences appear, the quantitative differences of physiological activity arise (energetic differences in our model). The areas with the highest activity predominate over other parts of embryo. He showed that morphological differentiation occurs on the basis of physiological gradients. Child explained the nature of the presence of dominant areas in the embryo by the external factors (differences in a media, illumination, polarity, etc.). In our model the inductor of energetic differences is Earth's gravitational field. It was found that the coincidence of polar axis and the plane division of maternal cells with gravity direction was the cause of the unequal distribution of DNA-energy between the daughter cells as well as varied metabolic activity, division ability and destiny (Kuznetsova, 2013).

Once again, we will remind you that all cells of undifferentiated multicellular embryo have the same DNA-particle, DNA-information, and are distinguished only by the DNA-energy. The differentiation degree of embryonic cells, excluding the root cap zone, is not so large; therefore, the cells still have the ultrastructure of meristematic type. This is the initial stage of the emergence of heterogeneous systems in the 4 local zones of the embryo (Figure 61). Embryonic cells in these zones are characterized by the different metabolic activity, division rates, and morphogenesis potency. Early embryogenesis differentiation is a reversible process. If the cells from different zones of undifferentiated embryo (except root cap) are put in the regulated conditions, *in vitro*, then a plant from the callus tissue can be obtained. This has been confirmed experimentally.

To summarize the above, it should be noted that the undifferentiated multicellular embryo is the straight in form, occupies a central position within the endosperm, and is characterized by a radial symmetry. Its development occurs by the certain plan. First the energetic matrix is created. The matrix is formed on the basis of the principle of bipolarity, in accordance with the direction of gravity and law of energy conservation. Apical meristem of root and shoot are it's a two centers of energetically strong cells. Chalazal (distal) meristematic center of an embryo originates from the apical cells of embryonic tier and its development occurs by the downward energy vector. Micropilar (proximal) meristematic center comes from the cells of embryonic suspensor of second order. The energy motion in this meristematic zone goes from lowest to higher level of biological energy. Taking into account that the seed formation is characterized by the high morphogenetic correlations, both cell lines come to one and the same energy level in the early embryogenesis end. Formation of an energetic matrix of embryo is the strongest genetically determined process. Sporophyte initial for its creation, should pass a certain number of cell divisions of each type (mitosis, asymmetric mitosis and activating mitosis), which follow each other in the species-specific sequence. This process is associated with the intensive embryo growth through increasing the number of its suspensor-embryonic cells.

Development of embryo occurs in close contact with the nourishing tissue of endosperm. Because the two heterogeneous systems by the DNA-particle in this process are involved, diploid tissue of embryo and haploid tissue of endosperm, begins to act the processes of incompatibility. A place of its function and morphological manifestation is the outer surface of embryonic complex, where the cells are characterized by the semi-permeable envelope. During embryogenesis, the thickness of insulating layer around the embryo gradually increases and gives rise to the chamber of embryo. The majority of outbred embryos have the optimal thickness of this layer, which does not prevent their normal growth and development.

LATE EMBRYOGENESIS

An embryo is characterized not only by the orderliness, but also species-specific organization: a certain shape, size, and internal structure (Buchholtz, 1918, 1846; Chowdhary, 1962; Dogra, 1978; Singh, 1978; Owens et al., 1982; Williams, 2009; etc.). The spatial arrangement of cells is laid in the embryonic complex, is preserved in the plant body, and is subordinated to a certain structural and functional organization. As known, any organization connects with the distribution and redistribution of material-energetic processes in a certain direction. The morphogenetic processes are superimposed on the energetic matrix of embryo and have another nature.

We think that during late embryogenesis the formation of the informational matrix of future plant or the morphogenetic field according to Weiss (1939) takes place. These processes initiate the transition to the histogenesis and an undifferentiated embryo gets the specific phenotype, which connects with the functioning of DNA-information. At this time a passage from the state of differentiation to the specialization of cells and tissues happens. Specialization is the unidirectional form-building process. It continues during all life of plant. All specialized cells lose the quality of totypotency. Transition toward the specialization is possible only in certain physical conditions, which are created inside the seed during late embryogenesis. DNA-information gives rise a few groups of cells with the certain phenotypic organization and functional purpose. This leads to the emergence of internal heterogeneity, development of tissue systems inside the embryo, which has a species specific form and size.

DNA-information, if we proceed from the corpuscular-wave nature of DNA, is a modulated wave which carries a species-specific information signal. The DNA-information of a sporophyte initial was laid during meiosis and fertilization, as a first informational signal of a future plant. During cell divisions, this information is transmitted from initial to embryonic cells. Therefore most embryonic cells are the equipotential during early embryogenesis. With regards to the energetic plan, the part of cells from two polar meristematic zones of the shoot and root are in early stages of differentiation (the zones of differentiated meristems). Now it seems obvious that before stage of late embryogenesis, embryonic cells are the totipotent by DNA-information and the high DNA-energy provides a full development of a plant. Exceptions are the root cap cells. They, after losing of biological energy during activating mitosis, are the differentiated cells with large amounts of starch granules and lipid droplets (Trenin, 1988). An analogous situation occurs in animals. It is experimentally proven that the embryonic cells on the blastula and early gastrula stages are not yet differentiated. Each is a totipotent and can provide a normal development of embryos (Tokin, 1987).

The beginning of embryonic organ formation occurs after an embryo reaches a certain size or "critical mass." In early July, the cell divisions inside embryonic complex become less random. Divisions are more often observed in the cells of proximal zone, which form a group of orderly arranged cells (Trenin, 1988). They become a root cap cells in the future. Three meristematic regions and hypocotyl-shoot axis appear (Owens, 1982). Less vacuolated meristematic tissue of hypocotyls and cotyledons develop in the center of embryo. Surface cells form a single layer of epidermis, then the cotyledons are differentiated. All this time the embryos increase in the size. During maturation, cell divisions progressively cease, but the process of their elongation for some time continues. Simultaneously, the accumulation of starch takes place. The embryo goes over into a resting state. By this time, its cells are completely filled with the protein bodies and lipid droplets and the amyloplasts absent.

A mature embryo is characterized by radial symmetry. It consists from a few tissues of different degree of differentiation and specialization (Figure 62a). The cells of the apical meristem of root and shoot have the same ultrastructure: Low-vacuolization cytoplasm, large nuclei, and isodiametric shape. Below cotyledons in the third zone of meristematic cells, a hypocotyl (hypocotyl-shoot axis of embryo) is placed. Less vacuolated cells are arranged in the center of embryonic complex and are the meristem of procambium. The meristematic tissue gives rise to the conducting system of a future plant.

A mature embryo of Scots pine is straight, elongated, occupies a central position inside the female gametophyte, and its cells are the heterogeneous by the morphology and functional purpose (Figure 62a,b). It has from 4 to 18 cotyledons and the two polar meristems of shoot and root. Between the size of the female gametophyte and the embryo, there is a positive correlation which increases during late embryogenesis. The correlation coefficient in the period of hypocotele differentiation is $r = 0.14$; during differentiation of cotyledon primordial is $r = 0.48$ ($P < 0.05$); in the beginning of cotyledon is $r = 0.51$ ($P < 0.05$); and in the end of cotyledon growth is $r = 0.77$ ($P < 0.01$) (Svintsova, 2002).

As mentioned above, the embryo consists of the embryonic cells of distal (micropillar) zone, developing by the principle of  stages, and the proximal (chalazal) zone, formed by the principle  stages. The processes of histogenesis also follow a definite plan. Morphogenesis is genetically determined process, which is carried out only within the local complex () of cells of distal zone. At the end of early embryogenesis, all its cells are identical by the DNA-particle, DNA-energy, and DNA-information. Further development of embryo is primarily related with a change of positions of separate cellular groups within this zone. A large part of embryonic cells during these changes loses the property of totipotency by two components of genome: DNA-energy and DNA-information. The formation of the information matrix of a future plant also occurs within this period. Its creation leads to a regular distribution of cells and tissues within the plant body (in accordance with the species information matrix). Furthermore, this causes irreversible changes to initial DNA-information which were laid in the genome of sporophyte initials during its reproductive development.

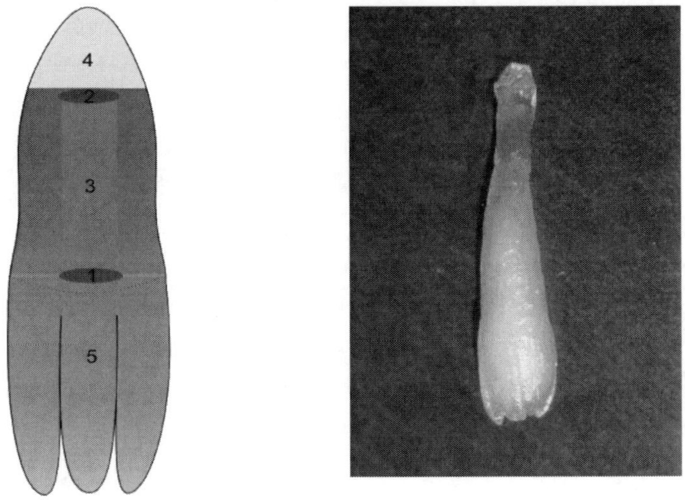

Figure 62. Schema of informational matrix of *Pinus sylvestris* embryo: 1 – apical meristem shoot; 2 – apical meristem root; 3 – hypocotyl; 4 – root cap; 5 – cotyledons.

The distal zone of embryo presents a cellular local complex which grow uniformly in length and width. When the embryo reaches the maximal width of corrosion chamber, the direction of growth begins to change. Sometimes the embryo continues to grow, due to the increasing diameter of the corrosion chamber. Due to compression, the inner and outer cellular layers of embryonic complex are in unequal position. After each division, the outer layers occupy the living space of nearby endosperm cells. The following layers are shifted in the direction of the outer and central layers of cells. This includes the cortex tissue, whose cells continue to divide the periclinal manner. The central layers have the bilateral pressure from the endosperm cells and an external cover of the embryo (cortex). Therefore, they are forced to change the growth direction, shape, and division type. Central cells are moving from the ordinary to the asymmetric mitosis and become more elongated shape. Increased intracellular pressure leads to a decrease in size of embryonic cells and the fixation of their direction growth.

Once more, important consequences follow from this physical situation. We think that during longitudinal compression and intracellular pressure growth, the conditions for the local distortion (irreversible modulation) of frequency and amplitude oscillation of the first DNA-information are created, resulting in the new information signal in the genome of this group of cells. These cells give rise to the hypocotyls tissue, which is disposed in the center of the embryo. It forms the hypocotyls-shoot axis and extends from the shoot meristem to the root meristem. The formation of the hypocotyls-shoot axis of embryo is the first step towards the creation of the information matrix of future plants and the meristem of procambium or the meristem of the primary conducting system.

It should be noted that the hypocotyl tissue is formed in the result of longitudinal compression of the meristematic cells of the local complex (), cell longitudinal divisions, and their growth in length. Therefore, all pro- and cambial cells have the same narrow elongated shape. The local complex originates from the energetically strong embryonic cell, appearing after the 2-3 activating mitosis of sporothyte initial. All the cells of complex are the totipotent by their genetic characteristics. The cells of the hypocotyle are identical with them by the DNA-energy and DNA-particle, but differ by the DNA-information. Therefore, they differ from the meristem tissue of root and shoot apex by shape and ultrastructure, but the energy potential (DNA-energy) of these three meristems is the same level. Therefore, the procambium tissue presents the 3rd zone of energetically strong totipotent meristematic cells. In the plant, a cambium tissue is the lateral meristem by the location, functions throughout the tree life, and forms the secondary conductiving tissues. In ontogenesis, these meristematic zones (apical shoot meristem, apical root meristem, and lateral cambium meristem) present the unified coordinated system, which have key importance in the life of plants.

All form-building processes touch only cells which belong to the differentiation zone of energetic matrix (Figure 61). With regard to the apical meristem, cambial tissue has a unique feature due to the asymmetric mitosis: The ability to unlimited proliferation and morphogenesis. Modification of information signal causes the directional change of information (morphogenetic) fields of the embryo as the whole. In turn, the morphogenetic field is the basis for the further development of the specialized cellular systems and initiation of the deeper processes of histogenesis. After the change of initial informational signal and the emergence of the morphogenetic field of embryo, the cells of hypocotyle develop by the specialization path and irreversibly lose the totipotency. This tissue is organized by the radial rows (Esau, 1980) and initially each layer of hypocotyle cells presents the radial row of procambial initials.

The embryo continues to grow longitudinally after the transverse growth ends. It also occurs in a limited space. It is limited by the size of corrosion chamber with the distal end; with the proximal it is limited by the micropillar end of seed peel. The embryo rapidly increases in size in both directions - the micropillar and chalazal zone of endosperm. Hypocotyle, on its micropillar end, carries the apical meristem root and the apical meristem shoot is located on its halazal end. Its cells intensively divide and lengthen along entire length of tissue.

Then the embryo enters into the next stage of morphogenesis and information matrix formation. During elongation in the micropillar direction, the embryo grows along the wide part of the corrosion chamber and occupies the space, where the cells of the primary suspension, the embryo-outsiders and the ovicell, are located. The apical root meristem is protected from the outside by the root cap cells. A growth of embryo in the halazal direction occurs along the corrosion chamber, which has the shape of a cone, with the protrusions of endosperm cells inside it. Chamber narrowing prevents the penetration of the embryo into the female gametophyte. The embryo is characterized by a clavate shape with a semicircular top at its distal end (Figure 61). The largest diameter is below the top in the region of the shoulders placement. Therefore, when the embryo grows in the halazal direction, its top, with the apical meristem, freely passes deep into the chamber and the shoulders get stuck. When the embryo is retained by the “funnel,” the surface cells by the shoulder perimeter are compressed simultaneously in several places. Inside the corrosion chamber, the protrusions of endosperm cells are produced with the greatest compression. Under this pressure, the DNA-wave of cells of the local centers comes down to a lower energy level and triggers for the new cell divisions. Intense divisions occur simultaneously with cell growth. The change of the wave parameters of initial DNA-information and formation of a new modulated signal takes place simultaneously in the genome of meristematic cells, which have entered into the zone of higher compression. The change of the rhythm and intensity of mitotic proliferation leads to the inception of cotyledon rudiment and formation of vertical "leaf-like outgrowths" – cotyledons.

Consequently, the informational matrix of embryo changes again and gives rise to the next morphogenetic field. This morphogenetic field initiates a series of form-building processes: Cotyledons become a species-specific shape and size. After their growth termination, the apical meristem shoot finds itself between the cotyledons, tightly pressed together. Due to their vertical position, these leaf-like morphostructures will protect the apical meristem shoot from damages during seed germination and initial stages of ontogenesis.

Quantitative traits such as the beginning of hypocotyle differentiation and cotyledone inception are characterized by the low variability: Coefficients of variation did not exceed 16% (Svintsova, 2002). The norm is the synchronous cotyledone inception and their uniform growth. In Scots pine the average number of cotyledons is quite constant and ranges from 5.99 ± 0.03 to 6.06 ± 0.03 (Vidyakin, 2013). It is established that during self-pollination and open-pollination, the number of seeds in which the embryos reached the stage of their differentiation and the seed production of cones by their value are approximately the same. The conclusion is made that the process of successful cotyledone differentiation can be considered as a criterion for the formation of viable seeds.

Morphogenetic correlations in the processes of seed formation are manifested in the coordinated development of the zygote, embryo, endosperm, and surrounding structures of seed (Batygina, Vasilieva, 2000). A relation between the volume of the endosperm and the embryo in seeds is defined for every species. The endosperm also contains a specific quantity of cells and nutrients, which will be used during embryogenesis and seed germination. Concerning the quantity of endosperm into a seed (its volume), relates to the value of energy gradient between the tissues of the embryo and endosperm, while also considering the length of the restraining stage during embryogenesis, which in Scots pine lasts ≈ 2 months. As the analysis showed, the growth of embryo and endosperm, concerning the transverse and longitudinal (chalazic-micropyle) axes, the achievement of species sizes and proportions (shapes) occurs with low variability of allometry parameters, indicating the stability and genetic conditionality of seed proportions and ways of their achievement (Svintsova, 2002). Between the size of female gametophyte and embryo, there is a positive relationship increasing during embryogenesis.

The increasing of embryo length is not the same in different plants. The regression coefficient varies from 0.90 to 1.84 (Svintsova, 2002). The differences between the embryo sizes are aligned in a result of compensatory growth. The same regularity takes place during development of female gametophyte: The significant differences in the beginning of its logarithmic growth are substantially reduced in the end of megagametophytogenesis. The same rhythm of growth processes is observed during cotyledone formation. Three types of mother trees, depending on the algorithm of growth processes in the embryogenesis, have been revealed: 1) The phase of intensive growth begins and ends early. 2) The later beginning of growth processes, which is compensated by their intensification. 3) The intermediate type. In the maternal trees this algorithm of growth processes is preserved in different years by a weather conditions.

It was found that the contribution of genotypic component in the total variability of trait, "embryo length," is increased: A coefficient of heritability in the wide meaning (H^2) in the beginning and end of intensive growth is 0.22 and 0.55 ($P < 0.05$ and 0.01, respectively). Development of female gametophyte is largely due to the genotypic control: $H^2 = 0.88$ ($P < 0.01$). Influence of genotypic features of father and mother plants on the variability of quantitative traits of seed is not the same. According the data of dispersion analysis, the force of influence of the maternal plants on this trait is 28-62%, depending on the diallel crossing variant. The force of influence of father's components is much lower – 0.2-10%.

The variability of quantitative parameters takes place at the individual level and within the maternal plant. The beginning of embryo intensive growth is characterized by high variability of quantitative traits (Svintsova, 2002). The coefficients of variation of linear sizes of the embryo and cotyledone length vary in the range of 35 – 54%. Growth processes in the embryogenesis are finished in the 2nd decade of August with the sum of effective temperatures of 1500-1590°. During seed formation, the variability of traits is significantly reduced and after their maturation is estimated as the low. It should be noted that these general regularities are confirmed on the organism and population levels. They testify about a strict genetic control of embryogenesis and its specificity.

As already mentioned, the female gametophyte originates from a megaspore, which is characterized by a haploid set of chromosomes and individual DNA-particle. A sporoderm limits its growth from the outside. Cell divisions in limited space lead to a pressure increase. The growth pressure is the main reason for the change of division types – from free-nuclear mitosis to the usual mitosis, and then from mitosis to the endomitosis.

During endomitosis, the chief events are the formation of polytene chromosomes. They are formed due to a multiple replication of DNA-particle without its subsequent divergence. The genome of mature endosperm may be in the different state: Polytene chromosomes without increasing their number, with multiplying ploidy levels, nuclear fusion, incomplete division, etc. (Nagl, 1978; Pichot & Maätaoui, 1997). The data testifies that the different violations of haploid cells take place during endospermogenesis.

During embryo development, we have another situation. The formation of a diploid genome of sporophyte initial is going inside the ovicell in the conditions of higher intracellular pressure too. Embryogenesis, in comparison with endospermogenesis, begins and proceeds inside limited volume within the corrosion chamber of endosperm and seed. Embryo development begins from the sporophyte initial and ends by the formation of multicellular embryo. Destruction of genetic system of its embryonic cells does not occur. Study of DNA content in nuclei of different gymnosperm species showed that the DNA amount in the embryo genome is 2C (Konar & Moitra, 1980). The exception is the energetically weak cells of embryonic suspension, where the DNA content in the nuclei reaches to 8C according to Nagl (1978). Meanwhile the DNA amount in the cells of apical meristem root is also equal to 2C. It becomes obvious that the diploid principle of genome organization is more resistant to increased pressure, compared to the haploid ones. Hence it follows that the spin-spin connections between homologous chromosomes give the diploid genome of embryo and plant additional stability.

The biological energy of sporothyte initial is a maximum level and an embryo has the great potency for growth and development. However, these potencies are realized only partially on the mother plant and never go beyond the seed. Therefore the embryogenesis, as well as the megagametophytogenesis, we consider as the genetically determined restriction stage (Kuznetsova, 2013). According to the gas law, the direct consequences of continuation of cell divisions and embryo growth inside a fixed volume are a further rising of pressure into the female gametophyte, an increased density of nucleoplasm, cytoplasm, an accumulation of a large number of biomolecules with a low speed, and a change of the metabolic activity of embryonic cells. Since the speed of biomolecules is directly connected with their ability to enter into a chemical reaction, the synthesis of chemically inert substances such as storage starch, proteins, and lipids begins. The analysis showed that the sequence of transition to the accumulation of nutrients in embryonic cells is the same as in endosperm cells. Initially the starch in plastids appears and size of its granules is increased. Then the biosynthesis of proteins and lipids begins and the starch disappears. Gradually the embryonic cells are completely filled with the protein bodies and lipid droplets. In the mature seeds, the lipid content is $\approx$ 1/3 of the total weight: 31% in *Pinus roxburghi*, 36% in *Pseudotsuga menziesei* and 30% in *Welwitschia mirabilis* (Konar & Moitra, 1980). Glycerides account for 86% of the total amount of fats. By the ultrastructure a cells of mature embryo and endosperm are not differ significantly. However, the cells of the cortex and cotyledons contain more protein bodies and the lipid droplets are the main inclusions in the cells of shoot apex and procambium (Trenin, 1988).

An embryo and maternal plant are separated from each other by the principle of a double protection (Kuznetsova, 2013). A new equilibrium state between them is reached when the more energetically strong embryo has isolated from a sporophyte by means of energetically strong tissue of endosperm and the two abscission layers – the embryonic chamber and seed peel. The embryos are considered an equilibrium state at critical point, where they lack the freedom degree. After seed maturation, the change of entropy, pressure, and volume are so decreased that they begin to head towards zero. The processes of dissimilation and assimilation are minimized (Tokin, 1987). Therefore, the embryos cannot go out from the equilibrium state by themselves. Without external influence they, as the pollen grains and ovules, are doomed to expire.

Summary

According to the corpuscular-wave conception, each seed has three interrelated and independent species-specific programs – the corpuscular on the basis of DNA-particle, the energetic on the basis of DNA-energy, and the informational on the basis of DNA-information. An embryo's body represents three interrelated and independent matrices – corpuscular, energetic, and informational, in accordance with which the plant formation in a space and in a time will occur. Each genotype creates three interrelated and independent biofields – the corpuscular, energetic, and informational, on the basis of which the exchange of matter, energy, and information is implemented and regulated. All these phenomena are manifestations of one indivisible substance – a DNA (particle and wave) in its three hypostases: DNA-particles, DNA-energy and DNA-information.

The corpuscular component of the genome is formed in the three stages of the reproductive cycle: Male meiosis, female meiosis, and fertilization. It identifies the same protein composition of the cell lines of one genotype, the incompatibility (interspecific, interspecies) of pollen, and tissues during vaccinations causes. The diploid genome of each initial sprophyte has an individual species-specific DNA-particle. Its primary structure presents a code form for recording corpuscular information on the basis of the certain nucleotide sequence, which serves as a matrix for the biosynthesis of proteins. Action of DNA-particle is restricted within the life of a single genotype in the whole expression: Pollen, ovicell, sporophyte initial, embryo, or plant. Accordingly, a corpuscular field is the body of these genotypes. A corpuscular matrix or individual diploid DNA-particle of a future plant is created during fertilization. Then, in the course of ontogenesis, it reproduces many times with the help of replication, transcription, and translation mechanisms.

The DNA-energy (or energetic information of genome) is formed during gametogenesis and proembryogenesis on the 6 activation stages of the energy line of reproductive cycle by means of transition of DNA-wave on the higher energy levels. The energetic matrix of a future plant is created during early embryogenesis. Therefore, each mature embryo is characterized both by a certain DNA-energy and by an energetic matrix or energetic plan of plant organization. The energetic field covers the embryo body and goes beyond it. Intraspecific differentiation of trees by the DNA-energy determines their competitiveness, division on the plus, minus, and average by the growth, and vital state genotypes. The energetic matrix of mature embryo consists from 5 zones: Three zones of energetically strong cells (shoot apical meristem, root apical meristem and procambium tissue), the differentiated meristematic cells (future specialized cells and tissues), and the more energetically weak cells of root cap. Apical meristem of shoot and root arise during early embryogenesis. A meristem of procambium is formed during late embryogenesis. This plan differentiation of a plant on the zones depends on DNA-energy levels: Three energetically strong, differentiated, and energetically weak (then the dead) cells will remain throughout all its life.

Only after creation of energetic plans form the information matrix and begin the specialization processes. Phenomena of differentiation and specialization have a wave, but of different nature. The first operates on the basis of DNA-energy, the second on the basis of DNA-information. The formation of the informational component of genome as well as the DNA-particle occurs during meiosis (male and female) and fertilization. The initial DNA-information in the form of a primary species-specific cosmic signal is laid. Therefore, until late embryogenesis, all embryonic cells are totipotent and the embryo is characterized as an undifferentiated informational matrix. Then, the genetically determine process of formation of a modulated informational signal occurs. It touches the strictly defined groups of differentiated cells of embryo and leads to the creation of non-uniform informational field. This informational matrix initiates the certain form-building processes, during which the transition of embryonic cells on the path of specialization takes place. The cellular specialization and histogenesis begin during late embryogenesis and are fully realized during ontogenesis. Cells undergo deep changes depending on the function, which they discharge inside the plant body. Embryonic cells lose the property of totipotency. The processes of morphogenesis lead to the formation of different specialized tissues: Epidermis, parenchyma, sclerenchyma, xylem, phloem etc. These perform various functions to aid in the survival of species within the range territory.

Embryonic Selection

The syngamy of gametes gives rise to many new combinations of genes, energetic states, and informational modifications; whereupon, genotypic variability of embryos increases significantly (Table 4). Embryonic selection regulates the genotypic composition from the initial diversity of embryos. The surviving genotypes are moving in the ontogenesis and participate in the formation of populations. The remainder die off (Stockwell, 1939; Barnes et al., 1962; Kuznetsova, 2003; Filonova et al., 2002; etc.). The selection can only exist if there is an alternative. In embryogenesis, a surplus embryo number is created and their competitive interaction occurs: Directly with each other and indirectly through contact with their environment. In this case, a huge number of random combinations of competitive interaction arise and, in each combination, the success of the most qualitative component is ensured.

Male and female gametes are haploid in higher plants. The end result of sexual reproduction is the creation of a diploid genome of zygote or sporothyte initial. Diploidity increases the homeostasis and stability of the genotype on the cellular, organism, population, and species level. An important factor of increased viability is the emergence of a diploid heterozygous genome during cross-pollination, which is more resistant to varying environmental conditions (Goldstein, 1992).

We are guided by the concept that fertilization is a conservative mechanism and a way of preserving the species identity. The main purpose is the maintenance of the corpuscular-wave norm of species in the alternation of generations. Conservative function of sexual reproduction was discussed in the works of Bell (1988), Eshell (1991), Shcherbakov (2005), etc. Therefore, if the purpose of meiosis is the enlargement of genotypic variability, then the task of fertilization causes it to narrow. Embryonic selection is carried out in two stages: 1) A competition and 2) a stabilizing selection. Both types occur by the three lines – a DNA-particle, DNA-energy, and DNA-information.

We think that the competition of embryos reflects the compatibility degree of the genomes of male and female gametes during formation of spin-spin connections between their homologous chromosomes. In the course of its action, the genotypes with mutational load, energetically weak genotypes-outsiders, unadapted with the environment, incompatible outbred, and inbreed genotypes are removed. Once more importance process occurs during the formation of spin-spin connections. A genetic system in the equilibrium state is returned. In result of this competition, the population gene pool pays a certain price in the form of a same empty seed and non-viable progeny. The average level of pine seed reduction is 13-21% (open pollination, optimal years, Stupinsky test-object). The quantity and quality of seeds are maintained at a relatively high level due to the presence of the polyarchegonial system. The selective advantages have the energetically strong genetically normal outbred genotypes, which are the compatible with the regional climate and the endosperm of seed.

On the embryonic stage a spectrum of genotypes of the embryo serves as a new population gene pool and the mother population as its adaptive habitat. According to Buchholtz (1920), from 1 to 16 embryos (polyarchegonial system, cleavage polyembryony) can go out into corrosion camera, but usually their number is not more than 1-3 (Figure 63). It was found that the average number of zygotic (genotypically different) embryos in the analyzed population is 2.1±0.06 per a female gametophyte (Svintsova, 2002). This parameter is close to the value, obtained on other populations, and is considered the optimal for Scots pine (Sarvas, 1962; Romanovsky, 1997).

Embryos compete with each other directly and indirectly through their interaction with the endosperm. Competition and selection among the zygotic embryos ends earlier than among the cleavage embryos (Sarvas, 1962; Svintsova, 2002). During embryonic selection, only one of them survives and the rest die within one month after fertilization (Dogra, 1967). In the early embryogenesis middle, the proportion of seeds, which is retains several zygotic embryos, does not exceed 2%. Cleavage embryos die at the end of early embryogenesis and also before the beginning of its organ differentiation. At the end of late embryogenesis, the frequency of the occurrence of seeds with additional embryos is about 1%.

The leading position occupies an energetically strong embryo, the genotype of which is compatible with the endosperm tissue. The leadership of embryos is determined early as the result of different extensions of the suspensory cells, which bear the embryonic cells into the corrosion camera at different lengths (Figure 63b). The order of embryo placement during embryogenesis never changes, which testifies that the nature of leadership is embedded in the genotype of the dominant embryo. The following paragraph discusses the cause of this phenomenon.

After activating mitosis, the cells of suspensor lose the ability to divide, but have some amount of biological energy. Therefore, they develop by the principle of differentiated cell (1 DNA particle – 1 DNA-wave, without cell division). Their DNA-energy, in accordance with the 2nd law of thermodynamics, is descended from level to level, up to zero. The nucleus (size, form, and chromosome number) does not change because their biological harmonic oscillators are not destroyed. On the contrary, the volume of the cytoplasm which is determined by surplus kinetic energy, significantly increases. The free energy, which enters into suspensor cells after each DNA-wave radiation, causes more rapid growth. Suspensor cells begin to vacuolizate and increase in size. During growth, the suspensors fulfill an important function – the embryonic cells travel from the ovicell into corrosion chamber of endosperm.

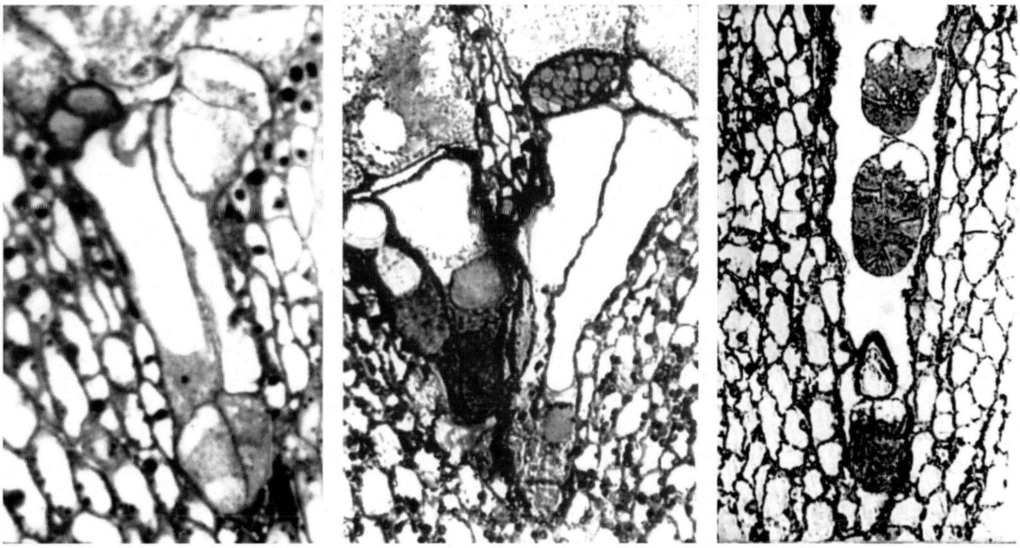

Figure 63. Development into a corrosion chamber one (a), two (b) and three (c) embryos.

As a rule, all primary suspensor cells a part of 16-cellular embryo grow at the same speed. This indicates that they received the equal amount of biological energy during proembryogenesis. In the competition between zygotic embryos, the leading position is provided by the suspensor of energetically strong embryos due to their higher growth rate. Approximately half of the genotype-outsiders are removed during embryonic selection since the population number of zygotic embryos is equal to the average two, according to Svintsova (2002).

When only one embryo comes out into a corrosion chamber, its genome can be of any genetic constitution. If two or more embryos develop in the chamber, many random variants of competitive interaction are created between them. For example, the competing embryos can be all strong, weak, or different by the DNA-energy level; all inbred, outbred, or different by originization; they also can have different degree violations. The most energetically strong embryo in every variant of interaction, or the embryo that enters into the corrosion chamber earlier, has the competitive advantage. Leadership of an embryo, established at the moment of its transition from the ovicell into the endosperm, during embryogenesis does not change. At the end of the early embryogenesis, all lagging embryos die. Therefore, if the energy potential of the gene pool of surviving embryos is decreased during fertilization; during competition, its level rises again.

The embryonic competition is important for the species survival because the deliverance of the population gene pool comes from a genetic load. Note that this selection type removes a significant portion, but not all low-quality defective genotypes. Therefore, the gene pool of the survivor embryos is presented by the genotypes of all categories by the line of DNA-energy, DNA-particle, and DNA-information. The process of genotypic selection continues during embryogenesis.

It has been established that embryos at the stages of proembryogenesis and late embryogenesis are destroyed very rarely - no more than 1-5% (Dogra, 1967). The "critical" point of embryogenesis is at early embryogenesis. In optimal years, the death of 20-25% of ovules in the cones happens in this stage (Sarvas, 1962; Dogra, 1967; Kuznetsova, 2003). Since this type of selection eliminates the embryos occupying dominant position into the endosperm, plump seeds emerge.

Early embryogenesis is a critical point in the development of embryos and occurs at the time of stabilizing selection. As a rule, it takes place from mid-June to mid-July, when stably warm weather with an average daily temperature above +10°C is established. It is shown that the intensification of embryonic development occurs at a more dynamic accumulation of heat and sufficient amount of rainfall. In the beginning of early embryogenesis, embryo sizes are characterized by a high level of variability. Embryo growth is limited by two factors: The intensity of the processes of incompatibility with an endosperm and the volume of the corrosion chamber, which has a species-specific volume and size.

In optimal years (80% of vegetation seasons in the research region), the process of embryogenesis is carried out in relatively normalized microclimate conditions (Table 4). In May, the cones grow rapidly and the covering scales are thickened. At the same time, the shoots and needles of the current year are growing. Embryogenesis proceeds inside rather big cones, which protect it from direct influence of the external environment. Normalized microclimatic conditions optimize the rates of embryonic development. They help to lower the levels of embryo variability and the formation of their morphogenetic correlations.

The internal environment of embryo is the female gametophyte, which isolate it from the maternal plant. Gametophyte is characterized by the species-specific form, size, and specific physical conditions (light absence, raise pressure, etc.). Endosperm tissue provides the embryo by nutrients during embryogenesis and seed germination. The endosperm of different female gametophytes of the same mother tree is variable formation for the following reason (Table 4). The gametophyte originates from a functional megaspore and a megaspore arises during female meiosis. Therefore, the endosperms of different ovules are not identical by their genetic nature. They, as well as the pollen, carry the information about the weather conditions during meiosis. It follows that the development of embryos within a mother tree occurs on the varying substrate (Table 4).

Stabilizing selection occurs simultaneously with processes of early embryogenesis and energetic matrix formation. Its development begins with the first division of sporophyte initial and ends the creation of multicellular embryonic complex with two energetically strong zones of apical meristems (poles of shoot and root), differentiating meristematic zone, and the zone of more energetically weak cells of root cap cells. The formation of bipolar matrix is a genetically determined process, which occurs in accordance with the law of energy conservation on the basis of DNA-energy. The bipolarity is the basic principle of inner organization of the embryo and plant, which is the main factor that regulates the processes of their differentiation and development during embryogenesis and ontogenesis. Early embryogenesis, as a meiosis, is a very vulnerable and ecologically dependent stage of the reproductive cycle.

Embryos are distinguished by the genetic constitution, the date of transition into the corrosion chamber, the time of "inclusion" of the new hereditary program that mostly defines the different intensity of their development. In early embryogenesis the morphometric parameters of embryos are characterized by a high level of variability. The result of the stabilizing selection decreases its level significantly and in late embryogenesis, it corresponds to the species norm (Svintsova, 2002). Variability of undifferentiated embryos is limited by several factors such as the genetic system of incompatibility between embryo and endosperm, and the volume of corrosion chamber and gametophyte, which have a fixed size, shape, and internal structure. The most important and vulnerable point is the embryo transition from transverse to longitudinal growth and process of energetic matrix formation.

The most norm-forming genotypes of embryos are able to form a bipolar energetic matrix within the gametophytes, which are the typical by the size and form. We think that the effectiveness of stabilizing a selection is provided by means of the female gametophytes, the morphometric parameters are under genotypic control of the mother trees. During early embryogenesis, the gametophyte has reached its maximum size; the cells of endosperm are in a dormant state and do not depend from the environmental conditions. It was found that during embryogenesis, the increase in correlation occurs between the linear dimensions of gametophytes and embryos. The proportions between the size of the endosperm and the embryo achieves are also at a stable level and correspond to the species norm.

Rapid growth of ovules becomes slower before fertilization (Sarvas, 1962). The number of dead embryos can be judged by the number of empty capsules whose sizes are equal to or greater than the size of smallest seed of its cone. Smaller capsules testify that the ovules are degraded before fertilization. Analysis showed that in seed production of each tree always meet the big, average, and small empty capsules of the typical and abnormal form. However, the proportion of big and anomalous capsules among the empty seeds is higher than among the plump ones. The probability of embryo death is higher in the seeds where the morphogenetic correlations between its length and width are changed. This takes place in seeds of all categories. The big ovules, which were the leaders of the previous stage, come out more often from the reproductive process. Perhaps this is due to the fact that the vector of selection changes its direction.

The big female gametophytes have a larger number of archegonia, volume of endosperm, and corrosion chamber. Apparently, the embryos of only certain genetic construction (energetically strong, without genetic violations, outbred) inside a wide chamber can reach the optimal size and pass through the process of formation of the energetic matrix into the undifferentiated embryo. Their number after fertilization has significantly decreased. All other less energetically strong genotypes will die because the relationship between the volume of the corrosion chamber and growth rate of an embryo is disturbed. Therefore, the proportion of large empty capsules among dead seeds is the highest. For this reason, an energetically weak embryo will be unable to proceed to the internal differentiation into gametophytes of average size. In the small gametophytes, only embryos with a lower level of biological energy are able to develop. Most of them have been removed during embryo competition and some would die during early embryogenesis. The stabilizing selection is removed from the reproductive process the more energetically weak and energetically strong genotypes, and the number of norm-forming genotypes in seed production is increased. The stability of a population gene pool of seeds becomes higher. These data are completely consistent with the theory of stabilizing selection of Shmalgauzen (1968), according to which the species norm of variability is constantly maintained and preserved in the alteration of generations.

As already mentioned, the endosperms of one tree represents a variable formation and carries the information about the climatic conditions of the year. During embryogenesis, a huge number of combinations of genotype-environmental interactions (from self-incompatibility to cross-incompatibility) are created. After open-pollination, most inbred and some outbred embryos are degraded as a result of self- and cross-incompatibility. They die because of the endosperm since they are unable to adapt to the new source of nutrition. Stabilizing selections of embryos on the varying substrate intensifies the differentiation of genotypes by the degree of compatibility with the source of nutrition, promotes to the selection of the most adaptive their components, and ensures success of the formation of genotype-environment relationships during embryogenesis. A more powerful insulating layer around non-adapted embryos (due to their tissues incompatibility) is formed afterwards. This layer prevents the transference of nutrients. Therefore, the probability of the energetic matrix's formation is significantly reduced. If the internal differentiation is absent, the processes of embryo destruction and death begins.

Stabilizing selection is an environmentally dependent process. Normalized microclimatic conditions inside cones play an important role in their realization.

However, these conditions provide only a partial independence of reproductive processes from the external environment. Stronger fluctuations of climatic characteristics from the regional norm change the course of early embryogenesis, increase the level of mitosis pathology, portion of inbred progeny, and reduce the yield of seeds significantly.

In a drought, a high temperature regime combined with a moisture deficit destabilizes and accelerates the processes of cell divisions and growth rates of embryos. In cold years, on the contrary, the main reason of embryonic death is the low rates of their development. The stronger the weather stress, the higher the degree of destructive changes. In abnormal years, survival of those embryos whose genotypes are more stress resistant and whose growth rate are correlated with the size of corrosion chamber and seed. They are able to switch over to the process of differentiation and form a normal energetic matrix. This is confirmed by data about the type of growth processes of embryos, which does not change in mother trees in different weather conditions: The Spearman rank correlation coefficient is 0.91 (Svintsova, 2002). Embryos, whose genotypes and morphometric correlation is severely disturbed, die before late embryogenesis.

INBREEDING DEPRESSION DURING EMBRYOGENESIS

In the second half of the twentieth century, the theory of embryonic lethal in the conifers was widespread (Koski, 1973). It is considered that many inbred embryos never developed because of the recessive genetic load (lethal and sublethal genes) resulting in a high frequency of empty seeds. Main arguments for a negation of genetic incompatibility system in them are the germination of inbred pollen and fertilization after self-pollination (Lewis, 1949, Bateman, 1952, Koski, 1973; Kosinski, 1981; Olds, 1986; Willims & Savolainen, 1996; et al.). We adhere to the concept of the genetic incompatibility system. In comparison with angiosperms, the gymnosperm's system is less developed and specialized, and also differs by several morphological and genetic properties. This is an earlier stage of its evolutionary development. However, the origin of the incompatibility of angiosperms from gymnosperms is not in doubt.

During proembryogenesis, the death of inbred and outbred embryos happens very rarely and does not depend from the pollination type. The process's incompatibilities are not detected. This indicates that a diploid embryo and a haploid endosperm can be considered as two heterogeneous systems, only after the embryo leaves the ovicell and will come in contact with the endosperm tissue.

Self-pollination has a negative effect on the development of reproductive structures. It increases the variability of many quantitative traits. However, the level of variability is not the same and depends on the trait nature, the weather conditions, and self-fertility of the mother tree. Table 8 shows an influence of inbreeding on different morphometric traits of pine (1-4 signs is the data of Svintsova, 2002). Some signs, such as the archegonium number and seed weight (1 and 7 traits), do not depend on pollination type.

In sexual reproduction, they reflect the development of female gametophytes - its energy potential and degree of incompatibility with the maternal tree. The influence of self-pollination on the fertilization (2nd trait) is uncertain, since the syngamy occurs in almost all the ovules. The increase of embryonic mortality (3th trait), which characterizes the intensity of inbreeding depression in early embryogenesis, is ≈ 1/2. Self-pollination has a great influence on embryogenesis rates (4th trait). The 6th trait shows the level of inbreeding depression in the population. Such a high degree of reduction of inbred seeds testifies about the high efficiency of natural selection in relation to inbreeding.

The "critical point" of embryogenesis is the stage of early embryogenesis (Sarvas, 1962). In open-pollination, the embryo death in 25% of ovules takes place (Dogra, 1967). After the transition to the corrosive chamber, the embryo passes to heterotrophic nutrition. In this time, the cells in the central part of the female gametophyte form a cavity. Its nearby cells contain many grains of starch (Figure 64a). Embryogenesis will be carried out in this nutrient-rich region. Between an embryo and an endosperm, as between two cellular systems with an energetically strong and an energetically weak energy potential, arise a centripetal streams of nutrients which provide further growth and development of embryo.

Table 8. Influence of self-pollination on the traits of the reproductive sphere of *Pinus sylvestris*

	Traits	Ratio of trait values at the inbreeding and outbreeding
1.	Archegonium number	0.99
2.	Ovicell fertilization	0.79
3.	Embryonic mortality in early embryogenesis	0.50
4.	Embryo growth in late embryogenesis	0.47
5.	Total seed number	0.67
6.	Plump seed number	0.07-0.08
7.	Seed weight	0.95

Figure 38a shows how the relationship between an embryo and an endosperm develops during a normal course of early embryogenesis. The trophic function of embryo provides several layers of endosperm tissue. After embryo appear in the corrosion chamber, the nearest energetically weak cells of endosperm turn and are attracted to it. Metabolism and planes of cell divisions are then changed. The adjacent layer is transformed into the secretory cells, which have a specific ultrastructure and high synthetic activity. The new synthesized biomolecules are not used for their own metabolism. They begin to move by the vector of the stronger biofield and are used for the nutrition of embryos.

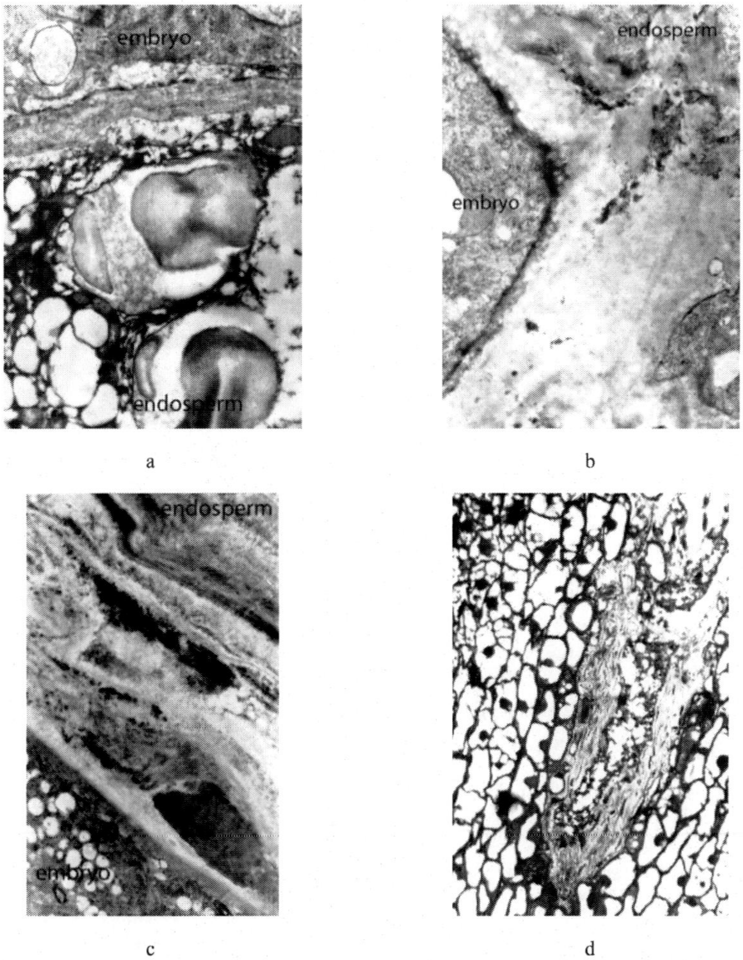

Figure 64. Relations between embryo and endosperm in the self-sterile form after open- (a) and self-pollination: beginning (b) and end (c, d) of early embryogenesis.

After embryos transition to the heterotrophic type of nutrition, they begin to grow rapidly (undifferentiated growth period). The nearby endosperm cells are rapidly destroyed in the result of excessive synthetic activity. They are replaced by the new cells. Their volume occupies the growing embryos. As known, a haploid endosperm and diploid embryo are differed by the DNA-particle. Therefore between them, as between the two heterogeneous systems, the self-incompatibility processes arise. Their appearance is due to the fact that the embryo can be used for its nutrition only in part of the substances of endosperm. The other biomolecules are detained by a semi-permeable cellular membrane which is located on the path of a centripetal stream of metabolites. They progressively accumulate around the embryo in the form of a "cover," giving rise to an isolating layer of the embryonic chamber. In open-pollination, the thickness of isolating layer is optimal and does not prevent the growth of embryo and its trophic function. While thickening, this layer initiates a transition of embryo to the inner differentiation and then to the resting state. In the mature seed, it represents the substance of the embryonic chamber, which fills the space between an embryo and an endosperm.

Morphological structures, which prevent the self-pollination in a coniferous species, are poorly developed. The intensity of self-incompatibility processes in the progamic period is low. Therefore, the proportion of inbred genotypes in the pool of zygotes should be significant. Most of them are removed when they are placed in direct and indirect (in system of "diploid embryo – haploid endosperm") competition between inbred and outbred embryos. Experience by artificial self-pollination is shown that in optimal years, the cone yield after self-pollination is reduced from 15-37%, and the plump seeds in the cones to 11-15 times in comparison with open-pollination (Table 8). The inbreeding depression in pine takes place during gametogenesis and embryogenesis, but the intensity of self-incompatibility reactions is much higher during embryogenesis.

By now it is proved that the inbreeding depression in coniferous is realized after fertilization: *Pseudotsuga menziesii,* Orr-Ewing, 1957; *Pinus pence,* Hagman & Mikkola, 1963; *P. glauca,* Mergen et al., 1965; *P. sylvestris,* Koski, 1971; *P. radiate,* Griffin & Lindgren, 1985; *P. contorta,* Sorensen, 1987; *P. taeda,* Williams, 2008; *Larix deciduas,* Kosinski, 1981; *Picea glauca,* Mergen at al., 1965; *P. abies,* Muller, 1977; *P. mariana,* Park & Fowler, 1984; *P. omorica,* Geburec, 1986; *Abies alba,* Kormutac, 1996; etc. It is shown that 64-83% of the inbred embryos of *Pinus taeda* die during early embryogenesis in 23 - 36 days after fertilization (Williams, 2008). In *Pinus sylvestris*, they eliminate within a month after fertilization, and the peak of their death also falls on the early embryogenesis (Kuznetsova, 2012). By the end of this stage, the ovules consist from the endosperm tissue with the normal ultrastructure and completely destroy the embryo, often surrounded by a thick layer of abscission layer (Figure 64-b,d). After the embryo degeneration, the destruction of its endosperm follows.

Embryos, in comparison with the endosperm, are characterized by a higher level of DNA-energy and a more powerful biofield. Its energetic potential is the maximal species value, as it was formed at the 6 activation stages of reproductive cycle. The 1st cell of endosperm undergoes only 2 activation stages (Table 3, Figure 27). In addition, the DNA-energy of endosperm is descended down to the 12 levels during the formation of female gametophyte. The powerful biofield of outbred embryo creates a centripetal stream of nutrients through the plasmatic membrane, which possess a selective permeability. Since the embryo and endosperm are partially compatible tissues, not all molecules will be used for embryo metabolism. Part of itis detained and settles around embryonic complex. They then accumulate gradually around the embryo. The self-incompatibility processes between inbred embryo and endosperm after self-pollination has the same cytological signs as in the open-pollination, but they proceed more intensively. Furthermore, their intensity is much higher than between the inbred pollen tube and nucellus.

Inbreeding depression in the self-sterile (*SS*) trees of pine is characterized by the highest intensity. An electron-microscopic study of early embryogenesis showed that shortly after the transition to a heterotrophic manner of nutrition in the contact zone with an endosperm, processes of incompatibility between inbred embryos and endosperm begin and accelerate. Figure 64b illustrates how the abscission layer between the embryo and the endosperm expands rapidly. Apparently this is caused by the inability of inbred embryo to use in its metabolism in a significant part of biomolecules of its own endosperm. It increasingly restricts the stream of metabolites towards the embryo and inhibits its growth while thickening. The width of abscission layer to the early embryogenesis end often reaches a size of 1-2 cells (Figure 64c). After reaching the threshold concentration, it becomes a mechanical barrier that completely separates the embryo from the nutrients of endosperm. Then the irreversible degradation changing of embryonic cells begin, embryos die, and all that remains is an empty capsule Figure 64d). After some time, the endosperm is destroyed. Therefore, the degradation of the embryo and endosperm are secondary events, with respect to the abscission layer. A very small number of seeds are able to overcome the barrier of incompatibility (average values of *SS* trees vary within – 0-20%). In this connection it is necessary to consider two important factors, firstly, the endosperm of mother tree represents a variable formation, and secondly, the intensity of incompatibility processes directly depends on the weather conditions. Therefore, the number of plump seeds in different years and in the cones of the same tree vary in more wide limits.

Inbred embryos are characterized by high variability of morphometric parameters during early embryogenesis. It was found that during self-pollination, the variation coefficients by the trait of "embryo length" can range from 45-60%, during open-pollination, respectively, from 15-38% (Svintsova, 2002). Differences of variability level testify about instability increasing of embryonic development after self-pollination. In turn, the destabilization of embryogenesis leads to the death of the majority of inbred embryos which do not correspond to the variability diapason of species by the growth speed and the development rate. Genotypes, which go out beyond the norm reaction, will not be able to develop inside a fixed volume of gametophyte and pass in time to the internal differentiation. All of them will be removed from the reproductive process in the course of stabilizing selection. The reasons of destabilization could be the following:

1) There is the dependence of embryo mortality from the self-fertility level of a tree. All conifers are cross-pollinated species, but they are not strictly cross-pollinated. Their distinctive feature is the differentiation by the types of seed reproduction systems which is realized at the intraspecific level, in the form of intra-population variability. Numerous self-pollination experiments on a different species have shown that they are characterized by high polymorphism by this trait and a wide intraspecific norm reaction to the inbreeding (Barnes et al., 1962, Park & Fowler, 1984; Kuznetsova, 2012, et al.). The

diapason of variability by the self-fertility level varies from 0 to 1 and more (Kuznetsova, Isakov, 1996).

Self-fertility is a genetically determined trait. The average level of sample, determined by the average K_{SF} and the intra-population distribution of trees by the classes – self-fertility (*SF*), partly self-fertility (*PSF*), and self-sterile (SS) did not significantly change in the optimal years (classes K_{SF} are given in Chapter 2). Structure of Stupinsky population by this trait fluctuates around the ratio – 8*SS* : 4*PSF* : 1*SF*. 2/3 of this sample are presented by the auto-sterile trees. The portion of *SF* trees is 7-9% and the high *SF* is even lower. Efficiency of inbreeding depression of trees, which are distinguished by the self-fertility level, is different. By the embryogenesis end, almost all the inbred embryos (95-100%) of high *SS* forms with the K_{SF} – 0-0,03 die; *SS* – 80-90%; *PSF* – 50-80%; *SF* – 50-10%.

Analysis of inbreeding influence on the embryonic mortality showed that the main cause of inbred embryo death of the *SS* forms is its incompatibility with the endosperm. In the open-pollination, the study of the frequency and spectrum of pathological mitosis in the seed progeny of the *SS* and *SF* forms do not reveal any significant differences between them. Electron microscopic investigations of embryogenesis, after self-pollination, are illustrated if the thickness of abscission layer is greater in a *SS* forms in comparison with the *SF*. This layer in the *SS* forms reduces the stream of metabolites and inhibits the embryo growth. In self-pollination, the thickness of abscission layer acts as a main factor that increases embryo differentiation by growth and development. In SF forms, the differences of embryogenesis between types of pollination are absent and correspond to the species norm. It should be noted that the intensity of self-incompatibility reactions depend not only from the self-fertility level of a tree, but also from embryo status by the DNA-energy. Embryos with lower energy potential and more intensive processes of self-incompatibility have less chance for survival. In the *SS* forms, only the single most energetically strong embryos overcome the barrier of self-incompatibility, which negatively affects their viability. Genotypes, which diverge by the morphometric parameters from the species norm, are removed during stabilizing selection. Consequently, the proportion of outbred progeny in the gene pool of surviving seeds is increased.

2) Dependence of embryonic mortality from the violation degree of DNA particle, DNA-energy, and DNA-information. The gametophytes in the fixations are sometimes found, which have also destroyed endosperm and embryo with the normal ultrastructure. Or, for example, in *SS* forms, the dying embryo can be surrounded by an abscission layer of insignificant width. This indicates that the death of these embryos is due to another reason.

The embryo viability depends on the number and spectrum of genomic and chromosomal abnormalities, which are enhanced as a result of transition in a homozygous state during self-fertilization. If we proceed from the fact that most mutations worsen its vital state and their viability is a function of DNA-energy, self-pollination, on the one hand, leads to the elimination of genotypes with lethal and semi-lethal genes, and on the other, increases embryo differentiation by the rate of growth and development.

Disturbances of embryos by the DNA-energy and DNA-information are manifested in the form of various morphoses, impaired development rates, which are the cause of embryonic lethality. In the norm, the resultant DNA-wave, arising during gamete syngamy, is a wave packet from two waves close by the amplitude, frequency, and species-specific modulation when the envelope amplitude follows the form of initial cosmic signal. Violations of the wave characteristics of the genome of a sperm or the ovicell, lead to a mismatch between the amplitude and frequency oscillations during formation of resultant DNA-energy; causing the distortion of the common information signal as a result of noise and hindrances of DNA-information of gametes. The sum of such waves creates a huge variety of wave packets with the same length L. Accordingly, the stronger their components (separate DNA-waves of sperm and ovicell) deviate from the species norm, the DNA-energy is more reduced during formation of resulting wave. In self-pollination, the identical disturbances are superimposed and enhanced, which result in greater variability of genome, destabilization of the rate of embryo development, and the appearance of different morphological abnormalities.

3) There is a dependence of inbreeding depression from the environmental conditions. The relationship between a self-sterile level and a temperature in the flowering plants was first established by East (1929). He found that the self-incompatibility reactions are characterized by a maximum intensity in the zone of optimal temperatures. Any divergence from the optimum, towards higher or lower temperatures, reduces the barrier-incompatibility in plants. This property of genetic incompatibility gives the possibility of obtaining viable inbred progeny from SI forms and is widely used now in the breeding practices of angiosperms. Therefore, the genetic incompatibility system not only regulates the close relationships, but also contributes to the species survival in adverse environment.

We received similar data about the dependence of the self-fertility level of *Pinus sylvestris* from the weather conditions. An experiment of self-pollination was performed on the same 50 random trees of the Stupinsky population. The years by climatic conditions were as follows: 5 optimal (1988, 1989, 1992, 1993 and 1996), 2 cold (1990, 1994), 2 warm (1996, 1998), 1995 - moderate drought, 1991 - strong drought, and 1999 had an intermediate position between the optimal and warm. In the course of the experiment, it was established that the self-fertility was a stable trait in the optimal years (R_{SF} = 0.13-14), and a labile trait in the abnormal years. Inbreeding is very sensitive to the environment conditions; the sensitivity threshold is 1999, slightly anomalous year (R_{SF} = 0.16). In cold and droughty years there is a significant extension of the norm reaction: from 0.00 – 0.75 (optimal) to 0.00 – 0.92 (1994). In the fertilization year, a direct connection between the average R_{SF} of population sample and the weather stress is proved. The common trait for the pine and the flowering plants is that in the cold, warm, and droughty years, it is changed only in the direction of greater self-fertility. Moreover, if a shift of norm reaction and a portion of viable inbred seeds are greater, the vegetation conditions differ more from the regional optimum (R_{SF} = 0.38 and 0.58, respectively, in drought 1995 and 1991). The ecological dependence of self-fertility trait is a confirmation that the nature of inbreeding depression of conifers and flowering plants is the same - a genetic incompatibility system.

Figure 65 shows how the self-fertility trait changes by the self-fertility groups in the optimal (1992, 1993) and stress gradient of 3 years: A warm (1998), a moderate drought (1995) and a strong (1991) drought. The trees of first (I) group, which is 2/3 of the sample, includes only *SS* forms, and the second (II) group unites *PSF* and *SF* forms. Then the average R_{SF} for the groups was calculated and took it for 100%. In optimal years, it is 0.056 and 0.362 for I and II groups, respectively. As the stress on weather strengthens, the difference between the variants increases. Even in the warm 1998 year, the response of *SS* genotypes is 1.5 times higher in comparison with the trees of group II. This indicates that the stress resistance of inbreed *SF* genotypes is higher. SC species of flowering plants are also more adapted to the unfavorable weather conditions than the SI species (de Nettancourt, 1977; Surikov, 1991; Agadzhanyan, 1992; etc.). I.e., there is an analogy between the reaction of contrasting species of pine trees and species of angiosperms by this trait on the weather stress.

For the practical selection, the highly *SF* trees are the initial selective material for obtaining valuable forms on the basis of inbred lines. Their study revealed a similar characteristic of generative sphere function in the both variants of pollination (Svintsova, 2002). A dispersion analysis confirmed the absence of a significant influence of self-pollination on the reproductive structure development (potential seed productivity, seed scales proportion in the fertile zone, reduction of potential seed production), in a pollination year.

Analysis of late embryogenesis of inbred and outbred embryos does not reveal the significant differences by the next traits: Embryo and endosperm sizes; number and size of cotyledons; morphological violations (Svintsova, 2002). The coefficients of variation between types of pollination do not exceed 10%. Variability of embryos by the number of cotyledons is also within the species norm. Developmental anomalies occur with a frequency of $\approx$ 1%. The parameters of seed productivity of the mother trees and their inbred and outbred progeny have a similar value. According to dispersion analysis, the influence of the factor "pollination type" on these traits is insignificant. The influence of the mother trees on the traits of seed productivity is significant: The seed total number is 40% ($P \leq 0.05$) and the plump seed number is 43% ($P \leq 0.05$). This indicates a certain tendency of maternal inheritance and a sufficiently high viability of their inbred progeny. In late embryogenesis, a low level of variability of the quantitative and morphological characters of inbred embryos is confirmed by the data on the selective advantage of inbred progeny of highly self-fertile forms and is obtained by their growth and resistance (Isakov, 1999).

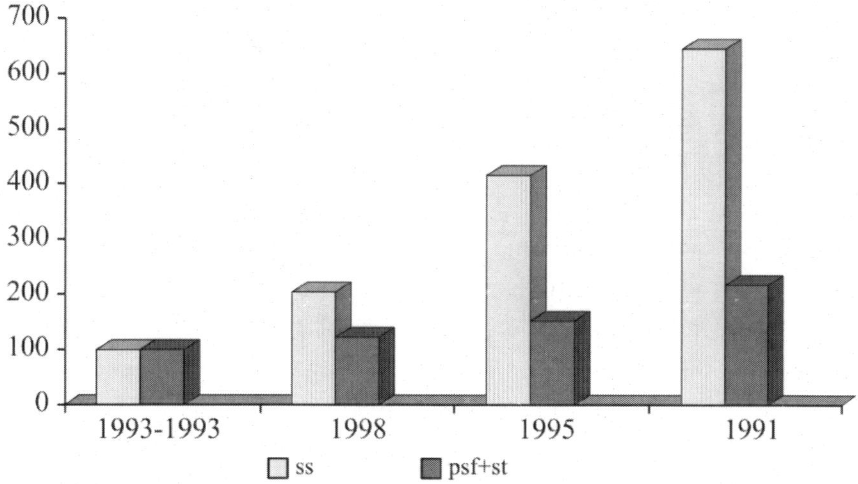

Figure 65. Change of self-fertility level at the *SS* and *PSF* + *SF* groups of *Pinus sylvestris* in the optimal, warm 1998, moderate 1995 and strong 1991 years.

The genotypic composition of seeds is formed every year in accordance with year climatic conditions. In higher plants, a genetic incompatibility system and cross-pollination regulate the closely related relationships through the selective advantage of the outbreeding and the exchange of genetic information between plants in populations. In conifers, its action is stretched to the entire reproductive cycle with the maximum realization in the end of early embryogenesis. Despite the specificity of system functioning, it is not inferior a flowering plants in the efficiency. In optimal years, a proportion of inbred progeny in seed gene pool of *Pinus sylvestris* does not exceed 7-9%. In abnormal years, seeds are adapted to unfavorable environmental conditions through the increase in the gene pool of the amount of inbred and resistance to weather stress genotypes.

Summary

Embryogenesis includes competitive and stabilizing selection. The result is a narrowing of the genotypic diversity of population gene pool, improving quality, sustainability, and adaptability. The main tasks of embryonic competition is the release of a significant part of a genetic load (energetically weak, genetically, and informational disturbed, inbred, non-adapted genotypes). The creation of a diploid genome should be associated with the formation of the spin-spin connections between homologous chromosomes and returns the genetic system to the equilibrium. Any defects of genetic material of the sperms and/or ovicells from the species norm (by the line of DNA-energy, DNA-particle, DNA-information) break this process. The main task is the raising of gene pool stability by means of increasing of the number of norm-forming and outbred genotypes. This type of selection can be compared with the "Procrustean bed" since the embryo should be embedded into the fixed volume of gametophyte. This takes place during early embryogenesis. In the main, inbred and "non-standard" embryos from the reproductive process are removed. In optimal years, their proportion is ≈ 1/5 of seed production in year; in anomalous years - from 20 to 100%, depending on the weather stress.

During meiosis and fertilization, a huge number of genetically non-identical genotypes of pollen, ovicells, and embryos arise. Their spectrum is much wider in comparison with the mother population. The set of discrete genotypes of this spectrum completely reflects the information accumulated by this population about various environmental factors. This genetic material undergoes a selection, initially on the base of different mother trees and then with the endosperm, which carry information about the current state of the ecological niches of every tree. Therefore, during fertilization and embryogenesis, a great number of genotype-environmental information is superimposed in different combinations and creates the resulting information in the form of a set of common norm reactions of the seed gene pool. Since an embryogenesis occurs in the normalized conditions, the genetic material of the seeds is only partially adapted to regional climatic conditions rather than closely linked to them; such seeds have more degrees of freedom. This is a necessary requirement for their development of different new habitats.

Chapter 6

ONTOGENESIS AND MAIN FUNCTIONAL SYSTEMS OF THE FOREST TREE SPECIES

The organic world, by its origin, is common. Evolution is one of the main forms of the development of living matter. Its evolutional development is constantly moving forward from the lowest primary prokaryotic forms to the highest multicellular plants and animals. The species are never completely repeated in the course of speciation. Separation of the organic world on two branches, animals and plants, happens at a certain stage of evolution. This process is due to the internal and external reasons while quickly moving in search of new food sources and mastering the land.

HYPOTHESIS OF PLANT ORIGIN

The common ancestor of plants are the unicellular prokaryotic organisms. In the previous work, we showed that the organic world was divided into the plant and animal world in accordance with the law of linear momentum conservation (Kuznetsova, 2013). The process of directed speciation consists deals with the appearance of a new species, which are a reflection of the higher type of organization of unicellular eukaryotic organisms. Eukaryotic type is characterized by a higher level of orderliness, the complication of internal organization, and progressive changes of the reproductive cycles of multicellular species.

The immobile lifestyle has a significant influence on all traits of the vegetative and generative spheres while manifesting itself in the form of species-specific particularities of development and organization of plants. Plant immobility is due to the subordination of living matter evolution to the fundamental laws of physics. In physics, the momentum ***P*** of a material object is called the product of the body mass on its velocity:

$P = Mv.$

According to the law of linear momentum conservation, the sum of the momentum of two particles, in the absence of external forces, remains unchanged. Therefore after its division on the plants and animals, the total angular momentum of living matter during their further evolutionary development should not change, i.e.

$$Mv_{enimal}+Mv_{plant}=const;$$
$$Mv_{enimal}=Mv_{plant}.$$

The vital strategy of the species of Metazoa is aimed at the intensification of the operative function of mobility (***v***). The strategy of the species of Metaphyta occurs in the opposite direction because they pass to a way of lifestyle attached to the soil (or another substrate). As for the mass (***M***), the total it is much higher in the plant species than that in the animal species. These distinctions between kingdoms are particularly evident for the less evolutionary advanced systematic groups of animals and plants (Protazoa and Phaeophyta). In angiosperms, the differences in total mass are less pronounced due to the unequal distribution of biological energy between their vegetative and generative spheres, in favor of the latter.

Ontogenesis

Ontogenesis begins with seed germination and continues until the end of life. All forest tree species are characterized by an attached lifestyle. Their development occurs within the species range and ecological niches through populations. Therefore, ontogenesis is sometimes called the population-ontogenetic stage. This stage divides in the following age-related and structural-physiological periods: The embryonic, juvenile, generative, and post-generative (reproduction and aging). During the embryonic period the energetic, corpuscular, and informational matrixes of the future plant are formed. Therefore, the embryonic period we have to pick out in a separate stage (Chapter 3). During ontogenesis, the growth associated with the increase of tree sizes, the formation of the structure elements of its body, and the development leading to qualitative changes in the structure and functions of the plant and its parts, are carried out.

As mentioned above, a gametogenesis is the genetically determined process of independent and interconnected implementation of the three programs of individual development of future plants – the energetic, corpuscular, and informational. Corpuscular information is laid in the primary structure of DNA-particle as a certain nucleotide sequence. During embryogenesis and ontogenesis, its chemical formula is not changed.

The number of cells increases by division of diploid initial (a cell from a cell) and all cells of one plant receive the same individual and species-specific DNA-particle.

Energetic information of plant is laid in the form of a set of individual and species-specific energy levels of the DNA-wave. In the course of ontogenesis, the energetic potential of plants is constantly changing. Vector energy moves downward. It is directed from the highest level (sporophyte initial) to zero (natural death level). Biological energy is distributed in the plant body is also uneven: From the maximum level (meristematic tissue) to zero (dead cells). We think that the tree age is defined by the average level of the total biological energy of all its cells (living and dead). Cell differentiation and creation of energetic matrix, occurs during early embryogenesis. The energetic matrix is formed by means of three types of divisions: Usual, asymmetric, and activating mitosis by the principle of bipolarity and in the accordance with the energy conservation law. An energetic matrix of all higher plants contains two centers of energetically strong cells of apical meristem (shoot and root), zone of differentiation, and energetically more weak cells of root cap.

Division directions and tree phenotype are genetically determined in space and time processes which controls the DNA-information and informational matrix. Informational component of genome, as the DNA-energy, has a wave nature. DNA-information by the physical nature represents the species cosmic informational signal and the modulated individual species-specific informational signal (after creation of information matrix). An information matrix of all higher plants has 5 zones which differ by their DNA-energy and DNA-information: Apical meristem shoot, apical meristem root, hypocotyls, root cap, and cotyledons. This matrix is created during late embryogenesis. As it has been shown, a new modulated informational signal only carries a DNA-wave of cells in the zone of differentiation of an energetic matrix. Therefore, the biological harmonic oscillators of different embryonic cells generate a non-uniform biofield, which acts within cell and beyond its borders. In embryology, the notion of biofield was given by Gurvich (1977), and a morphogenetic field by Weiss (1939). By Weiss, a morphogenetic field is the genetically determined information of an organism as a whole. During ontogenesis, a biofields of cells are superimposed on each other, creating a series of constantly changing morphogenetic fields. Therefore, a DNA-information is the dynamic component of a genome; its changes have not a chaotic, but a genetically determined character in accordance with its information matrix. DNA-information defines the processes of tissue specialization, a species-specific phenotype of plant, and its age-related changes.

Ontogenesis Beginning

Ontogenesis begins from embryo exit from the resting state. Seeds of higher plants vary by form, differentiation degree, and sizes. In pine, the end of embryo growth and its transition to the dormancy occurs in August. All metabolic processes are reduced to a minimum. The seeds will be inside closed cones until the spring of next year. They represent an autonomous formation, which is not connected trophycally with the mother tree and is isolated from it structurally. However, a tree continues to bear a protective function with respect to the progeny, promoting to its better preservation during the autumn-winter period. In March – April, cones open and the seeds leave mother trees. The seeds are spread by the wind for a distance up to 1 km. Then the seeds germinate and give rise to the plants of new generation. They are doomed to perish if the cones do not open and seeds are remained within them.

The common for all species is the uniform principle of internal organization and the fact that the root tip should be directed towards micropilar end of seed. Biology of seed germination is covered in the works of Bewley & Black, 1982, 1994; Black et al., 2006. Numerous experiments have shown that seeds for germination require water, air, and heat. Apparently, the water in combination with the optimal temperature (+22 ± 10°C), is the trigger that outputs the embryo from the dormancy. The income of water takes place by means of an absorption process and the seeds gradually swell. During swelling, their volume increases, the micropilar channel opens, air begins to come in through it, and a connection is established between embryo and environment. Pressure inside embryonic chamber changes and cell divisions are initiated.

Let's look at the physical nature of this process. The analysis will begin with the fact that during late embryogenesis, the biological harmonic oscillators of embryonic and endosperm cells are stopped, the wave function on the DNA disappears, and the seed passes into a dormant state. However, the oscillator, as a physical system, is not destroyed. In physics, this state of system is described as the equilibrium at the critical point. We think that pressure is the main factor that is marking a smooth transition of embryo from the active proliferation to the dormancy. Since the mature seed represents a closed system, due to the impermeability of seed coat and the absence of metabolic processes, the possibility of the activation of biological harmonic oscillator from the inside is excluded. Therefore, it can leave this state only under influence of external forces.

We think that the external factor which outputs the embryo from the dormancy is the water and pressure. Pressure in the embryonic chamber begins to decrease during the swelling of seeds, after the micropilar channel opens and expands. Through this hole, the seed peel connection is established between the embryo and the external environment. The optimal temperature and decreased pressure are an external force that activates a biological harmonic oscillator of one or more nearby embryonic cells and it again begins to oscillate. Embryos from this push comes into resonance, the biological harmonic oscillators of all cells begin to oscillate, and the embryo outputs from the dormant state. During resonance, it experiences a significant stress, which may be the cause of its death.

The embryos are characterized by the highest stability and viability if the corpuscular-wave parameters of their genome and the energetic, corpuscular, and informational matrix correspond to the species norm. After activation, the following chain of intracellular reactions and main processes of metabolic activity comes into action: Respiration, activation of enzymes and their biosynthesis, fermentation of nutrients, their movement from endosperm to development zones, cell divisions, and their growth.

The biological harmonic oscillator of each embryonic cell, after activation, is characterized by a certain energetic level, which depends on the energetic potential of the embryo, and its position within the energetic matrix. Inside an embryonic complex, the apical meristemes of shoot and root have a maximal level of DNA-energy. This tissue presents a complex of totipotent meristematic cells that retain the morphogenetic potency throughout all life of plant. It is the source for origination of other cells and tissues.

In essence, nature has created the model of "perpetum mobile" on the example of the apical meristem (shoot, root) and energetic matrix. The cells of apical meristem maintain their genetic constitution and the ability to proliferation during ontogenesis. Totipotency of meristematic cells provide a special type of cell division – an asymmetric mitosis, which depends from gravity direction and law of energy conservation. During asymmetric mitosis, two different cells by DNA-energy are formed (Figure 66). One daughter cell, which is the identical with the mother cell by the biological energy level, remains in the meristem tissue. The second daughter cell, which has lost two quants of biological energy, enters on the path of differentiation and specialization. It gives rise to various tissues and organs of plant in accordance with its spatial location inside the informational complex of an embryo and the direction changes of its cellular lines by the energetic and informational component of genome.

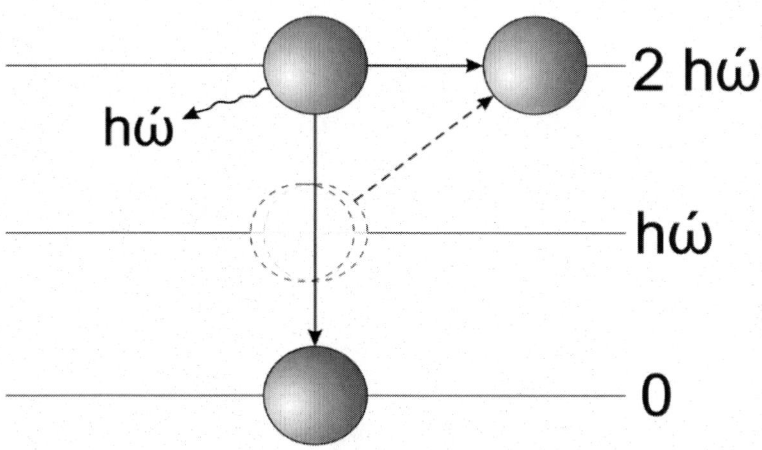

Figure 66. Scheme of change of the DNA-energy during asymmetric mitosis.

Higher plants are characterized by an immobile lifestyle. Its main task at the beginning is the formation of connection with soil substrate, according to the principle of "genotype-environment" interaction. Therefore, each seed is provided with a special samara and has a fixed position during flight – with a pointed (micropilar) end down. It ensures a better attachment to the substrate in the moment of landing. The subsequent step of attachment to the substrate is carried out with the help of primary root, which goes out first during germination of seeds. The process of germination and transformation of the embryo into a sprout occur in two stages: First the radicle and then the shoot when cotyledons appear (Figure 67-a). This order is disturbed by mechanical damage of the chalazal pole of a seed peel. In this case, the shoot with cotyledons appears first; the radicle remains inside the seed that leads to the death of sprout (Figure 67-b).

a

b c d

Figure 67. Germination seeds in norm (a) and after breach of peel (b); seedlings in the beginning (c) and in the end (d) of 1st vegetative season.

An order of seed germination is laid during embryogenesis and is the common feature of all seed plants. At the proembryogenesis end, the suspensor carries the embryo from the ovicell into the female gametophyte, becoming closer to the chalazal tapering end of the corrosion chamber. Its longitudinal growth occurs along the corrosion chamber and mainly in the direction of the micropilar end of gametophyte because the growth from the opposite (chalazal) side limits by the layer of compact endosperm cells which are completely filled with protein and lipids. In addition, the region of cotyledons is the widest place of the embryo. Wide cotyledons and endosperm fix the embryo position in the chalazal part of seed. The embryo length begins to increase during seed germination. Region of embryonic root is narrower and has a pointed tip. Therefore, the embryo growth after activation goes only in one direction – to the micropilar end of seed.

Swelling of pine seeds takes about two days. In optimal conditions, they begin to germinate within 4-5 days. During this time, the embryo exits from the dormancy. The main intracellular metabolic processes are activated. Biological harmonic oscillators of embryonic cells, in accordance with the 2nd law of thermodynamic, move down to the lower energy level and emit a free energy quantum. This energy initiates the first cell divisions. It should be noted that cells of all zones divide. However, the intensity of cell divisions is different depending on the zone. Cell development occurs by means of division and elongation. The transition of the embryo to the proliferation process occurs at the beginning of ontogenesis.

The embryo grows rapidly at this time. However, the embryonic root grows more quickly due to the higher proliferation activity of its cells. It is established that the intensity of divisions in the cells of root and shoot apical meristem is different (Ivanov, 2011; Ivanov & Dubrovsky, 2013). In the root meristem, a large part of embryonic cells participates in the proliferation process; the relative division rate is greater than in the cells of hypocotyls. A cell cycle does not exceed more than several days. After the beginning of the division process, the embryonic root grows rapidly, comes near the micropilar end of seed and abuts against it. Under its pressure, the seed peel in the region of the micropilar canal is destroyed, the embryonic root comes out, and gives rise to the primary root. It grows very quickly and the next day its length reaches ≈ 1 cm. After seed germination and root attaches to the soil substrate, a developing embryo begins to absorb water, carbon, oxygen, and ash elements from it.

The second stage of germination is the exiting of the cotyledons with the terminal bud from the seed. Until now, the trophic function of embryo was provided only by the cells of endosperm. Therefore, the most part of endosperm nutrients are completely utilized and only one peel remains from the seed. The cotyledons go out from the seeds due to the intensive growth of hypocotyl. It is bent into an arc and they gradually walk out from the peel. The enlargement of the gap peel, due to the growth of cotyledons in the width, helps the sprout exit from the seed. Then the hypocotyl of sprout straightens. Its cotyledons become green and begin to participate in photosynthesis.

During the 1st vegetative season, the aboveground part of a seedling successively replaces two types of morphogenesis: An escape with the cotyledons and terminal bud (Figure 57c); and an escape with single juvenile needles and terminal bud (Figure 57d). The first morphogenesis relates to the formation of a new type of information matrix – the homodynamic metameric matrix. This is the development plan of a plant during ontogenesis. The principle organization of the energetic matrix and corpuscular matrix are not changed.

New types of informational matrix is formed on the basis of meristematic cell layers of the hypocotyls tissue. Cells of hypocotyl have an elongated shape and are arranged by the parallel rows (layers) along its longitudinal axis. Before embryo exit from dormancy, its cells are totipotent by the DNA-energy and are characterized by the same maximal level of biological energy. In the course of series of asymmetric mitosis, the growth of embryo length occurs a specific way. Each cellular layer of hypocotyl gives rise to one metamer of primary stem, within which the derived cells are located, one above the other, in according with the descending energy vector. Its upper layer has a higher energy level (embryonic cells of hypocotyl), the last layer of metamer - the lower level. Biological energy of other cells inside metamer is intermediate between them.

In the course of embryo formation, the number of horizontal layers inside hypocotyl is determined. The length of embryonic hypocotyl is increased ≈ 12-15 times during seed germination. Meristematic cells in the process of elongating a primary stem of escape lose the property of totipotency. After asymmetric mitosis and the separation of its tissue into metamers, they become different by the DNA-energy. Their distribution by the energetic level is formed in a certain way. Figure 68 shows how the DNA-energy of initial hypocotyl cells change after four divisions by the usual and asymmetric mitosis.

Let us assume that the initial potential of hypocotyl tissue is equal to 8 energetic levels. Afterwards, each mitotic division becomes lower by one level. After usual mitotic divisions, 16 meristematic cells with the identical DNA-particles and DNA-energy are formed. Their energetic potential has become lower by 4 levels compared to the embryonal cell of hypocotyl. After asymmetric mitosis, 16 derivative cells arise, which are the identical in DNA-particles and differ by the DNA-energy. The scheme of this type of division is shown on Figure 56 and 58. After 1st division, two daughter cells arise (Figure 58). One of them, identical with the mother cell, remains the initial, and the second cell descends lower on the two energetic levels – from 8th to 6th. The formula of 1st division is: 1:1. During 2nd division, the initial again gives rise to cells of 8th and 6th energetic levels. Two cells from an energetically weak derivate appear: one is its copy (6th level), and the other of the 4th level. The formula of genotype distribution after two divisions is: 1:2:1. After the third division, two cells from initial are formed again – the 8th and 6th energetic level. Two derivates of the 6th level give rise to two cells of the 6th level and two cells of the 4th level. From the last energetically weak cell of the 4th level, originate cells, respectively, the 4th and 2nd levels. The cell distribution according to the

DNA-energy is as follows: 1:3:3:1. The fourth asymmetric mitosis is ended by the formation of 16 cells; one of which remains at the highest 8th energetic level and the second daughter cell descends on two levels below. Three derived cells of the 6th and 4th levels give rise to three cells with the same, respectively, the 6th and 4th levels; and three cells of lower, respectively, the 4th and 2nd levels. Two cells of the 2nd and the lowest level are formed from the energetically weak cell of the 2nd level. The formula of the frequency distribution of cells is: 1:4:6:4:1. Now we determine the sum of their energetic levels: 8+(6x4)+(4x6)+(2x4)+0=64. The sum is divided by the number of cells – 64:16=4. The energetic potential of metamer corresponds to the 4th level. Hence it follows that the average level is the same after usual and asymmetric mitosis.

The variability range of derived cells after asymmetric mitosis expands by the descending energetic vector on two levels and the level of initial cell remains the same. The frequency distribution of cells by DNA-energy obeys the law of normal Haus-Laplace distribution: The smallest and largest value is represented by one cell, with the mode in the zone of average energetic state of the metamer. Since cells do not occupy all energetic levels (through one level), the cellular complexes with identical DNA-energy within a metamers are created. These meristematic cells are totipotent only within the complex and differ from the cells of other complex not only by DNA-energy, but also by DNA-information. This may be a prerequisite for their isolation within the metamer and initiation of the development of specialized cells, tissues, and organs.

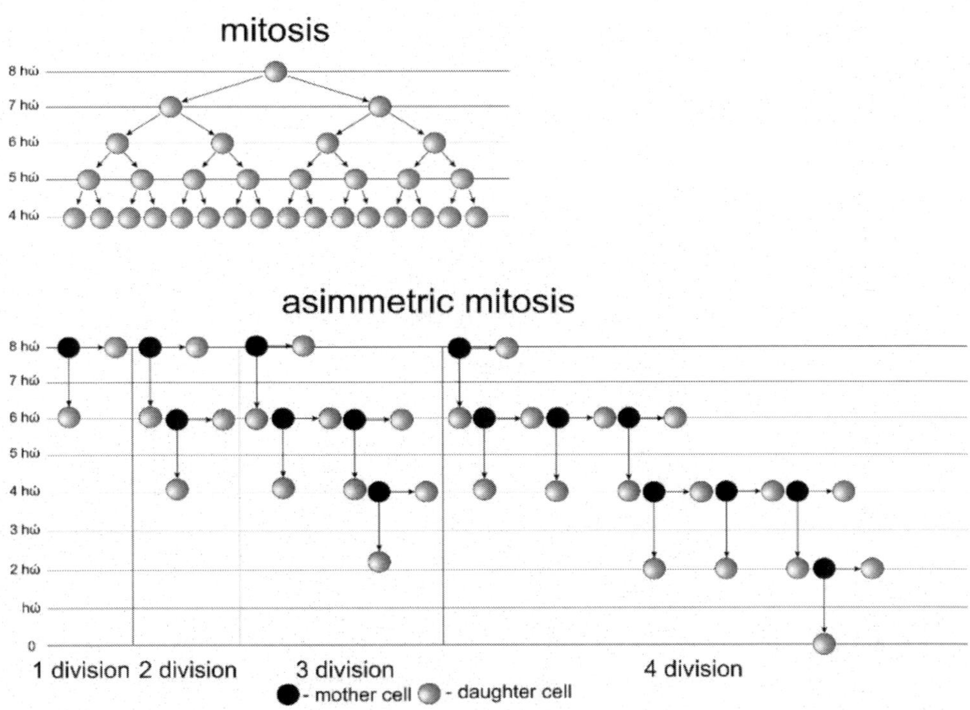

Figure 68. Schema of the distribution of cells by the energetic levels during usual and asymmetric mitosis.

The upper, intermediate, and lower layers within one metamer are meristematic cells. This is due to the fact that the meristematic zones of embryo are characterized by a very high energetic potential. Asymmetric mitosis lowers it, but not by much. This level of reduction is not sufficient for their transition to a state of differentiated cells. During seed germination, all metamers of developing stem consist only from the meristematic cells. Therefore, it is not possible to distinguish the division of the 1-year seedling into metamers. By the end of the formation of metamers, the cells begin to pass from division to the process of elongation and seedling continues to grow rapidly.

All cells of one-year seedlings are identical by DNA-particle and their differentiation by DNA-energy has the same principle of distribution as in the energetic matrix of a mature embryo. In regard to the DNA-information, the formation of informational matrix is not ended during late embryogenesis and its development continues during the 1st vegetative season.

The change of pine sprout occurs within a month after germination, when the main photosynthetic organ, the cotyledons, are replaced by the 1-needle shoot. In July, the embryonic bud of escape begins to grow and form a 1-year seedling. Cells of meristem and its derivatives, during regular sequence of asymmetric and usual mitoses, form the identical metamers which are located one above the other inside a terminal bud. Further growth of the first shoot is carried out by means of cell divisions and elongations of metamers. A specificity of this informational matrix is that its formation is accompanied by significant morphological changes of the needle. In the 1st vegetative season, seedlings form the cotyledons, then the juvenile 1- needle shoot; in 2nd season, the typical 2-needle shoot is formed.

The underground part of the seedling is also changed. The root system and tree crown originate from the apical meristems, respectively, a root and a shoot. Both these meristems have the disk-shape and represent two interconnected gyroscopic platforms. However, the principle of root and stem development is essentially different by nature. The primary root develops from the apical root meristem located at the hypocotyl end. All its cells are the derivatives of the apical meristem and are not divided into metamers.

It should be noted that the gyroscopic platform is not all meristematic tissue, but only its resting center in the apical meristem. Its cells are rarely divided (Clowes, 1967; Ezau, 1980; Evert, 2006, etc.). They have approximately the same size and low rates of synthesis of nucleic acids and protein. It is established that a sharp decrease of the division rate is associated with a lengthening of the G_1 period (Clowes, 1967). Apparently, the low proliferating activity of meristematic cells is due to a fixed spatial position of the resting center, as a gyroscopic platform. Proliferating activity of meristematic cells near the resting center is $\approx$ 10 times higher. The zone of intensively dividing cells occupies a considerable area from the apex to the older part of root. In pine sprouts, the zone of dividing cells is 10 mm from the tip, where the cell divisions are partially combined with the elongation and with the processes of differentiation. All newly formed cells of root, depending on their spatial position within the information matrix, are differentiated in the three meristematic tissues - the epidermis, the cortex, and the central cylinder. Further growth of root in length are provided mainly by the elongation and specialization of differentiated cells. Consequently, the initially totipotent cells of apex become the differentiated and specialized tissues with a specific ultrastructure and functional purpose.

The main root and lateral roots of pine represent the taproot system, which penetrates deep into the soil. Therefore, the apical meristem of root after seed germination forms a long primary root (Figure 67c). Then lateral roots begin to develop and seedlings develop the branched root system to the 1st vegetative season end (Figure 67d). For the reason of apical dominance, the differences between length of the main and lateral roots of seedlings initially is very large ≈ 4-6 times (Veselkin, 2013). The tips of roots are covered with a root cap, which protects the apical meristem from damage. It helps to move the roots through the soil during growth. The cap consists of the parenchyma cells, which originate from the root apical meristem (Esau, 1980). These are energetically weak cells, so the duration of their life is not great.

We think that the histological distinctions between stem and root are more determined by the differences of the aboveground and underground informational matrix of seedlings and in lesser extent by the external conditions when their development occurs. The aboveground part of plant presents the system of metameres with the structure of nodes and internodes. The underground part is organized easier. The primary root originates from the cells of apical root meristem, with the combination of the two types of divisions – asymmetric mitosis and usual mitosis which are the processes of cell divisions and their elongation. Therefore, the location of tissues into the root from level to level varies insignificantly. Underground and aboveground parts of plants are connected in a certain part of the seedlings which is called the transition zone. Within this zone, the root structure is gradually transformed into the stem structure.

Embryos and seedlings of all species of *Pinaceae* have the same principle of inner organization. The aboveground part of the seedling consists of a hypocotyl and cotyledons, the underground part of a root. They differ only by the size and number of cotyledons, whose number in pine is 3-18 (Mirov, 1967). Moreover, the principle of organizing the embryo and the tree is similar (Mirov, 1967, Lanner, 1976; Bewley & Black, 1982, 1994; Black et al., 2006). Apparently, species and plants differ by the corpuscular-wave parameters of their genome, but all are characterized by the one-type principle of matrix organization (corpuscular, energetic and informational).

Indeed, the diploid state of the genome is the characteristic of all species of higher plants. In ontogenesis, the genetic identity of plants by the DNA-particle provides the DNA replication, according to the complementarity principle. The bipolar principle of embryo and plant organization is the universal feature of plants. Their energetic matrix has the axial structure coinciding with the gravity direction and two poles (apical root meristem and apical shoot meristem), which energetically dominate over its other parts. In ontogenesis, the number of meristematic cells or is constant or decreases. The maintenance of their energy potential and the development of plants occur in accordance with the energy conservation law. A plant body consists of the cells of different degrees of differentiation, including dead ones. The older the tree, the more its body consists of differentiated and dead cells. The informational matrix is the basis on which the processes of specialization of cells and tissues will occur. In the center of a mature embryo, it is possible to distinguish the stem-like axis of weakly vacuolated meristematic cells of procambium – the hypocotyl and in the apical part – the cotyledons with the bud of the shout. In the opposite pole, there are the signs of cellular specialization which are the characteristic for the roots. Meristematic cells and tissues of differentiated embryos by their location and cytological characteristics have a similarity with tissues, which then develop in the seedling (Esau, 1980). The future epidermis is located outside the layer of protoderma. The layer of more vacuolated cells of the main meristem is under it. It should be noted that in embryonic cells, the property of totypotency is lost in the process of differentiation and specialization. The exception is 3 energetically strong zones of apical meristem of root shoot and procambium meristem.

The plan of the plant body is created in a certain sequence and the matrices are superimposed on each other: A corpuscular - during proembryogenesis; an energetic - during early embryogenesis; an informational - during late embryogenesis. The diploid state of a genome and the bipolar principle of plant organization are the common features for all higher plants. During late embryogenesis, the traits of class, family, species, and individual level of plant organization appear. Now it seems obvious that a genotype and a matrix of each mature embryo contains species and individual information on the internal organization and program development of the future plant.

The ability of plants has a certain orientation in space and grow in accordance with the gravity direction is a well-known fact. It is described as the phenomenon of geotropism. Geotropism can be positive or negative. The root is characterized by the positive geotropism as it grows in gravity's direction. The stem has the negative geotropism because its development occurs in the opposite direction. The process of geotropic turn demonstrates the seed germination on filter paper when the seeds have a horizontal position. Figure 67a shows that the geotropic turn is carried out in the region of the growing root segment and happens soon after its exit from the seed.

Seeds fly out from the cones in the spring when the soil is most saturated with water and rapidly warms up, which helps their rapid germination. During flight, the pointed end of the seed is directed downward. This increases the probability of deepening and linkage with the substrate after its fall to the ground. In this case, the embryonic root grows directly downward. However, as a rule, the seeds have a different position during germination (horizontal, inclined, etc.). In this case, the root tip shortly after seed germination bends and begins to grow vertically downwards, in accordance with gravity.

It is believed that a root cap is considered the formation which protects the apical meristem, helps the root motion in the soil, and controls the processes of geo-reception (Wilkins, 1966; Esau, 1980). The place of gravity reception is the central column of root tip cells, in which amyloplasts act as the statolits. According to our views, the cells of root cap can't be the system of gravity reception. These are energetically weak cells and last link in the cell line of differentiation. The process of root cap formation is related with the action of energy conservation law. After their DNA-wave reaches the zero energy level, they die. Therefore, the function of root cap cells is the partial removal of dead cells from the tree body, which are continuously formed during ontogenesis. We assume that the geotropism mechanisms must have another nature.

We have noticed that the biological harmonic oscillator of the meristematic cell structurally resembles a wave gyroscope. In plants, the principle of its action is based on the effect of a standing DNA-wave inside the nucleus. The nucleus of meristematic cells has a spherical shape. In this case, the nucleus can act as a resonator of an axisymmetric form. This form ensures a high stability of cellular gyroscopes to the beats. The time of its work is very large. The gyro will work as long as the meristematic cell exists; theoretically up to infinity. Hence it follows that the apical meristems of root and shoot represent two gyrostabilized platforms, which are held in the certain position by a system of cellular gyroscopes and do not participate in the vibrational processes of the plant. The property of such platforms is the preservation of conditions for an asymmetric mitosis, which reproduces the genome of initial cells of the embryonic meristem in their cellular lines.

If we proceed from the wave nature of DNA, 3 meristems of stem, root and cambium represent a device responsible for the geotropic reaction of the whole tree, for the unity of underground, and aboveground parts. This aids the processes of stabilization and orientation in space. These are two gyrostabilized platforms interconnected by means of hypocotyl tissue that operate as a one system. Their interaction is based on the effect of the independence of elastic waves of resonator oscillations from the vibrations of external environment. This system has a high accuracy of mutual binding of the axes of sensitivity of gyroscopes. This quality is achieved by the fact that the genomes of cells are characterized by close and unchanging corpuscular-wave characteristics.

MAIN FUNCTIONAL SYSTEMS OF THE FOREST TREE SPECIES

Let's consider some of traits connected with immobile lifestyle of plants on the example of forest tree species. A large part of the analyzed particularities are common to the entire plant world; some of them belong only to the higher plants. Analysis of the functional systems of forest species are carried out based on morpho-geographical, anatomical, embryological, ecological, physiological, etc. investigations (Pravdin, 1964;

Esau, 1980; Levitt, 1980; Kozubov et al., 1982; Raven et al., 2005; Evert, 2006; Williams, 2009; Nobel, 2009; etc.).

ENERGY SOURCES

Life is a highly ordered process that requires a large amount of energy. In previous work we have shown that plants and animals can exist, develop, and perform work by receiving energy from internal and external independent sources (Kuznetsova, 2013). The inflow of energy from outside is provided by a continuous exchange of substances between the individual and the external environment. Animals have three independent sources of energy: Two inner (biological harmonic oscillator and metaphase chromosomes) and one outer (respiration). Due to the fact that the plants are unable to actively move in search of food, they must have such a source which is capable of providing their trophic function at a fixed lifestyle.

Plants spend energy with a high degree of efficiency. However, the nutrients are contained in a limited volume of soil substrate which is not enough to provide the tree trophy during many years in one place. Therefore, the energetics of plants is supplied by four energy sources. One more source assists in providing the plant with energy – photosynthesis. The energy sources of plants are shown in the Figure 69.

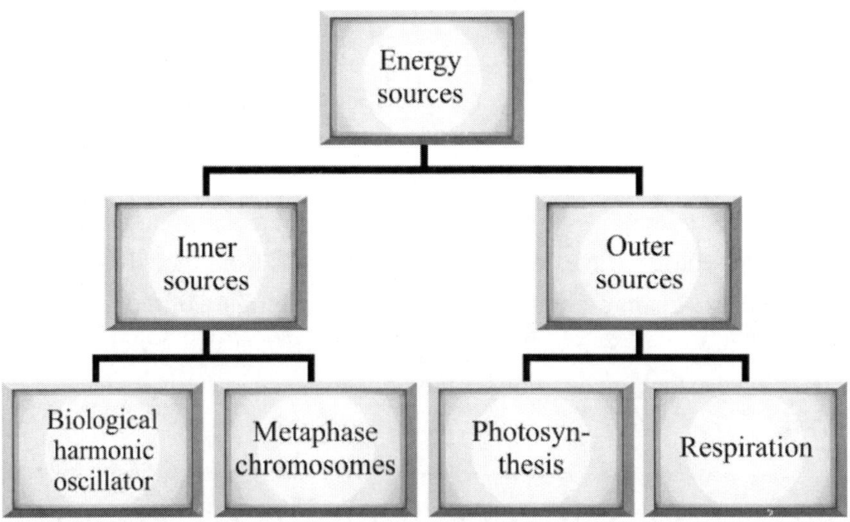

Figure 69. The inner and outer sources of free energy in the plant cell.

The biological harmonic oscillator or DNA-wave energy is the 1st source of inner energy, which is present in all living cells of plants, animals, and microorganisms. A detailed description of this source is given in Chapter 1. Here we list only its main characteristics:

DNA-energy is the energy of quantum harmonic oscillator. The nucleus together with the oscillating DNA, function in the cells as the physical system – the biological harmonic oscillator. It as any oscillatory system, has energy (potential, free), a biofield, and a set of common quantum-mechanical properties. Metabolites flowing in the cells occurs with the participation of biofield forces. Free energy is released once in the cycle - during the G_2 interphase period when the oscillator descends to the lower energy level. This energy initiates new cell divisions and is spent on energy supply of such intracellular processes as replication of ATP from ADP, cellular growth, etc. The harmonic oscillator can be in the operating (interphase cell) and in the non-operating (dividing cell and resting cell) state. The transition of DNA-energy to higher energy levels is a strictly genetically determined process. It occurs during sexual reproduction on the basis of species reproductive cycle which leads to the rejuvenation of generative cells. In ontogenesis, the DNA-energy of vegetative cells and plants tends to the equilibrium (0th energy level or natural death level), in accordance with the 2nd thermodynamics law. The cells inside the body of embryos and plants are heterogeneous by the DNA-energy levels and are characterized by the different degrees of differentiation. The highest level of energy has the meristematic cells of shoot and root apex and cambium. It is significantly lower in differentiated cells. In all dead cells, the biological harmonic oscillator is destroyed.

The 2nd inner energy source is the metaphase chromosomes. This source originates with the appearance of mitosis and occurs only in the proliferating cells of animals and plants. Its physical nature is the elastic energy of highly compressed DNA molecules. Elastic energy is accumulated while chromosomes coil during the G_2 period and prophase of mitosis. In telophase, energy is released and enters into two daughter cells during chromosome despiralization. The energy of metaphase chromosomes is used in the energetic supply of such processes as a transcription, a cell growth, and a creation of centripetal and centrifugal energetic streams. The centrifugal stream functions on the basis of processing or the ordered exponentional disintegration of RNA molecules, the process of which occurs with the liberation of free energy. The centripetal stream acts in the ATP – ADP system. ATP molecules are the universal source of energy for many biochemical processes, which proceed in the cell. Both streams provide the energetics of interphase cells: The maintenance of their metastable state, biosynthesis of proteins, nucleic acids, lipids, etc., are the making of different works.

External energy sources are respiration and photosynthesis. Respiration is available in all species of plants, animals, and microorganisms. Respiration occurs throughout their life and operates based on chemical energy. In plants and animals, respiration, as a biological phenomenon, represents a combination of redox reactions by the splitting of energy-rich carbohydrates and other organic substances. It also involves the transformation of liberated free energy to the chemical bonds of ATP molecules. These reactions occur with the absorption of oxygen and give off carbon dioxide. The simple energy-poor carbon complexes (carbon dioxide, etc.) are used as the building blocks for various intracellular metabolic processes - the synthesis of amino acids, sugars, etc. Respiration refers to the external sources since the nutrients in the body of animals and microorganisms come from the environment. In plants, we see another situation. There are no specialized organs for respiration. The nutrients are created by the plants themselves during photosynthesis and they are the initial substrate for respiration. I.e., the molecules of energy-rich substrate are synthesized and utilized within the body of a single plant and the main components for these redox reactions (O_2, ions, etc.) come from the environment. By this reason the respiration, in relation to the plants, is nominally an external energy source. In plant and animal cells, the specialized organelle for respiration is the mitochondria.

Photosynthesis is a complex and surprisingly perfect physical and biochemical mechanism for the capturing of solar energy and its transformation into the energy of ATP chemical bonds. It occurs with the participation of chlorophyll and other light-absorbing pigments which are present in chloroplasts of cells and consists from a light-dependent and light-independent phases. Plants during photosynthesis give off O_2 and absorb CO_2. A photosynthetic reaction is usually expressed as:

$$6CO_2 + 6H_2O \xrightarrow[\text{chlorophyll}]{h\upsilon} C_6H_{12}O_6 + 6O_2$$

Photosynthesis is an external source of energy and is found only in plants and photosynthetic microorganisms. Photon energy that comes with sunlight initiates and provides the work of this source. The leaves are the main morphological organs where this reaction is carried out. Specialized organelle of cell is the chloroplast. Photosynthesis leads to the formation of carbohydrates and other important nutrients that are necessary to maintain energy balance and trophic function of plants. Part of oxygen which is released during photosynthesis goes to their respiration and the rest goes to the atmosphere. Adding to photosynthetic activity of plants on the Earth, the circulation of oxygen and the optimal atmospheric composition are supported. Physical models of respiration and photosynthesis are described in more detail in this Chapter.

SPATIAL COORDINATION SYSTEM

Embryos and plants are characterized by the bipolar principle of organization and a certain orientation with respect to gravity direction. All plants have two poles of asymmetry (root pole and shoot pole) and three meristematic tissues (root apex, shoot apex and cambium/procambium). These tissues consist from the totipotent meristematic cells and their development is very weakly dependent on other parts of the plant. All cells of each meristem have the same DNA-energy. During ontogenesis, they, as more energetically strong tissues, must dominate other parts of the plant. Bipolarity and asymmetry are the main factors of spatial coordination of cellular systems.

The bipolar matrix is created during early embryogenesis during the first two activating mitosis of embryo and persists throughout plant life. According to the energy law conservation, the meristematic status of root apex and shoot apex is maintained by means of continuous formation of differentiated cells. In the ontogenesis, the genetic identity of meristematic cells is provided by the means of the asymmetric mitosis. The main requirement for their realization is the coincidence of the polar axis of mother cells and the planes of divisions with the gravity direction. After asymmetric mitosis, one daughter cell remains in the meristem and the other is added to differentiated cells which are developed in the various tissues of plant. Since more energetically weak daughter cells are also divided by the asymmetric mitosis, a gradual transition from one differentiated cellular group to another is observed in the interjacent zone between two apexes, which ensures the harmonious development of shoot and root (Esau, 1980; Evert, 2006).

All main processes of generative and vegetative sphere development are characterized by the strict orientation, with respect to gravity direction. The branches of seed plants are laid in the form of buds in the leaf axils (Esau, 1980; Evert, 2006). It has been established that the annual increment of branches is closely related to the bud length from which they develop (Powell, 1977). In the bud, the metamers number of future shoots and the intensity of growth by the elongation are defined (Vorobyev et al, 1989). In the next spring a male, female, and vegetative shoots develop from according buds. The shoot growth occurs primarily by means of cell divisions and then due to their elongation. It is shown that at the beginning of their growth, the meristematic cells of the columnar type are successively divided and undergo transverse divisions (Esau, 1980). In this period, all newly formed shoots grow only upwards (Figure 70a). This indicates that their cells are divided by asymmetric mitosis. Apparently, the vertical position of the shoots and asymmetric mitosis promote an orderly distribution of cells by the energy gradient along a branch. The female strobiles have only a vertical position (Figure 70b). As shown above, the reproductive cycle by the female generative line is arranged in accordance with the gravity direction. Another situation takes place with the male generative sphere. Male buds and microstrobiles can have different spatial positions inside the tree crown.

Microsporangiogenesis and microgametophytogenesis (activation stages, ᚦ) depend on the gravity direction (Figure 28). Initial microsporangiogenesis are oriented by superficial cells, which are located by the circle of primordium microsporophyll. Therefore, this reproductive stage can occur in the microstrobiles with different space positions. Microspores have a spherical shape and a certain place inside a tetrad of microspores and archesporium. This is an axisymmetric generative cell, so it can initiate the activating stage from any space position too. Three activating mitosis, which a microspore passes during microgametophytogenesis, has an exact spatial orientation. This is evidenced by the traces of two protallial cells and the vegetative cell position inside the pollen grain, which are placed one above the other. Intensive growth of male shoots begins after pollination. Figure 70c and 70d shows that they are oriented upwards, regardless of spatial position of their microstrobiles.

Figure 70. Vertical growth of young branches of main, lateral, female shoots (a, b), and shoots with a horizontal (c) and vertical directed microstrobiles.

Two zones of meristematic cells, which arise in the distal and apical pole of embryo, can be considered a complex of identical by the structure, spatial location, and synchronized by the frequency and amplitude of cellular gyroscopes. Its principle of operation is based on the wave nature of a cell, whose biological harmonic oscillator represents the set of standing chromosome-waves oscillating inside the nucleus or a closed resonator cavity of axisymmetric shape.

Two systems of cellular gyroscopes (root apex and shoot apex) are located at the opposite ends of the embryo and plant body. From the physics position, these meristematic tissues present the two gyroscopic platforms, the stability of which is provided by a system of meristematic cells or the cellular gyroscopes. They fix the spatial position of cells inside two meristems along the gravity direction. Therefore, during ontogenesis, the meristematic cells of root and shoot apex are divided mainly by means of activating mitosis, which is the chief factor of their differentiation and future specialization.

Inside the embryonic complex, the gyroscopic platforms are located at opposite ends of the hypocotyl: At the lower end (root pole) the meristem of the root rudiment is placed and at the upper end above the cotyledons (the escape pole), the meristem of the shoot rudiment. In the tree body, the shoot and root meristems function throughout its life. These meristematic tissues, or two gyroscopic platforms, should be viewed as a whole - the system of endogenous coordination of the growth and development of the aboveground and underground part of the tree. This system secures the same orientation of its main trunk and major root and stability of the vertical position of the tree during ontogenesis. If one of the gyroscopic platforms for some reason is destroyed and dies, its place is taken by the apical meristem of one of the lateral branches or the lateral root.

As already mentioned, all conifers are characterized by a monopodial type of branching. The initial period of growth of the shoots and needles is carried out by the divisions of cells and their derivatives within the metameres, which were laid during their development inside buds in the previous summer. At first, all young shoots grow vertically upwards (Figure 71a). So their cells divide mainly by the asymmetric mitosis.

Photosynthesis activity of a tree requires a well-developed crown. The formation of a new tier and the development of existing branches occur annually. Therefore, the crown represents a complex formation. The metameres, within one year's shoot, have the same structure, but the cells within the metameres are differentiated according to the descending energy gradient.

The vertical position of the young shoots is maintained by the small sized, weak vacuolization of cells, and elasticity of their cellular walls which give them a certain resistance to the external environment. In the second half of May, when the shoot reaches ≈ half of its annual increment, the branches begin to grow by means of elongation of the cells within the metameres. Under the weight of growing shoots and needles, the lateral branches are bent at the base of the shoot and gradually change the spatial position from the vertical to the more horizontal (Figure 71b). In June, the branches take up the more horizontal position (Figure 71c). The terminal shoot is directed only vertically upwards. Its vertical spatial orientation is supported by a system of cellular gyroscopes - the interconnected meristems of tree (apical meristem of shoot and root).

a b c

Figure 71. The change of spanial position of lateral branches during their formation: a – vertical young shoots; b, c – beginning and end of their bend.

AGE-RELATED CHANGES

The terminal shoot of a tree grows only vertically upwards. As mentioned above, asymmetric mitosis can theoretically be compared with a work of perpetuum mobile because one daughter cell is always an exact genetic copy of the mother cell and in the cell lines, it never gets old. The cells of apical meristem could be indefinitely divided by means of asymmetric mitosis if it were not the necessity increase and maintain the volume of tree photosynthetic surface. Therefore the process of inception and development of apical and lateral buds are realized with a participation of two division type. During ontogenesis, the asymmetric mitosis always occurs in combination with usual mitosis which leads to the aging of vegetative cells including energetically strong merictematic cells and tissues. The energetic potential of a tree gradually decreases from a maximum (embryo level) to the zero level (level of dead tree) in accordance with the 2nd law of thermodynamics. It follows that the life of individual trees is finite and always ends in death.

As known, photosynthesis is one of the basic sources of plant nutrition. Without access to light, they quickly weaken and die. In forest trees, the main organs implementing the functions of photosynthesis, transpiration, and gas exchange are the leaves or needles. After seed germination, these functions are performed by the cotyledons and then by the juvenile needles (single-leaf, crescent form). Obviously, the realization of photosynthetic reaction is impossible without abundant branching. Typical needles appear on the 2nd vegetative season, the lateral branches begin to develop on the 3rd year. A large size of photosynthetic surface is achieved due to the constant renewal of the crown: The formation of new branches, the annual growth of the existing, and the death of old branches and leaves (needles). The scheme of the distribution of branches by age and sex in the tree crown is shown on the Figure 72.

The tree crown is a morphologically complex, dynamic, and a multi-age integrated system. The age-related changes of tree crown have the complicated algorithm of development, which depends from many internal and external factors. The most significant endogenous factors are the initial level of tree biological energy, the age of the tree itself, and the age of its branches and needles. In regard to the exogenous factors, the annual increment of the shoot length depends mainly on the weather conditions of the previous season (Vorobyev et al., 1989).

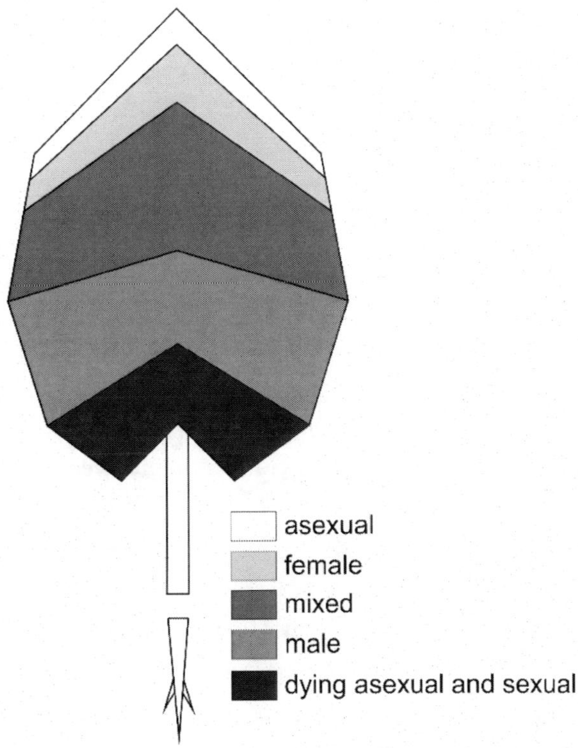

Figure 72. Scheme of distribution of tiers in the tree crown of Scots pine.

The cells of all vegetative and reproductive morphological structures of a tree crown originate from the apical meristem as the youngest tissues of the plant. It is shown that the size of any branch positively correlates with the volume of its resting (winter) apical meristem (Goroshkevich, 2016). The formation of main shoot and lateral branches of 1st order is due to the proliferating activity of the apical meristem, which is characterized by the highest level of DNA-energy within the crown of tree. Cells of the central zone of an apical meristem, as the part of tree gyroscopic system, are characterized by a lower meristematic activity and are the source of all cells of the plant crown. An important part of the vegetative season is when its meristematic cells are divided only asymmetrically. In August, their divisions are carried out by mitosis. The usual mitosis, on the one hand, increases the number of totipotent meristematic cells of the shoot apex, and on the other, lowers their energetic potential by one energy level. Consequently, the volume and number of cells of the apical meristem is doubled. The central and peripheral zones appear in the meristem tissue. The increase in volume above the "critical mass" leads to the weakening of vertical endogenous bonds between the meristematic cells as the cellular gyroscopes, which have located on the periphery of apex. Therefore, the partition of the peripheral layer of the apical meristem into several parts causes their isolation and subsequent separation from the central zone of the shoot meristem (zone of gyroscopic platform) becomes possible. The central part of meristematic tissue remains part of the common gyroscopic system which regulates the tree vertical position in space.

All changes into a meristematic tissue are strictly coordinated in the time and are carried out in a certain sequence. In August, when the rate of growth of branches slows down and the DNA-energy of all cells of central zone becomes on the level less, a transition of meristematic cells from the asymmetric to usual mitosis and the process of bud formation begins. We think that this transition is due to internal and external reasons. The external factors are the change in environmental conditions: Decreasing the length of daylight, lowering of night temperatures, etc. The internal reason is that the diameter of an apical meristem is limited by the width of the tip of its shoot and by this time it has reached this boundary. As known, the summer meristematic tissue has a larger volume than winter meristematic tissue (Curtis, Popham, 1972; Goroshkevich, 2016).

Before budding, the horizontal layers of energetically weak meristematic cells are formed during asymmetric mitosis and the peripheral meristem is produced during usual mitosis. These energetically weak derivative cells are characterized by high mitotic activity. Their development will follow by the decrease in energy potential and morphogenesis. During budding, the peripheral meristem gives rise to the several local cellular complexes or the primordium, where active morphogenetic processes are initiated. Part of their cells go by the path of differentiation and specialization.

The development of all vegetative (shoot and leaves) and generative (microsporangium and ovules) morphostructures begins with the formation of primordium tissue. A vegetative initial of future branch, needle, or generative initial of microsporangium, ovule, archegonium, etc. become one or more of the surface cells of the primordium and the longitudinal axis of symmetry which coincides with the direction of gravity. A significant difference between the initials is that the development of the vegetative morphostructures occurs outside of primordium, and the development of generative organs and cells inside a primordium tissue.

A group of meristematic cells of the peripheral zone, during development of shoot primordium, undergoes a number of periclinal mitotic divisions and gives rise to a local complex of totipotent cells. On the vegetative cone it has the form of a hillock. The external surface of the primordium consists of a zone of parallel arcuate cell layers. Cells of the surface layer primordium, whose longitudinal axis coincides with the gravity direction, become the initial of future lateral shoot. Relationship of different morphogenetic processes is revealed during formation of shoots. They also consist from the central and peripheral histological zones. The first gives rise to the core of the stem and the second, to the primary cortex and lateral organs (Sacher 1954).

It is necessary to consider the fact that the primordium is formed during usual mitosis. This means that the DNA-energy of each cell generation becomes lower by one energy level. Accordingly, the energetic potential of initial cells of the lateral shoots is less than the level of apical meristem of the main shoot. Consequently, the life expectancy of the lateral branches should be less than the tree itself and its main shoot. The duration of the lifetime of the branches of the 2nd, 3rd orders is even lower.

In the ontogenesis of forest tree species, the asexual period of development is always present and has a different duration. *Pinus sylvestris* usually ranges from 7 to 15 years. The juvenile period of development is characterized the rapid growth of a plant. It relates to the increase of height, sizes of crown, and root system. Growth in height depends from the combination of many genotype-environmental factors. Among the internal factors, we believe the initial level of its DNA-energy plays a leading role. The annual increment of plants is formed in the cell's growth, vegetative tissues, and organs. In the juvenile period, the following regularities are observed: An increase in the absolute sizes of increments over the years and a decrease in their relative values. The average increment of 1-year seedlings is 2-3 times higher than the average height of their sprouts; the annual increment of 2-year seedlings is equal to the height of a 1-year seedling; the increment of 3-year plants is only 50% from the height of 2-year seedlings, etc.

The 1st vegetative season has its own specifics. As it has been shown during seed germination, the aboveground part of 1-year seedlings are divided into metamers, the typical needles, lateral branches are absent, and the energy potential of apical shoot meristem corresponds to the level of embryo shoot meristem. This is the beginning of heterotrophic period and juvenile (asexual) stage of tree development. In August, when the linear growth of the seedling is coming to the end, the first apical bud is laid and are larger than the embryonic bud, which is located between the cotyledons. Apparently, the volume of embryonic meristems (shoot and root) is not sufficient to provide the morphogenetic processes. The volume of meristematic cells, by the number, are doubled after usual mitosis. As a result, the energetic potential of seedling is decreased by one level in comparison with the embryo. Therefore, the optimal volume ("critical mass") of meristematic cells is reached and the central zone of shoot apex meristem is formed and functions throughout tree life. It is established that with rising of apex diameter, the size of main meristem is changed more than in the peripheral meristem (Goroshkevich, 2016). On the next year, the stem diameter of the 2-years seedlings increases more than twofold.

In 2-year seedlings, the analogous processes occur. In August, when the linear growth of shoots becomes slower, the transition of meristematic cells from the asymmetric to the usual mitosis takes place. The cell number is doubled and the volume of apical meristem begins to exceed the "critical mass." DNA-energy of a meristematic tissue of apical bud decreases relative to the 1-year seedling and embryo, respectively, on one and two energetic levels. The size of the central zone increases and a peripheral zone appears with meristematic cells. The peripheral meristem is the basis for the formation of lateral buds. Then the process of apical and lateral bud inception takes place. Before their initiation, a decrease of DNA-energy takes place. In the lateral buds, this diminution equals the number of mitotic divisions during primordium formation. In August of the next year, the size of the central zone of meristematic tissue still slightly increases and its peripheral zone becomes much greater. In 2-year seedlings, as a rule, form 1-3 lateral buds. In subsequent years, the number and annual increment of lateral buds increase.

During the juvenile period, a system of branches and needles are formed, which are differentiated by their lifespan and, apparently, by the level of DNA-energy. In the tree crown, the duration of a needle's life is the lowest and equal, as rule, to 2-3 years. Apparently this is due to the fact that a leaf primordium originates from the deep cell layer which is located within a metamer, the lowest level of biological energy. The meristematic cells of this layer, after formation of leaf primordium, are divided by usual mitosis and their energy potential becomes even smaller. Therefore, if we proceed from the wave nature process, the lifetime of needles cannot be long.

Needles are the vegetative organ of the tree crown; their main function is photosynthesis. Their development begins with the formation of leaf primordium. Leaf protubercles are laid in the peripheral zone of the apical meristem as a result of periclinal divisions of cells, which are characterized by higher proliferous activity (Esau, 1980). The leaf initial is the group of the superficial cells of primordium. The initial passes to the anticlinal divisions (asymmetric mitosis) and the lateral protuberance on the shoot apex begins to form. As the derivatives of initial cells continue to divide by asymmetric mitosis, the leaf begins to grow vertically upwards from the lateral protuberance in the form of needle-like outgrowth. The needles are not divided on the metameres and made up of the epidermal, vascular, and main (parenchyma) tissue systems. Each tissue consists of specialized differentiated cells. The peculiarity of differentiated cells is the transition of biological harmonic oscillator to the lower energy levels does not lead to cell division. After reaching the zero level, they die.

The processes of branch formation, needle formation, and the decreasing of DNA-energy within a tree will continue up to its death. As mentioned above, the plant transition to sexual reproduction is possible only with a certain diapason of DNA-energy levels. In gymnosperms, the apical meristem of main shoot is the part of the gyroscopic system of tree. Therefore, it cannot initiate the reproductive process. The initiation of the reproductive process begins only when the DNA-energy of the apical and lateral buds of lateral braches will reach the level from which the plant's transition from vegetative to sexual development occurs. As a rule, the sexual reproduction of Scots pine begins in 9-15 years and this process continues until old age. Development of megastrobiles comes on 2-3 years earlier than microstrobiles. The formation of megastrobiles occurs mainly in the upper part of crown and the microstrobiles - in the lower part. This indicates that the initiation of reproductive processes by the female generative line requires a higher level of DNA-energy than for its initiation by a male generative line.

With increasing age, the spatial distribution of generative organs directly depends from own age of branches and change their position in the tree crown. In pine, unlike cedar, the localization of branches by the sex is poorly expressed. Apparently, the lower levels of female floral meristem and the upper male are overlapped; therefore, in the center of crown there is always a wide area with microstrobiles and megastrobiles (Figure 70). Inception of first megastrobiles occurs on the shoots of the 1st or 2nd order, mainly in the upper and middle part of crown. In subsequent years, the number of shoots of the 1st, 2nd, and 3rd orders, whose floral meristem has reached the DNA-energy necessary for megastrobiles inception, becomes larger. With increasing age, the number of female shoots begin to decrease in the lower part of the tree crown since the DNA-energy of their floral meristem descends below the minimum level. For example, a female strobiles of a 40-year cedar had 97.8% shoots of the upper tier; the 200-year cedar – 37.5% (Vorobyov et al., 1989). The first microstrobiles appear on the relatively weakly-growing shoots of the 2nd and 3rd order. With the age, their number of shoots, progressively increases in old trees and reaches a maximum value.

In the first decades after the beginning of sexual reproduction, the trees continue to grow rapidly, whereby, the position of the branches relative to the main shoot rapidly changes. Now the inverse relationship between the growth rate of tree and its age is established (Vorob'ev et al, 1989). It is shown that in early ontogenesis, the linear growth of a plant is characterized by an abrupt age curve; thereafter, the stabilization phase of growth takes place and it is replaced by a phase of growth activity reduction. This algorithm change of age curve is explained by the fact that the tree rate of vegetative growth increases more quickly than the energy potential of apical meristem decreases during early ontogenesis. By the age of 40-50 years, growth stabilizes and then slows down. In 200-year-old cedars, the length of annual female branches remains unchanged during 35 years and the growth intensity of leading and lateral shoots becomes the same (Vorob'ev et al, 1989). A decrease in DNA-energy meristematic cells of the tree crown leads to irreversible age-related changes, top drying, and plant death.

PHYSIOLOGICAL HOMEOSTASIS

The life of plants and animals occur in a constantly fluctuating environment. Their evolution proceeds in the direction of greater independence of species from the external environment by creating an internal, more controlled medium (Lehninger, 1985; Rubin, 1999; Alyokhina et al., 2005; Costanzo, 2017, etc.). However, the separation of the internal from the external environment and the homeostasis types in them, are different. In an evolutionarily advanced animal species (mammals, birds), there are three types of homeostatic systems: 1) Isovolemy (volume constancy of blood, lymph, etc.); 2) isotony (composition constancy of pH, proteins, lipids, etc.); 3) isothermy (body temperature regulation). In higher plants, the isotony type of homeostasis was developed.

Undoubtedly, plant and animal organisms require additional energy for maintenance of a homeostatic state. During animal evolution, the surface of a body's contact with the environment is gradually decreased, which leads to a reduction of energetic expenditures required for the exchange of water, salts, and gases. At the same time, the specialized organs, which directly interact with environment, have received a powerful development. In plants, on the contrary, the surface of body's contact with the environment, by reason of a fixed lifestyle, constantly increased. The specialized organs, such as the leaves and root system, make up their main volume. They provide the maximal access of plant to the light and soil substrate. A tree grows throughout life due to the bipolar organization of its energetic matrix and the presence of two meristems. Because the volume of body is not fixed, the type of regulatory system as the isovolemy, is excluded. There is also no isothermy in plants and their life directly depends on the conditions of the environment. Apparently the main reason for the inability to regulate the temperature is due to a lack of energy resources and fixed lifestyle. The ability to cross into a dormant state is another property of trees. In the resting state, the biological harmonic oscillator is stopped and the intensity of respiration and metabolism is sharply decreased. This allows them to survive the unfavorable vegetative conditions during winter or summer seasons.

Isotony functioning on the basis of continuous water-salt exchange in system of soil – plant – atmosphere is one of main regulatory system of plants. The principle of its operation is connected with development of an internal, constantly circulating medium and specialized systems that ensure the processes of absorption, distribution, assimilation of water, nutrients, and transpiration. All cells of plants are in close contact with its internal medium, which provides their supply with nutrients. Water reaches up to 80-95% of their mass. This is the medium for the biochemical reactions. It takes part in the photosynthesis reaction. The preservation of composition constancy of the circulating fluid inside a plant, when an environmental condition fluctuates within fairly wide limits, is the most important of its functions. Water, ions, and microelements, which it brings to the cells, directly participate in the functioning and character of enzyme activity.

Physiological homeostasis is a main requirement for the normal intracellular metabolism. A stable internal circulating medium increases the efficiency utilization of energy from the external sources (photosynthesis and respiration). In hot weather, it controls the temperature regime of the plant and prevents the overheating of leaves.

Intracellular pressure and osmotic concentration of internal medium differ substantially from their parameters in the external environment (Sliacher, 1970; Tselnicker et al, 1993; Tinker & Nye, 2000; etc.). Internal mediums (xylem juice) and outer solutions have a different ionic consistency. A significant difference of electric potentials can be detected between them (Clarkson, 1978). Transition of ions from cells to xylem can occur in the conditions in which their entrance from an environment is excluded. The hydrostatic pressure on the envelope's inside cell is always greater than the pressure on it from outer solution. This difference arises as a result of exceeding the osmotic pressure in the root cells over the osmotic pressure in the internal solution of xylem. In plants, the osmotic pressure of root meristem varies within wide limits (1-3 and up to 10 atm.). Therefore, the ions and water in the suction zone is actively absorbed by the cells and are retained in them.

At the cellular level, the physiological homeostasis is created during absorption of water and nutrients from the soil substrate by the cells of root meristem. Their biological harmonic oscillator is in working state during interphase. DNA-waves oscillate inside nucleus and is the source of biofield, which acts inside and outside of the cells. Undoubtedly, its forces should take part in the creation and regulation of the processes of physiological homeostasis. We think that for the appearance of intracellular medium, the leading role has two factors: The semi-permeability plasmatic membrane and the working state of biological harmonic oscillator of root cells are used as a biofield source.

Depending on the stage of cell cycle, the water and nutrients can enter or leave the cell by means of passive or active transport. During interphase, the duration of which is equal $\approx$ 90% from cell cycle; the oscillator is in the working state. Therefore, a biofield forces help in the active transport of water and nutrients into a cell from the soil substrate. Semi-permeability of plasmatic membranes also regulate the penetration of ions and molecules, depending on their size and charge. The entrance of ions K^+, inorganic, and organic molecules inside a cell, occurs simultaneously with the processes of the extraction of ions HCO_3^-, Na^+ and Cl^- from cell. Absorption of water creates a turgor inside the cells, initiates their growth by elongation, adds to the elasticity of the tissues and organs of plant. Active transport continues until an equilibrium state is reached between soil substrate and root meristem. Then these processes are stabilized at the level of a definite osmotic pressure and certain osmotic concentration of molecules and ions.

Wave function on the DNA and the biofield disappear during mitosis. At this time, the excess water and K^+ ions leave the cells by a concentration gradient or by a passive transport. They pass to an intercellular space. Conversely, the ions H^+, Na^+, Cl^-, and Ca^{2+} enter inside the cell. The molecules of proteins, polysaccharides, nucleic acids, and other large biomolecules does not pass through a semi-permeable membrane. In the telophase, the biological harmonic oscillator and biofield are again restored. In the interphase, cells actively absorb the water and nutrients from the soil against concentration gradient and the osmotic pressure begins to rise until the equilibrium state; streams of water and ions will be balanced with the removal of these particles beyond cells.

Water, carbon, oxygen, and ash elements are transported from the intercellular space of the root to the leaf surface by the gradient of water potential. For their motion, the trees have a special water-conducting system, which begins with the energetically strong meristematic cells of root, passes into the xylem tissue of root and trunk, and then into a network of branches and leaves. Backflow of intracellular liquid from root is impeded by the layer of the endoderm, the cells of which regulate the movement of substances within the water-conducting system of the plants. Water streams move along the intercellular spaces of the mesophyll of leaves, fall into the stomatal cavities, and evaporates through the stomata into the atmosphere. Only small portion of the intracellular liquid during mitosis is excreted to the outside in the form of mucus, which promotes the movement of roots in the soil substrate. Their cellular walls have thickenings: Casparian strips, which contain a suberin and lignin, making this part of root an impermeable for the water and nutrients.

In conclusion, it should be noted that in plants, nature has created a reliable type of physiological homeostasis. The isotony system is well adapted to the biology of forest tree species and their immobile lifestyle. It acts as a pump that a soil solution absorbs (sucks). Since just after the internal medium was created it moves through plant body. During life, a huge amount of water and nutrients pass through it. All living cells of tree are in direct contact with the water stream, which continuously circulates. Xylem sap represents a solution consisting of water and a complex mixture of different substances (ions, microelements, salts, sugars, amino acids, metabolites, etc.). Its molecular composition is initially formed by the cells of the root and contains the entire set of substances for maintenance viability of cell systems and state of their physiological homeostasis. Since the intensity of absorption and transpiration depends from the species, temperature, humidity, and light, the composition of xylem sap varies depending on the stage of development, vital state, time of day, and season. These exogenous and endogenous changes are well coordinated with the biological rhythms of a plant, which act at all levels of its organization. The main role in these processes, apparently, belongs to the biological harmonic oscillator. Its work, as a physical system, can be carried out only within certain limits of the temperature and osmotic pressure: For the most part of plants it is 0-40°C and 5-30 bar.

GAS EXCHANGE SYSTEMS

In animals, the gas exchange between an organism and environment is carried out only owing to one energy source – respiration; in plants, with the participation of two sources – photosynthesis and respiration. Both of these external sources, together with inner sources (biological harmonic oscillator and metaphase chromosomes), present the main mechanisms of transformation and accumulation of energy in the plant cells. At the present time, the processes of gas exchange, respiration, and photosynthesis are widely covered in the literature (Slater, 1970; Kursanov, 1976; Swanson & Webster, 1977; Lehninger, 1985; Nobel, 1993; Tselnicker et al., 1993; Rubin, 1999; Singhal et al., 1999; Alberts et al., 2002; Chentsov, 2004; Alyokhina et al., 2005, Tripathy & Pattanayak, 2012; Taiz et al., 2014; Costanzo, 2017; etc.). Photosynthesis represents the totality of complex, biochemical, and biophysical processes, in which, the oxygen is released and the carbon dioxide absorbed. During respiration, on the contrary, the carbon dioxide is released, and the oxygen absorbed. In plants, the high efficiency of gas exchange is provided due to the huge number of leaves in the crown, a large area of the leaf surface, the presence of cracks in the bark of trees, lentils, great number of stomata, and intercellular spaces. It is established that during the process of respiration and photosynthesis, the oxygen and carbon dioxide comes easier in the cells and is secreted from them through stomata and intercellular spaces.

The essence of gas exchange is concluded in the diffusion of gases, mainly oxygen, and carbon dioxide, through the outer surface of plants and animals. In animals, the depth of gas diffusion depends on the size of molecules, but as a rule, it does not exceed one millimeter (Riegel, 1967). Another situation happens in plants. It is known that all their body is permeated with the intercellular spaces; through the system of cavities and passages relate to the external environment. The mesophyll cells interact directly with the environment through the stomata. Stomata represent the openings in the epidermis and limited by two cells with a gap inbetween. In the leaves are located either on both its sides, but more often, on the lower side only (Ezau, 1980). In conifers, the stomata are distributed by the parallel rows along the long axis of the needles. It leads us to believe that in plants, the exchange of gases with the external environment takes place directly through the intercellular spaces and stomata.

RESPIRATION

Respiration is a basic vital function that provides animals and plants with energy. At present, comprehensive material on the biochemistry of these processes has been accumulated (Slater, 1970; Kursanov, 1976; Lehninger, 1985; Alberts et al., 2002; Gifford, 2003; Alyokhina et al., 2005; Bouta, 2005; Plaxton & Podesta, 2006; O'Leary & Plaxton, 2016; Costanzo, 2017, etc.). Therefore, the main emphasis in this analysis, we will focus on the physics of some respiration mechanisms.

In plants, the specialized organs for respiration are absent. This function is performed by each cell independently. In the cellular level, respiration mechanism in all eukaryotic organisms are similar. Respiration can be described, firstly, as the gas exchange with the environment, by means of O_2 involvement and, secondly, a significant source of external energy for intracellular processes and functioning. Its main function deals with metabolism, by converting the chemical energy of nutrients into the energy of ATP chemical bonds. In the course of respiration, ATP molecules are constantly created, which, as a centrifugal energy stream, is necessary for the energy supply of a cell's vital activity.

All living organisms create the orderliness of structure by means of energy of internal sources and the energy which they extract from the environment. Channels of the energy movement within the cells relate to a certain highly ordered processes. Respiration is the complex, multi-step process of biological oxidation of organic substances, with the release of free energy. In its initial stage – a partial splitting of sugars into two molecules of pyruvic acid – oxygen is not required. This part of the reaction takes place in the cytoplasm and the energetic output is insignificant. Then in the cycle of tricarboxylic acids, which end the main paths of oxidative splitting of carbohydrates in the presence of oxygen, a pyruvic acid is oxidized by the CO_2. This reaction occurs inside the mitochondria.

Oxygen enters into the cells through the intercellular spaces from the environment. In the mitochondria, the main reactions of energy conversion connected with the Krebs cycle and the electron transfer chain are realized. The electrons move along the electron transfer chain, are gradually descended from level to level, and emit the quanta of free energy. This energy is stored in the chemical bonds of ATP. Electrons, in the final stage of transport, are attached to oxygen molecules which are restored to the water.

Mitochondria are present in all living tissues of plants and animals, although in plant cells, the number of organelles is less than in the animal cells. It consists of a matrix, smooth outer membrane, intermembrane space, and an internal membrane with the cristae. Cristae mitochondriales form the protrusions deep into the matrix, which significantly increases the inner surface of organelle. The mitochondria do not originate de novo. They multiply during division by means of constriction or fragmentation. Their life cycle is usually several days. This is a very labile organelle. They constantly move and change shape. By its shape, they are the spherical, elongated, branched, and twisted. The number of mitochondria in the cell varies greatly (from 1 to 100 000) and depends on differentiation degree and functional activity of cells. The cristae amount in them is different too.

The intensity of respiration depends on a number of internal and external factors: Time of year, functional loading, age, etc. In relation to the environment, the cells within a tree have different locations: Inside the plant body (trunk, generative organs), inside the soil substrate and plant body (root), and in direct contact with the air (leaves). It is the highest in young rapidly growing cells and tissues, then in leaves, and even lower in the stem and root tissues. Most parts of plants can exist and develop in a limited temperature diapason - 0-40°C. Cell respiration sharply decreases in winter and summer when the temperature becomes lower than +5°C and higher than 45°C. In the cells of secretory type, the mitochondria are especially numerous. In jacket cells, they reach huge sizes that sometimes form concentric formations that occupy a significant part of cytoplasm (Figure 73-a). Without fertilization, the ovicell dies and the jacket cells gradually pass to the inactive state. They are vacuolated, number of organelles decreases, and their ultrastructure undergoes the significant changes (Figure 73-b).

Inner and outer membranes of mitochondria are characterized by the different structures and functional purpose. An outer membrane has a thickness of 6-7 nm, and consists mainly of phospholipids (85%) and proteins (15%). It separates the organelle from cytoplasm and is characterized by the nonspecific permeability. Small molecules and ions can penetrate through pores of this membrane. Most enzymes that catalyze the reactions of Krebs cycle are into a mitochondrial matrix. An inner membrane also consists of proteins and phospholipids, but a ratio between them is the direct opposite: Protein/lipid = 3: 1. It has a specific permeability. The electron transporters and set of enzymes, which carry out the ATP synthesis, are localized in the inner membrane. The same principle of the organization of the respiratory mechanisms implies the existence of specialized intracellular mechanisms and special systems, which ensure the delivery of oxygen and the emission of carbon dioxide, regardless of cell location.

Mitochondrial DNA, specific mRNA, tRNA, rRNA, and small ribosomes of prokaryotic type are found into the organelle matrix. DNA is a double-stranded circular molecule, whose size is much smaller than the nuclear DNA. In a living cell, the probability of its detection outside the mitochondria is equal to zero. Synthesis of mitochondrial DNA is not associated with the synthesis of the DNA in the nucleus. The rRNA, tRNA, and some proteins are synthesized with particioation of mitochondrial DNA. There is evidence to suggest that the chondriom, like the cell nuclei, represents the oscillatory system, which acts in the cytoplasm as a set of autonomous harmonic oscillators.

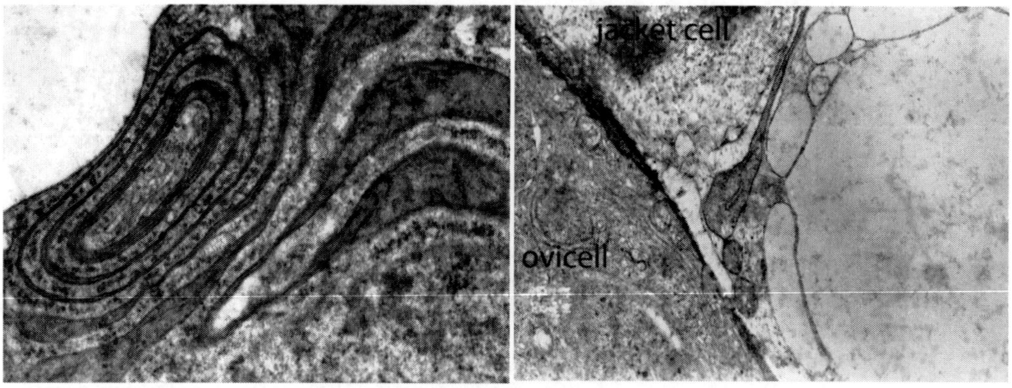

a b

Figure 73. Jacket cells during oogenesis (a) and 10 days after fertilization (b).

Indeed, a mitochondrial genome is similar to the prokaryotic genome. A DNA-particle of prokaryotic cell is one small size (≈ 7 μm); connection with histones does not form and represents a closed circular double-stranded DNA molecule within a separate zone – a nucleoid. The size of a nucleoid is about 0.4 microns in diameter. Mitochondria, due to the specificity of inner membrane, also have limited ability for secretion and absorption of large molecules. Wave motion of DNA is limited by its inner membrane of length L. Therefore, a DNA-wave must behave like the standing wave and its operation principle should be the same as in the prokaryotic oscillator (a cell together with the oscillating DNA). If this is so, the mitochondria should have a number of common quantum mechanical properties. A mitochondrial oscillator as the physical system must possess a certain own energy and biofield, which is created by a great number of oscillating dipoles of DNA.

It is known that the molecules of water, carbon dioxide, and oxygen are particles, in which the passage through the plasmatic membrane is determined by the concentration gradient. Oxygen (O_2) and carbon dioxide (CO_2) belong to the symmetric nonpolar molecules. The main differences between them are as follows. The oxygen consists of two molecules, it is lighter, and possessed by paramagnetic properties. Carbon dioxide is diamagnetic and consists of three molecules; therefore, it is heavier. Now let us assume that the cell is in osmotic equilibrium with the external environment. In this case, the concentration of oxygen and carbon dioxide inside and outside the cell should be the same. The oscillating dipoles of mitochondrial DNA are the source of a non-uniform electromagnetic field, which exists within the organelle and beyond. Its largest intensity in the zone close to the DNA: The field changes as $1/r^3$, $1/r^2$. In the far zone, it decreases to $1/r$.

A distinctive feature of paramagnet is that it does not have its own magnetic moment. Such particles in a non-uniform magnetic field acquire an induced magnetic moment, which is directed along the field. Therefore, O_2, like the paramagnetic substance, hitting in the non-uniform magnetic field near mitochondria, begins to move to the region of space where the field is stronger. O_2 molecules, moving centripetally along field vector, enter into the inner space of organelle. As consequence, thanks to the mitochondrial field, the continuous entrance of O_2 inside mitochondria can be carried out in all cells of the plant.

Another situation takes place with the process of CO_2 exchange during respiration. CO_2 excretion occurs in the final stages of redox reactions connected with the Krebs cycle, which proceeds inside the mitochondria. It is the diamagnetic substance and its molecules have their own magnetic moment. The carbon dioxide, as a diamagnet and a larger molecule, cannot penetrate into the mitochondria by means of passive and active transfer. Inasmuch as the induced dipole moment of CO_2 in the non-uniform magnetic field is directed against the field, all its molecules formed during respiration immediately begin to move into the region of a less strong field. Initially, they pass from mitochondria into the cytoplasm, then from cells into the intercellular spaces along the concentration gradient.

PHOTOSYNTHESIS

The nature for better energetic supply of plant has created an additional and very effective energy source – photosynthesis. Its main purpose is the formation of organic substances from carbon dioxide and water with the participation of light energy. The various specialized plant systems are involved in the function of this source: Vegetative organs (leaves of species-specific shape and size), tissues (mesophyll), organelles (chloroplasts), pigments (chlorophyll, carotinoid) which give the leaf color, and complex of connected redox reactions.

The leaf is the main vegetative organ of higher plants for the maintenance of the functions of photosynthesis, gas exchange, and transpiration. As a rule, it has the form of a flattened plate. It provides a large photosynthetic surface and the maximum permeability for gases. The leaf surface of a tree reaches the gigantic sizes, which in many times, exceeds the area of the soil that it occupies. The leaves have a xeromorphic structure: Stiff, small, and with a small evaporating surface. They can survive the unfavorable periods of summer and winter with water deficit and variations in temperature.

Most parts of leaves occupy the mesophyll tissue (Figure 48-a). Mesophyll or chlorenchyma is the main chlorophyll-containing parenchyma of a leaf blade, which is enclosed between two epidermal layers and contains many chloroplasts. In conifers, the mesophyll tissue is represented relatively by the homogeneous and lobate cells or it is differentiated into the columnar and spongy parenchyma. The folds appear as a result of the ingrowth of inner layers of envelope into the intracellular cavity. These spaces are very small. Gas exchange occurs through the stomata and stomata slit, which deeply enters into tissue of mesophyll. The stomata are located on an entire needle surface. Differentiated types of mesophyll is found in plants with the plate leaf blades. In plants of temperate climatic zones, the columnar parenchyma, as a rule, is located on the upper (ventral) illuminated part of a leaf; the spongy parenchyma is located on its lower part (Figure 74-a). Such leaves are called the dorsoventral. The cells of columnar parenchyma are densely shifted into one or more layers and are stretched perpendicular to the plane of the leaf. Their location to the epidermis surface increases the number of cells that have access to light, not obscuring each other. This also facilitates the outflow of assimilates. The spongy parenchyma represents the inner surface of the leaf, which provides CO_2 absorbing from the atmosphere. It also includes a stomata and a system of intercellular spaces, which constitute 15-20% of total leaf volume. The cells of this type of parenchyma are smaller in a size and have a more rounded shape (Figure 74-b). In comparison with the columnar parenchyma, the number of chloroplasts is less by ≈ 1/3.

Plastids occupy an important place in the life of plant. A typical chloroplast has the shape of a biconvex lens, a length of 5-10 μm, a width of 2-4 μm. It is surrounded a 2-membrane envelope on the outside. The outer and internal thylacoid membranes have a thickness of about 7 μm and contain approximately the same amount of lipids and proteins. The inter-membrane space is ≈ 20-30 μm. As a rule, the increase of chloroplast number occurs through the precursor structures – a proplastides. Proplastide is a very small, colorless organelle, which is presented a microscopic harmonic oscillator: A spherical body with a double membrane and an oscillating DNA molecule inside it. Then, the proplastides begin to increase in size and the longitudinal membranes of thylakoids are formed (Figure 74-c). These longitudinal membranes give rise to the specific thylakoids of grain stocks and stromal lamellae. Chloroplasts reproduce, sometimes by means of usual division (Figure 74-d).

An envelope of chloroplast consists of outer and internal membranes. An outer membrane separates the organoid from cytoplasm, has more even contours, and is permeable for most organic and inorganic molecules. An internal membrane, like in the mitochondria, is differentiated into two structural components: The thylakoid system and the stroma, which represent the common functional space. The stroma is a weakly-structured hydrophilic plastid matrix that is isolated from the inter-membrane space by means of an internal thylakoid membrane. It is established that this membrane has another chemical composition: It is characterized by structural asymmetry and selective permeability (Swanson & Webster, 1977; Singhal et al., 1999; Albertsson, 2001; Alekhina et al., 2005; Nevo et al., 2012; etc.). It forms two types of membrane systems in the stroma – a long stromal lamellae and grana stocks. The specific thylakoid membranes are stacked in the piles in the form of granas. The main part of pigments (≈ 80%) is concentrated in the grana stacks and their content in the fraction of stromal lamellae is less (Albertsson, 2001). In chloroplasts, the amount of granas and thylakoids per the

grana varies greatly and depends on many factors such
is estimated that a grana number in chloroplasts can r
starch and lipid droplets which are formed during photos

In fact, the stromal lamellae are a network of long
go through all plastid body, are arranged parallel rows,
The thylakoids have the form of disks that are very clos
concepts, they are formed by the layering (folds) of
cavities of thylakoids communicate with each other and
In regard to the thylakoid membrane, it represents a bin
molecules embedded in it. Lipids are divided into two
The cyclic lipids, such as chlorophylls and caro
photosynthesis, account for 65% of pigment lipids of th
al., 2005). The non-cyclic lipids are represented by the
sulfolipids.

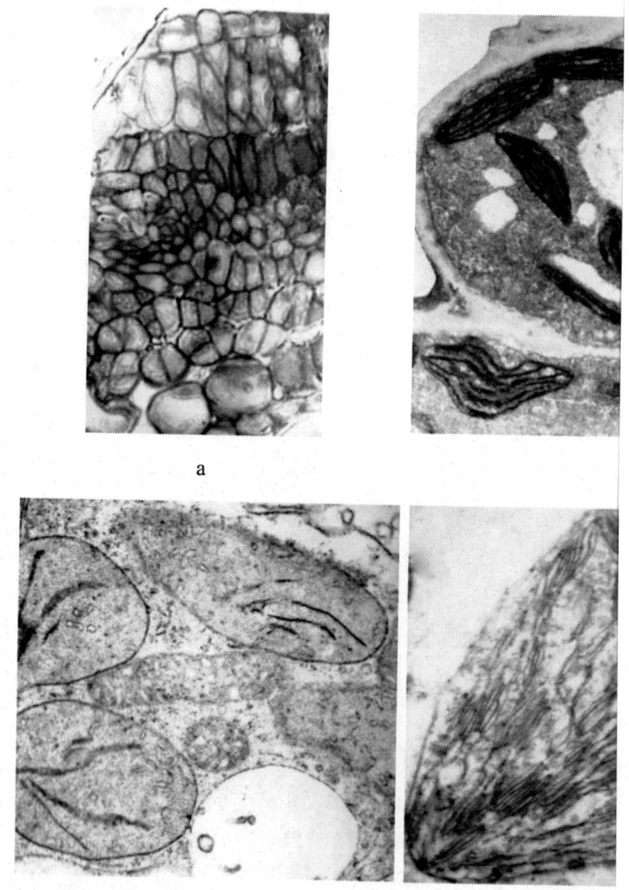

Figure 74. Structure of plant leaf (a), cell of spongy parenchyma (b
Populus balsamifera.

The stru
(transverse)
and the lum
the certain
classes in th
lumen space
these memb
Phospholipi
participate i
up to 70-80
the membra
all galactoli
lipid bilaye
sulfoquinov

The chl
plastid, can
composition
does not for
together wit
prokaryotic
synthesizing
synthesized
proteins, dif

The chl
division, de
for the dark
to half the
electron tra
chloroplasts
nucleus and
into the ch
through its
cytoplasmic
luminal spa

Now le
Analysis wi
the differen
must be re
Oxygen, as
fields and

diamagnetics, is popped into the region of weaker fields after the formation and leaves from mitochondria into the cytoplasm. In photosynthesis, the gas exchange proceeds in the opposite direction. Biosynthesis of carbohydrates with CO_2 participation (diamagnetics) is carried out inside the chloroplast, where it enters from the cytoplasm; i.e., CO_2 molecules are moved to a region of strong field. O_2 (paramagnetics) comes out from the chloroplast into the cytoplasm after secretion; i.e., O_2 molecules are moved towards a weaker field. This is the specificity and uniqueness of the photosynthetic gas exchange. Let us analyze these provisions in more detail.

It is known that photosynthesis is a complex chain of biochemical reactions and physical processes which proceed in two phases: The light-dependent and the light-independent. The chlDNA particle is localized within the organoid stroma. The chlDNA-energy is the source of an electromagnetic field only during light-dependent phase. In the light-independent phase, the harmonic oscillator of chloroplasts does not work. Chlorophyll and solar light are required for initiation of photosynthesis. Photosynthesis begins with the simultaneous excitation of the chlorophyll molecules in photosystem I and photosystem II.

The thylakoid system of a chloroplast participates in the light-dependent part of photosynthesis. Photosystem I, photosystem II, all pigment-proteins, electron transport, and ATP-synthase complexes; these are necessary for the realization of light-dependent reactions and are located in the grana stocks and stromal lamellae. The most intensive processes of photosynthesis take place in the thylakoid membrane of granas. High-efficiency light-collecting "shares" of light quanta is concentrated within the internal membrane of thylakoids (Tikhonov, 1996, 2012). These “light shares” represent the photosynthetic reaction center with 200-400 molecules of different pigments, mainly chlorophyll *a* and chlorophyll *b*, and partly carotenes and xanthophylls.

It is established that the thylakoid membrane during light phase is a bimolecular lipid layer which consists of different non-polar glycolipid molecules, mainly, galactolipids and phospholipids. Cyclic glycolipids are non-polar molecules and consist of a long hydrophobic tail and a non-polar head (Figure 75a). Molecules of pigments and proteins are neutral molecules and are embedded within this bimolecular layer. Galactolipids are concentrated on the external side of the thylakoid membrane. As a rule, the porphyrin nuclei of chlorophyll are localized between contacting membranes of grana thylakoids.

Now consider the biophysical model of gas exchange in the chloroplasts. Chlorophyll absorbs solar energy quantum and the excited electrons leave its molecule. Pigment in this moment becomes a short-term, positively charged molecule. In response to the appearance of a charged particle, the instantaneous dipoles with a continuously changing direction of poles, arise synchronously in all molecules of the lipid layer. Naturally, the placement of poles of neighboring molecules will be regular (Figure 75b). When these molecules are attracted, instantaneous dipoles are mutually oriented and their direction begins to change with the same phase. The appearance of intermolecular dispersion interaction causes the drawing closer of molecules of the cyclic glycolipids within the thylakoid system. Permeability of the bimolecular lipid layer of chloroplast becomes lower. Therefore, the intensity of light illumination, frequency of chlorophyll excitation, frequency of instantaneous dipoles origination, and intensity of photosynthesis should be interrelated and interdependent processes. It is now shown that the daily course of photosynthesis in the optimal weather conditions is approximately equivalent to the intensity of solar insolation.

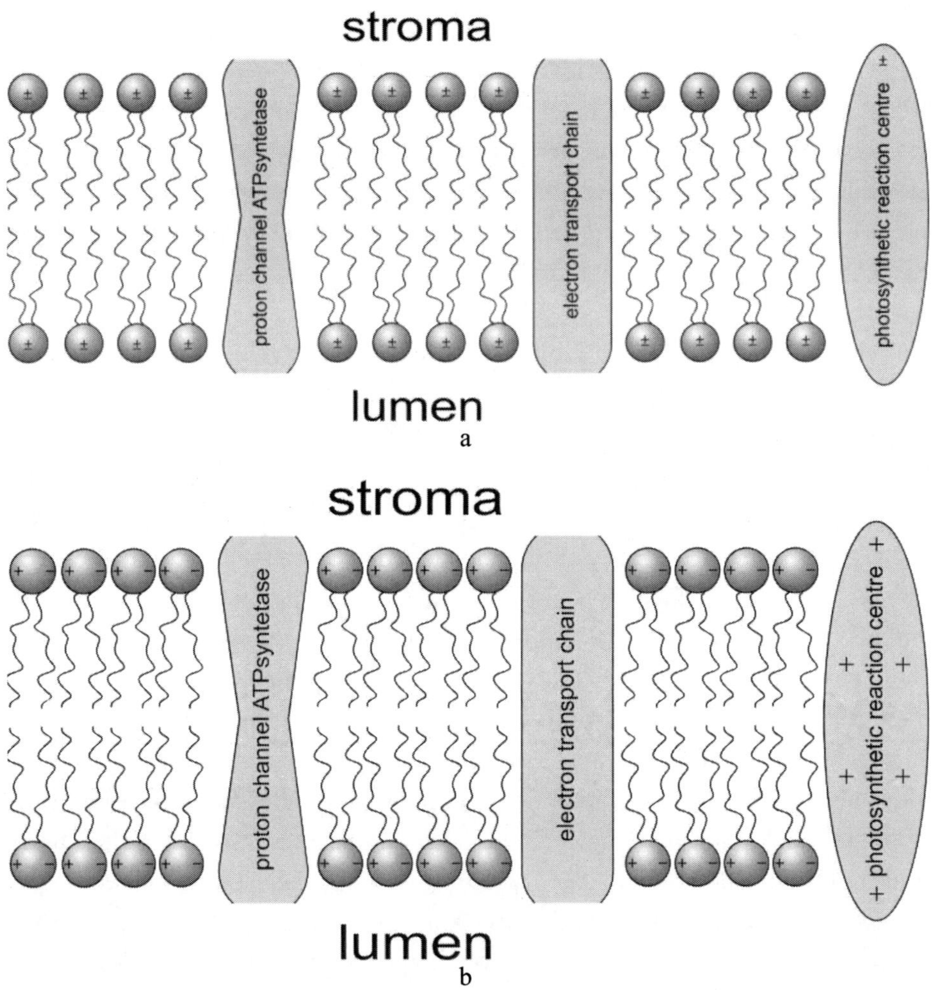

Figure 75. Structure of thylakoid membrane during light (a) and dark (b) phase.

The absorbed photon gives the chlorophyll certain energy: Chl + hυ → Cl*. The period of its excited state may reach 10^{-8} sec. Electrons that have absorbed the quantum energy passes from its main to an excited state; as a result, the pigment chlorophyll loses one electron. The electron itself has a magnetic moment and is the paramagnetic particle. Therefore, the exited electrons tend to go over to the space with a stronger field; i.e., into the stroma. They move inside thylakoid membrane towards the stroma by the certain channel because in this time, its membrane is impermeable. In photosystem I, their movement occurs along the electron acceptors of the electron transport chain, located on the proteins of photosynthetic membranes, to which they transfer some of its energy. The transfers of electrons requires that each carrier is restored and oxidized. The attachment of electrons to the carrier molecule occurs during reduction and the loss of the electron by the molecule takes place during oxidation. The energy that an excited electron emits during electron transport is captured and used for the restoration of the co-enzyme NADP in $NADP^+$ and the synthesis of ATP from ADP. Thus, the solar energy is converted into chemical energy and accumulates in the form of the bonds of two energy-rich compounds - ATP and NADPH.

In photosystem II, solar energy is used to split water: $2H_2O \rightarrow 4e^- + 4H^+ + O_2$. This is another light-dependent process that proceeds into granas of the thylakoid membrane (Hill's reaction). Initially, the surplus energy of electrons is transferred to the hydrogen ions which are present in the aqueous solution due to a dissociation of water molecules. The solar energy trapping occurs along the electron transfer chain through a series of intermediate states to the final electron acceptor in the photosystem I – a molecule $NADP^+$, in order to reduce $NADP^+$ to NADPH. In regards to the positively charged chlorophyll, it is neutralized by unexcited electron from water and pigment recovers its initial neutral state. Since a large number of chlorophyll molecules are simultaneously excited and neutralized in the daylight, the thylakoid membrane of chloroplast remain impenetrable all this time.

Because of the hydrolysis and ionization of water molecules, three paramagnetic particles are formed: Hydrogen atoms (protons), electrons, and molecular oxygen. According to the physics laws, they must move inside plastids in the direction of a stronger magnetic field. However, they cannot enter into the stromal space through the bimolecular layer of thylakoid membrane since it is impenetrable at this time. For this, protons and electrons in the membrane have two special channels. Electrons use a chain of electronic protein-carriers through which lumenal and stromal spaces, as well as a photosystem I and a photosystem II, are connected. The transition of protons, which are formed inside the thylacoid during water hydrolysis, occurs through the ATP-synthetase complex. It has been established that the synthesis of ATP is directly related to the transfer of protons through the enzyme ATP-synthetase. Molecular oxygen does not have such a special channel. Therefore, O_2 molecules are located initially closed from the stroma in luminal spaces. After accumulation in granas, the oxygen molecules pass to the stromal lamellae; then from lamellae to the cytoplasm, the intercellular spaces, and the atmosphere by the concentration gradient.

In the twilight period, there is a smooth transition of light from day to night. If we proceed from the corpuscular-wave nature of the photosynthetic process, there must be one more twilight intermediate phase between the light-dependent and light-independent phases. In this time, the cardinal reconstruction of plastid takes place on another principle of function without light. Indeed, all chlorophyll molecules, in the absence of light, return to the initial unexcited state and are neutral. During twilight, the intensity of solar illumination decreases, the frequency of chlorophyll excitation and the instantaneous dipoles origination become less frequent, and then disappear. Without the inducing factor (positively charged excited chlorophyll molecules), the dispersion interactions between cyclic molecules of glycolipids in the thylakoid membrane are decreased. The non-equilibrium distribution of charges and intermolecular attraction between glycolipids completely disappears and the bimolecular lipid layer of a thylakoid membrane becomes semi-permeable.

Harmonic oscillators of chloroplast is in working order when the thylakoid membrane becomes the semi-permeable. Therefore, the processes of active trans-membrane transport of different ions and molecules through the chloroplast membrane against concentration gradient occurs; the forces of electromagnetic field created by the chlDNA-wave takes part as well. Most ions and biomolecules, including the proteins (ribosomal, electron transport chain, key enzymes for CO_2 fixation, etc.), penetrate inside the plastid from the cytoplasm. Depending on the molecule size and its chemical constitution, the first of them are built into the outer membrane of the plastid, the second, into the outer or internal thylakoid membranes, the third are sent to the stroma. Conversely, the chloroplast proteins synthesized into the stroma or remain inside the chloroplast and participate in the light-independent reactions, or enter into the internal thylakoid membranes, where the photosynthetic light-dependent reactions are carried out. In addition, fatty acids, amino acids, sugars, and polysaccharides (starch) are formed here. Galactolipids from the stroma are embedded in the membranes, where the increase of sizes and further differentiation of the thylakoid system of chloroplast continues. Polysaccharides pass into the cytoplasm, where the synthesis of sucrose is completed. The intensity of the active transfer of particles (ions, molecules) from the cytoplasm to the plastid exceeds their centrifugal stream. In chloroplast, there are 80% Fe, 75% Mg, 70% Zn, 65% Ca, and 50% Cu in a leaf. The increase of molecule concentration leads to a rise of osmotic pressure. Chloroplasts begin to grow, while the differentiation of thylakoid system continues. Further increase of osmotic pressure leads to the cease of work, but not to the destruction of its harmonic oscillator. The physical situation inside chloroplast changes significantly.

The absence of a wave function on the chlDNA leads to the disappearance of the electromagnetic field or the driving force that ensured the movement of the particles against the concentration gradient during intermediate (twilight) phase. The active transport of ions and molecules through a semi-permeable membrane ceases and the light-independent phase of photosynthesis begins. Its distinctive feature is that the movement of molecules between cytoplasm and plastid is carried out by means of passive transfer by the concentration gradient. Molecules that easily pass through a semi-permeable membrane by passive transfer are characterized by small sizes. In the beginning and during light-independent phase, the CO_2 begin to enter from the cytoplasm into the stroma of chloroplast. Their molecules can't enter inside the stromatal space when a harmonic oscillator of chloroplast is in the functional state because of the diamagnetics and due to the impermeability of thylakoid membrane (light-dependent and twilight phase of photosynthesis). The main part of the related molecules of CO_2, which were formed inside the stroma during previous light-dependent phase, has been exhausted by this time. Proceeding from this, we assume that the entrance of CO_2 can be one of the inductors of light-independent reactions.

The cycle of light-dependent biochemical reactions, which consists of many stages and ends by the synthesis of energy-rich organic compounds such as glucose from CO_2 and water, is called the Calvin-Benson cycle. It includes three phases: 1) carboxylation, 2) recovery, and 3) regeneration. The carboxylation phase begins with the reaction of CO_2 fixation by the "acceptor" molecules by means of attaching to the ribulose bisphosphate (RuBP), which play an important role in the process of carbon fixation. The enzymes that participate in this reaction are contained in the water-soluble fractions of stroma. The key enzyme involved in the first main step of is carboxylation is RuBisCO and the reaction catalyzed by it is the slowest in the cycle of carbon fixation. In the stroma the concentration of CO_2, when its molecules are included in the cycle of Calvin-Benson, is constantly decreasing and at the same time, their quantity is replenished after CO_2 entrance in the chloroplast by the concentration gradient by means of passive transport: Initially from the cytoplasm into the plastids, then from the leaf intercellular spaces and the atmosphere into the cells. The reduction phase leads to the formation of a short-lived six-carbon compound, which then splits into two molecules of glyceraldehyde-3-phosphate. This process proceeds with the participation of energy of the ATP and NADPH, which were synthesized during the light phase. Then a glyceraldehyd-3-phosphate enters in the cytoplasm where it is used for further synthesis of monosaccharides and is transported to other cells. Part of its molecules, after passing through a series of reconstructions, turns again into RuBP, the primary CO_2 acceptor. These regeneration processes ensure the continuous operation of cycle. Starch in chloroplasts is synthesized from the remaining molecules of glyceraldehyde-3-phosphate.

During the twilight phase, the active transport of substances between the chloroplast and the cytoplasm occurs: The oxygen into plastids enters, just like the paramagnetic molecules, and the carbon dioxide from them is removed as the diamagnetic molecules. Therefore, the distinctive feature of the beginning of the light-independent phase is the complete absence of CO_2 and a large amount of O_2 in the stromal space. Superfluous oxygen is removed by means of simple diffusion and oxygenase activity of the RuBisCO enzyme. The products of this reaction are involved in the processes of photorespiration - a phenomenon that is observed in the first minutes after the displacement of plant from light to darkness. A phenomenon of photorespiration is found in many plants, including forest species.

At night, the stomata of leaves open so the contact of their cells with the external environment has the maximal potential. The content of CO_2 in the air is 0.03-0.3%. Inside the chloroplast, it is absent. Its molecules, after stopping of harmonic oscillator, begin to penetrate in the chloroplast by the simple diffusion. In stroma, all molecules of carbon dioxide are converted into a connected state. Then CO_2 undergoes the reactions of Calvin-Benson cycle and the part of fixed molecules begins to accumulate in the stroma. The content of the RuBisCO enzyme, which participates in the carboxylation reactions, is half of plastid protein. During night a rather large reserve of fixed CO_2 is created. Since there are no unfixed, its molecules in the stroma are a concentration gradient which is constantly maintained between the cytoplasm and the chloroplast; the night ensures the continuous movement of CO_2 molecules through the semi-permeable membrane. Carbon dioxide enters in the chloroplast by a concentration gradient from the air, roots, and mitochondria. The processes of gas exchange of the chloroplasts and mitochondria are interrelated, and organelles are often located nearby. During the day, the chloroplasts provide the mitochondria O_2, and during night, the mitochondria are a source of CO_2 for the chloroplasts. This goes to show that the plants must release very little carbon dioxide at night.

In the morning, the pigments of chlorophyll begin to absorb the light quanta and go over into an excited state, initiating the light-dependent reactions in photosystem I and photosystem II. Accordingly, positively charged molecules and instantaneous dipoles begin to appear in the thylakoid membrane. This leads to the attraction of cyclic lipids and the membrane becomes impenetrable for molecules and ions. The volume of membrane at the moment of its transition from semi-permeable to non-permeable state is somewhat compressed. The chloroplast oscillator comes into action from this push. Carbon dioxide, as a diamagnetic molecule, cannot longer get into the chloroplast and go out of it. During day, the light-independent reactions continue to work due to the accumulated carbon dioxide and ions, as well as the ATP and NADPH, which begin to enter into the stroma from light- dependent reactions, and a fixed carbon is consumed gradually. All CO_2 which is produced by the mitochondria during respiration is released into the atmosphere. Therefore, in the day time its volume increases sharply in comparison with the night.

To summarize the above, it should be noted that the importance of photosynthesis for the survival and evolution and life of plants on Earth is not possible to estimate. Photosynthesis represents the highly effective biochemical and biophysical nanotechnology, which is initiated and realized by means of solar energy in the micro-volume of chloroplast. Its final product is oxygen which is necessary for all higher forms of life, and the high-energy organic compounds (sugars), which provide a trophic function of plants, animals, and many microorganisms. In this chapter, we tried to describe the mechanism of gas exchange as a physical process which is carried out on the "reverse" principle: The oxygen, as a paramagnetics, is removed from chloroplasts and the carbon dioxide, as a diamagnetics, is fixed inside a stroma and stored in the form of different organic substances. It must be recognized that there are still many "dark spots." In plants, we are only beginning to understand the fundamental knowledge of many of mechanisms of their gas exchange.

SUPPORTING AND CONDUCTING FUNCTIONS

The main functional systems of animals are characterized by a higher level of specialization and differentiation. Their supporting-motor apparatus includes a complex of bones, cartilage, joints, ligaments, and muscles. In plants: A root, stem, and leaf belong to several vegetative organs that carry out the chief metabolic and energetic processes. They are multifunctional systems responsible for nutrient absorption and their reprocessing, transpiration, gas exchange, fluid circulation, and other important functions. Apparently, the system's multifunctionality is due to the fixed lifestyle. This kind of lifestyle is the energetically disadvantageous for the survival of any organism. The combination of different functions decreases energy losses when the independent cyclic processes are realized in the closed space of plant and defined as mutually conditioned and mutually related events. Biosynthesis of "nutrients" (photosynthesis) and their reprocessing (respiration, the accumulation and dissimilation of energetically rich organic substances) occur simultaneously within one organism.

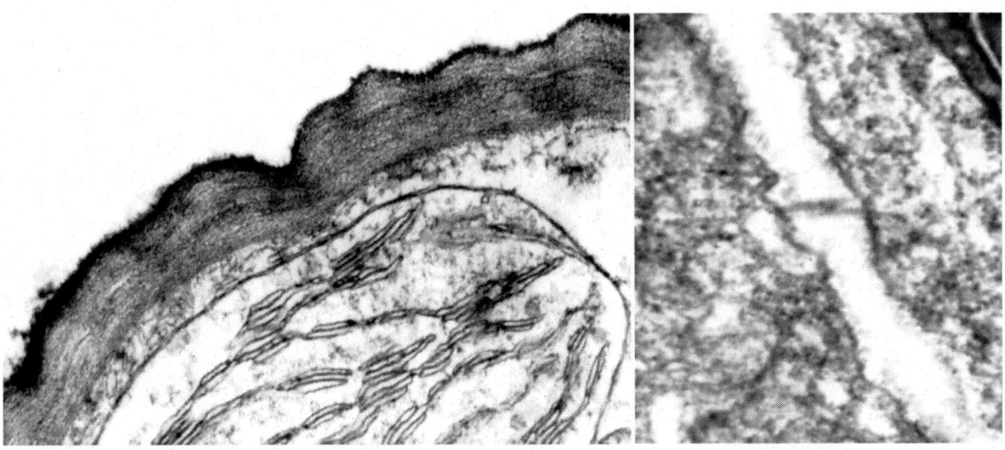

a b

Figure 76. Primary cell envelope (a) and pore (b) in the cells of young leaves.

In forest trees, the supporting function is performed by the different tissues and cells (a living and a dead), which take mechanical properties during stem development. The stem provides a link between the root and crown of tree. During ontogenesis, its crown reaches a considerable volume and weight. The main load falls on the trunk, or on the ground axial organ, which carries the system of lateral branches. The trunk must have a necessary safety factor, sufficient volume to withstand the mechanical overloads and must ensure the vertical position of tree in strong winds and snowfall. Moreover, the necessary durability and stability are given to the plant cells, which surround the outside by a thick envelope.

Most plant cells have a dense envelope. It gives them strong durability, regulates water flow, and maintains turgor pressure. The primary cell envelope is formed after division and initially has the property of extensibility. It contains cellulose, hemicellulose, and pectin and consists of the elongated microfibrils, which give it the look of a multilayer structure (Figure 76-a). Neighboring cells are connected through pores (Figure 76-b). Some cells (cortex, tracheids, etc.) have a secondary envelope which is formed from the inside. This envelope often contains a lignin which gives more rigidity and mechanical stability. They preserve their structural integrity and perform the supporting functions even after cell death.

A tree trunk begins its development as an axial shoot. Then apical and lateral buds on the shoots are laid, which ensures the plant's growth and branching. In forest tree plants, trunk thickening occurs annually after a shoot elongation. Its growth in the thickness provides the tissue of lateral meristem – the cambium. The cambium represents one layer of meristematic cells which is located between the tissue of xylem and the tissue of phloem in the stems and roots of the tree. The structure and peculiarities of its development in forest tree species of angiosperms and gymnosperms are given in the many works (Ezau, 1980; Evert, 2006; Rathgeber, 2011; Fromm, 2013; Koch & Schmitt, 2013; etc.). Procambium originates from a hypocotyl, the tissue of which passes through all hypocotyl-root axis of differentiated embryo.

Procambium cells are characterized a narrow-elongated shape and represent a continuous cylinder of hypocotyl tissue between the stem apical meristem and root apical meristem (Figure 62). After the seeds exit from dormancy, the meristematic cells of each horizontal layer of hypocotyl become the cambial initials and begin to divide mainly by asymmetric mitosis and sometimes by usual mitosis. At the sprout stage, the axial system formation of future plants occurs. In a result of active proliferation and longitudinal growth of cells, the sprout grows rapidly in length.

The sprout is longer than the embryo ≈ 12-15 times. The primary stem thickness compared to the primary root change slightly ≈ 2 times (Figure 67c). Totypotent tissue of embryonic hypocotyl, after asymmetric mitosis, becomes the heterogeneous ones along the longitudinal axis of plant. Its internal structure is undergoing significant changes. As it has been shown, after seed germination and up to beginning of embryonic bud development, many identical segments or the metamers inside primary stem and primary root, in a result of sprout polarity, are formed. The number of metamers in the sprout is equal to the number of horizontal layers in the hypocotyl tissue of embryo. Each metamer is the structural basis for the subsequent thickening and growth of shoot and root. Meristematic cells inside the primary stem are distributed in a certain manner: In the horizontal plane they are identical by the DNA-energy; in the vertical plane, their energetic level decreases in the gravity direction – from a top of metamer to its base. After usual mitosis, the DNA-energy of procambium cells is decreased on one level.

The embryonic buds and first escape with juvenile needles develop in summer. In this time, the length of a seedling and the thickness of its stem increase size by 3 times or more. Stem thickening and growth of a 1-year seedling occurs on the base of the horizontal and vertical growth of metamers. The inner stem structure is formed on the basis of metamers as well.

Cambial initials are the upper layer of a metamer, a cell that has the highest level of DNA-energy. During escape growth, these initials undergo the certain number of asymmetric mitosis and one division by usual mitosis. In the 1st vegetative season, the cambial initials are divided twice by usual mitosis. After asymmetric mitosis, one daughter cell remains a cambial initial and the other energetically weak cell develops along the path of differentiation and specialization. Then the cambial initial grows, reaches the certain size, and undergoes the next asymmetric mitosis. Consequently, the cambium is maintained as the meristematic tissue throughout the long life of a trunk. Energetically weak cell become the cambial initial of the 2nd order. Its first division occurs by usual mitosis. Derivatives of these divisions alternately depart either towards the outer or the inner part of stem; i.e., in the direction of future phloem or the future xylem. Then cell divisions occur in the horizontal and vertical layers of each metamer. The stem thickness of a 1-year seedling increases. The outer layer represents the radial cylinder of meristematic cells and the inner layer of the core tissue continues.

Cambial initials and their derivatives form a wide zone of meristematic cells inside the stem of a 1-year seedling – its core (Figure 77a, b). The core is made up of parenchyma tissues. It consists of the living undifferentiated cells, which initially are capable of growth and division. It displaces the cambium cells from the center to the periphery of the stem. Cambial tissue takes the form of a hollow cylinder, separating the outer and inner parts of the stem (Figure 77b). The wood parenchyma of a core is the place of deposition of the reserve nutrients of seedling. At the end of 1st vegetative season, one apical bud is laid on the top of shoot. The metameric organization of the future shoot is formed in the intrarenal period of its development.

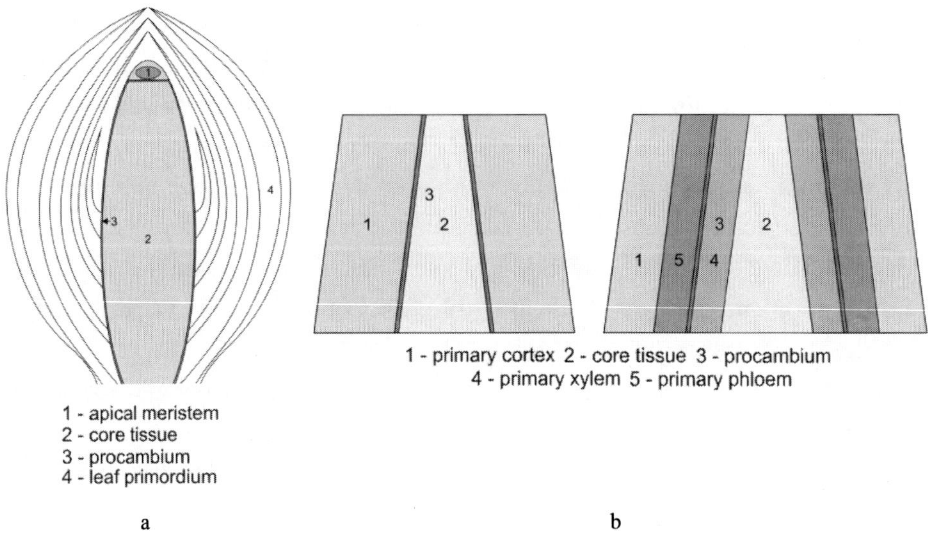

Figure 77. Scheme of primary stem of 1-year seedling (a); primary stem (b) and apical bud (c) of 2-year seedling.

There are two types of cambium cells. Most of the cells are spindle-shaped initials, having an elongated form with oblique pointed tips. They are extended along the axis of the trunk and their length is much greater than width. In coniferous, spindle-shaped cells reach a length of 2.5-3.5 mm, and in dicotyledons, they are shorter ≈ 1 mm (Raskatov, 1974). The cells of the 2nd type are the radial initials. They have thin cell walls with a slightly elongated or isodiametric form. These cells differ in the transverse section. Spindle-shaped initials and their derivatives make up the axial system of a stem and give rise to elements of bast and wood (vessels, tracheids, sieve tubes). Radial initials form its transversal system are the pith rays.

In the 2nd vegetative season, an escape with the typical pine needles begin to develop. Inside a terminal bud, the cambial tissue is laid in the form of a hollow cylinder in shoot apex. This zone is located below the level of leaves. Meristematic tissue above this level is not differentiated. Apparently, the scheme of metamer development is the same as in the 1-year plants. Namely, the cells of the lower layer of stem apical meristem, with a certain periodicity, are divided by the asymmetric mitosis. Their derivatives form a new metamer in the central part of which is disposed as a continuous layer of meristematic cells of the columnar shape. The upper cells of this metamer are the cambial initials. In comparison with the embryonic meristem, their DNA-energy is lowered by two energetic levels. Each cambial initial undergoes asymmetric mitosis, after which one daughter cell becomes an initial and the second energetically weak cell passes to the form-building process. Its derivatives alternately give rise to the cells of the primary cortex and primary core. Parenchymal cells of a core displace the cambial tissue from the central part of stem and it takes the cylindrical form. Simultaneously, the leaf primordia in the peripheral zone of the metamer are laid.

After some time, the lower cells of apical meristem undergo an asymmetric mitosis, during which, the next generation of cambial initials is formed. They initiate the development of the next metamer which occurs in the same sequence as the previous one. The divisions of apical meristem continue until the autumn, giving rise to new metamers. Overlapping each other, they then begin to form a rudimental escape inside terminal bud (Figure 77a). These intrarenal metamers differ from each other only by the inception time and by the stage of internal development. Their internal structure is analogous to the organization of metamers of a 1-year seedling by the following traits: 1) Cambial initials of a rudimental escape and shoots with juvenile needles within one seedling have the same level of DNA-energy. 2) Cambial tissue has the shape of a hollow cylinder and is in the center of metamer. Inside the cambial layer is the parenchyma tissue of the core, designed to perform different functions. At the juvenile stage of ontogenesis, the core implements the trophic, transport, mechanical, and other functions. The core tissue in the bud has the shape of a cone; the base of which is connected with the core of current year's escape. 3) The placement of the cytological zones of meristematic cells within intrarenal metamer, which are the identical by the DNA-energy and DNA-information, is similar to the location of the histological zones of tissue and organs within an escape metamer.

In 2-year seedlings, a process of the specialization and further differentiation of cells, tissues, and organs (branches, leaves) begins. Cambial initials are divided by asymmetric mitosis. Half of the daughter cells after asymmetric mitosis remain the same the cambial initials. The thickening of stem and root is connected with periclinal divisions of energetic weak cambial cells and the process of formation of primary xylem and primary phloem. Derivatives of these divisions alternately depart either towards the phloem or xylem. The derived cells of the outer layer are differentiated in the elements of secondary phloem (bast) and the derived cells of inner layer are developed in the elements of secondary xylem (wood) (Figure 77c). Horizontal growth of a stem is associated with an increase of the number of cambial initials. For these purposes, each cell of initial undergoes usual mitosis and energetic potential of cambial tissue is reduced annually by a level. At the end of the 2nd vegetative season, one apical and one-three lateral buds are laid on the top of shoot.

Cambial initiation of 3-year plants gives rise to a secondary xylem and a secondary phloem within each metamer along trunk length. However, strict alternation in the formation of their cells does not occur (Raskatov, 1974). After this age, a cambium lays aside more elements of the xylem than the phloem. Thanks to the annual resumption of cambium function, the share of wood accounts for 9/10 of the total trunk volume of an adult tree.

Xylem and phloem belong to complex tissues, which consist of conductor, mechanical, and storage elements. Cells of xylem live for a short time. The wood parenchyma and the pith rays also serve as storage tissue for a short time. All of these elements undergo certain structural and functional changes, die, and remain inside the tree body until the end of its life. Aging in the cells of xylem occur quickly because the DNA-energy of their cells is reduced after the each mitosis. The process of differentiation leads to the vacuolization of cells, as well as the thickening and lignification of the secondary envelopes. The secondary envelope covers the primary envelope as a spiral. Due to this, xylem cells continue to grow for some time. Cells reaching an energetic level of zero, die. After death, the shape and strength of their cellular wall has not change and they continue to carry water supply and support functions.

The circulation of liquid (water, carbon, oxygen and ash elements) is carried out by xylem tissue. It consists of dead cells. Tracheids and vessels of xylem are connected to each other by means of pores, which serve to transport the water solution. They keep the ability of water conduct only several years. As a rule, the loss of water conductivity takes place due to the clogging of pores. Lifting of liquid to a high altitude is mainly ensured by the capillary effect in narrow tracheid vessels, and by the high intracellular pressure that gets up to certain level in the meristematic cells of root with the participation of the bio-field created by the biological harmonic oscillator.

Cells of phloem carry out a descending stream of organic substances made during photosynthesis (sugars). This tissue consists of the sieve elements, parenchyma cells, specialized companion cells, and different unspecialized cells. All sieve cells have areas of pores at their ends. The highest concentration of sucrose is reached in the place of its formation - the leaves. Transport of soluble organic compounds represent a mechanical process carried out through the pores under the influence of gravitational forces and osmotic pressure.

Phloem cells function only when they are alive. They quickly become vacuolized cells, the cytoplasm takes a parietal position, freeing the way for organic substances. Nuclei are present only in the companion cells. These cytological signs show that the vital state and principle of the development of phloem cells correspond to the state of equilibrium cell. Equilibrium is a cell which has reached a zero energetic level (Kuznetsova, 2013). It does not exist below this quantum state. The cytoplasm and energetic system of sieve-tube cells, which previously provided their viability, begins to collapse. First of all, the nucleus envelope and the wave function on the DNA are destroyed. For some time, the sieve-tube cells function as non-nuclear systems. Submicroscopic structures of the parietal layer of their cytoplasm is provided by the energy of centripetal and centrifugal streams. This layer promotes the sliding of molecules and facilitates the movement of carbohydrates through the sieve tubes. After the protoplast is destroyed, the cell quickly loses the ability to conduct solutions and is replaced by the next-generation of sieve-tube cells.

Excretory System

Secretion or excretion regards to the basic functional systems. The secretion is usually understood as the process of removing various metabolic waste, toxic substances and dead cells from organisms. Animals and plants in the environment remove excess water, end products of metabolism, salts and toxic substances, entering into the organism or formed within it. The necessity of excretion is due to the fact that numerous biochemical reactions occur in the cells and are associated with the processes of transformation and distribution of energy. In the course of these reactions, different by-products are formed which do not participate in further metabolism. Therefore, they must be removed from the cells and organism. The toxic substances entering from outside or formed within organism can lead to cell and animal death. The majority of them are neutralized or isolated by means of special cells and cell systems and then turned out from organism in the external environment.

Carbon dioxide, water, and ammonia are the main end products of animal metabolism. For the removal of the organism to the external environment, animals have a special excretory system including specialized intracellular organelles, cells, tissues, and organs, which in the course of evolution were improved and became more complex. Animals carry on an active lifestyle by constantly moving in search of food within their territory of habitat; therefore, the metabolic wastes are distributed in different places of a species range.

Once more product the excretion of plants and animals is the dead cells. It is known that throughout the lives of individuals, a large number of cells are created and perish, which must be constantly excreted from the body. In animals, such organs and functional systems as the cardiovascular system, lymphatic systems, kidneys, liver, etc., take part in the process of dead cell removal. With aging, the number of dead cells increases, and leads to their accumulation and imbalance. However, in animals, their share of dead cells is not as great as in plants.

From the aforesaid, it becomes obvious that the excretory action of immovable plants must have a number of significant differences in organization and functioning (Roschina & Roschina, 1993, Roschina, 2014). In forest tree species, its specificity is caused by longstanding ontogenesis and a fixed lifestyle. The micro-niche, which occupies the tree, is quite small in comparison to its volume. Components of excretions are also the products of primary and secondary metabolism, toxic substances, and dead cells. If by-products and toxic substances are released and accumulated around the tree, they can cause it to die. Dead cells are no less important problem of plants, the number of which reaches up to 80-90% of the volume of the adult tree. Therefore, they must have the special channels for their isolation inside the plant body or for excretion in the external environment. As the analysis has shown, the plant excretory system is very effective and fully complies with plant biology. It is also multifunctional. Moreover, its spatial organization determines the histological organization of tissues within the tree body and is connected with the action of other functional systems (supporting, conducting, gas exchange, physiological homeostasis, etc.).

At the cellular level, the vacuoles are the site of concentration for the end products of metabolism. They are a part of the vacuolar system, which includes endoplasmic reticulum, Golgi apparatus, lysosomes, etc. The vacuolization degree of differentiated cells can reach 70-90% of cell volume. For comparison, vacuoles in the animal cells occupy only 5%, which is due to the movable lifestyle and high level of specialization of their excretory system.

Vacuoles are multifunctional organelles. Even Höfler (1945) showed that the vacuole in the cell works like an osmometer. Its volume in the hypertonic solution varies inversely with concentration. The excess water entering in the cell by the concentration gradient is continuously removed from the cell with the participation of biofield forces. Due to this, the osmotic pressure inside the cell is constantly maintained at a higher level than the pressure of outer medium (Troshin, 1985). Vacuoles, after isolation, take a spherical shape, which illustrates the influence of the biofield on their shape in the living cells. The distribution of toxic substances and secondary metabolites into simple components with the participation of special enzymes occurs inside the cavity of vacuoles. Then most of them are removed from the cell.

Vacuoles from the cytoplasm are separated by means of a single membrane – the tonoplast, which is characterized by a selective permeability. The cavity of vacuole is filled with a hypertonic aqueous solution containing various inorganic salts, sugars, organic acids, and organic salts. Therefore, the water molecules move by the concentration gradient to the vacuole and support the turgor in the cells. The inorganic substances, such as the phosphates of potassium, sodium and calcium are contained. Organic acids (oxalates, citrates) are often found here. Therefore, a vacuolar juice has an acid reaction (pH - from 2 to 5). Excretory products (tannin, anthocyanin pigments, alkaloids, etc.) and reserve compounds (sugars, proteins) accumulate in vacuoles too. They are deposited in the vacuoles because the ability to remove these substances outward from plant body is limited.

Vacuoles can also perform the autophagic function. Hydrolytic enzymes are found in the small and large central vacuoles. Their tonoplast forms the invaginations and a part of cytoplasm walks into their cavity. Mitochondria, plastids, elements of endoplasmic reticulum, ribosomes, and lysosomes within such an autolysosome may be contained. After lyses of content, the membrane of autolysosome disappears and the degree of vacuolization of cells increases. Autolysosomes are often found in the cells of secretoty type (Figure 78) and in the suspensor cells of 16-cellular proembryo (Figure 55-c).

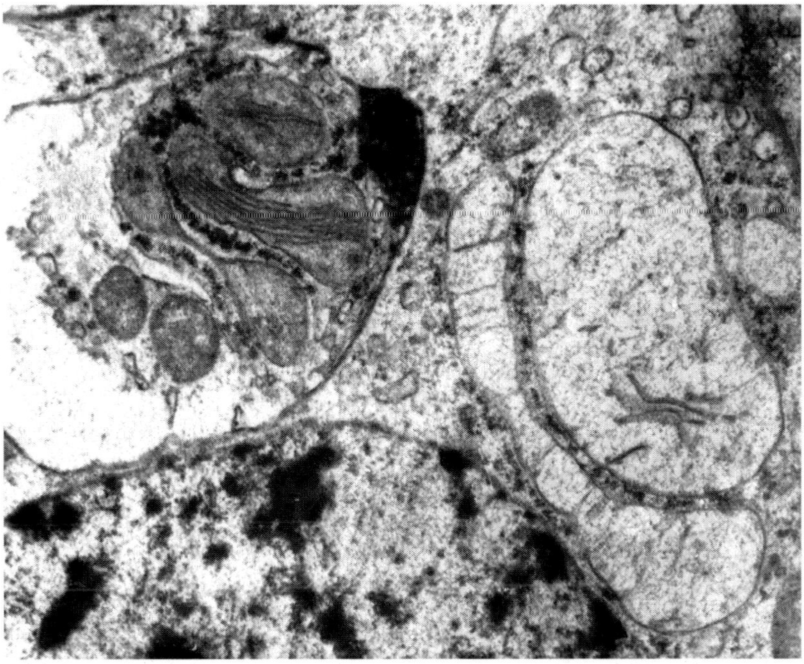

Figure 78. Autelysosomes of jacket cells of *Pinus sylvestris*.

The vascular system unites all cells of plant into one whole. A supply by water and different ions, which are necessary for photosynthesis, respiration, and other intracellular reactions, are provided from the soil substrate through the root cells; they are a continuously circulating internal medium by the ascending vector of xylem inside vacuoles and cytoplasm of cells. Carbohydrates, formed during photosynthesis, are transported from the chloroplasts into the cytoplasm and vacuoles, then by the descending vector with the phloem juice to the places of their utilization as the substrate for respiration or the substance for storage.

At the organism level, the by-products of metabolism are removed or localized in special storage tissues by various ways: With the withering away of roots and branches, the loss of leaves, the separation of the outer layers of cortex, through stomata, lenticels, resin ducts, lactation, airway system, etc. In multicellular plants, they accumulate in the tissues of the cortex, sometimes in the wood. The processes of secretion of the essential oils, carbohydrates, solutions during guttation, nectar, resins, phenols, alkaloids, proteins, etc. also belong to the excretory activity of plants. In conjunction, all these devices, together with the vacuolar system, represent the excretory system of a plant.

Plants are the main source of nutrients for many animals. This is due to the fact that the plants, with the participation of solar energy during photosynthesis, are able to synthesize the organic substances with a high reserve of internal energy from the inorganic substances. Their metabolism is carried out by the principle of a semi-closed cycle, which is a distinctive characteristic of the plant organism. The carbohydrates and other substances created during photosynthesis are a source of free energy and carbon skeletons for the biosynthesis of many organic substances in the cell. The main end product of this reaction is carbohydrates and by-product is oxygen. After complex processes of transformations and splitting, part of glucose molecules are used for the energetic needs of a plant. Starch, lipids, and proteins are stored in the form of energy-rich reserve substances. The volume of by-products formed during primary and secondary metabolism in the plants is significantly lower in comparison with animals. In addition, the excretory system main function is not the removal of nitrogen compounds from the plant body. Most of them, after re-utilization, take part again in the processes of intracellular metabolism.

The excretory system functions directly correlate with the circulation of internal medium: With the processes of its creation, the movement of xylem juice from the roots to the leaves, and the phloem juice from the leaves to the roots. It then delivers water and its nutrients dissolve in the cells. The biofield has the important role in the processes of active absorption of the soil solution by the root cells.

Forest trees absorb water mainly from the soil. Water in the water-permeable soil layer can be in the different conditions. Plants can use only the gravitational and capillary water. Gravitational water can move under the influence of gravity in a downward direction. After rain, it quickly evaporates and moves to the lower horizons of soils. Therefore, the water regime of plants between the rains depends on the capillary water and, accordingly, the water-holding capacity of the soil. This water rises through capillaries of soil particles from groundwater in different directions upwards. It is well absorbed by the roots and can provide the plant by water during a long, dry period.

The root grows and constantly shifts relatively to soil particles. The suction zone, with root rhizoids, is located behind the zones of division and elongation. The absorption of a water solution from the soil takes place by means of numerous rhizoids. Biofields and semi-permeable membranes take part in this process. Molecules of water are the molecular dipoles; ions are characterized by the charge. Therefore, the biofield induces their centripetal movement towards the root cells. The water molecules, under the action of biofield forces, begin to enter in them against the concentration gradient. The ions also come into motion. All this leads to increased intracellular pressure in the root cells. The number of molecules increases during their biosynthesis inside cells. The semi-permeable plasmatic membrane alters the molecular and ionic composition of a soil solution. The speed of their passage through a membrane depends on the size and charge of ions: K^+ - more easily, Na^+ - weaker, Fe^{3+} - practically do not pass. Large molecules and soil particles do not penetrate through the membrane. Therefore, the proportion of impurities in the root cells is much less than in the soil solution. All this leads to increased intracellular pressure in the root cells.

The suction zone of roots are differentiated, non-dividing cells with a constantly working biological harmonic oscillator. The oscillator, such as a pump, generates and continuously maintains high root pressure in the herbaceous plants, 100-300 kPa, and in the forest trees, 500-800 kPa. High root pressure and transpiration promote the active extraction of the intracellular solution in vessels by means of one-way diffusion of molecules through the membrane and their movement from roots to leaves. During diffusion, they again pass through the "sieve" of semi-permeable membrane and a special internal medium of tree is created.

Xylem juice represents a highly diluted aqueous solution of mineral and organic substances. It is created by double "purification": The soil solution through a cellular semi-permeable membrane, removing some particles, increasing the proportion of others, adding nitrogen compounds, enzymes, carbohydrates, and others biomolecules synthesizes in cells. As a result, the internal circulating medium differs substantially from the soil solution by the molecular and ionic composition. A large part of harmful contaminants contained in the soil solution remains on the surface or inside root rhizoids and together with the by-products of cellular metabolism accumulates in them. Therefore, the life duration of rhizoids is short. Depending on the species and intensity of root growth, it can be from 3-5, but not more than 20 days. The withering away of root rhizoids can be considered as a mechanism of excretion, as the process of purification of internal medium from heavy metals, poisons, and by-products of intracellular metabolism. As the root grows, new root rhizoids form, which move to the other parts of the soil and do not make contact with the old ones. For this reason, the by-products of plant metabolism do not re-enter into the root cells. For the above, it is necessary to add that the excretion system at this stage of function includes the processes of continuous root growth, selective absorption of soil solution, purification from impurities, and normalization of the molecular composition of xylem juice.

The next stage of purification occurs during movement of xylem juice from the root to the leaves. Xylem juice represents a homeostatic medium, which circulates by ascending vector and is characterized by a dynamic equilibrium of chemical composition and physical properties. Over 95% of water absorbed by the roots evaporates in the atmosphere and only a small part is used for the intracellular metabolism and plant growth. Excessive water is removed during transpiration. The negative pressure, which is created during evaporation, and high intra-root pressure makes the "water pillar" constantly move up. The main organ of transpiration is a leaf. Evaporation of water takes place via aerenchyma tissue of mesophyll and stomata chinks into the atmosphere. The stomata of most forest plants are located on the underside of the leaf blade. Transpiration is regulated by changing the stomata chinks and directly depends on the daytime and weather conditions.

The by-products are also removed by means of xylem juice. They move with the circulating medium and are deposited in the tissues of wood, bark, leaves, and fruits, which are periodically discarded. Many excretory substances (organic acids, mineral salts, calcium ions, etc.) are removed by the same way or preserved in the dead tissues of a tree. Organic acids are preliminarily neutralized and precipitated as insoluble crystals.

Calcium ions react with the pectin and oxalic acids, forming insoluble oxalate and calcium pectate. In this form, they cannot harm the living cells of trees.

Now we will move onto the analysis of another excretory function - the removal of dead cells during ontogenesis. It is especially important for multi-annual woody plants.

During ontogenesis, cell divisions occur in a certain sequence and are characterized by a definite orientation, with respect to gravity. Somatic cells are divided by means of two ways: Usual and asymmetric mitosis. Both of these types of mitosis' are carried out in accordance with the 2nd law of thermodynamics. The asymmetric mitosis preserves the totipotency of initial cells and leads to a rapid, deep differentiation of the derived cells. Suppose that the energetic potential of initial somatic cell corresponds to the 8th energetic level: Now let's see how the DNA-energy and the ratio of living and dead cells change during 9 divisions of asymmetric mitosis (Figure 79). The first dead cell appears after the fifth division and then their number begins to increase rapidly. If we continue this analysis, then, beginning with the 11th division, the number of dead cells will exceed the number of living cells.

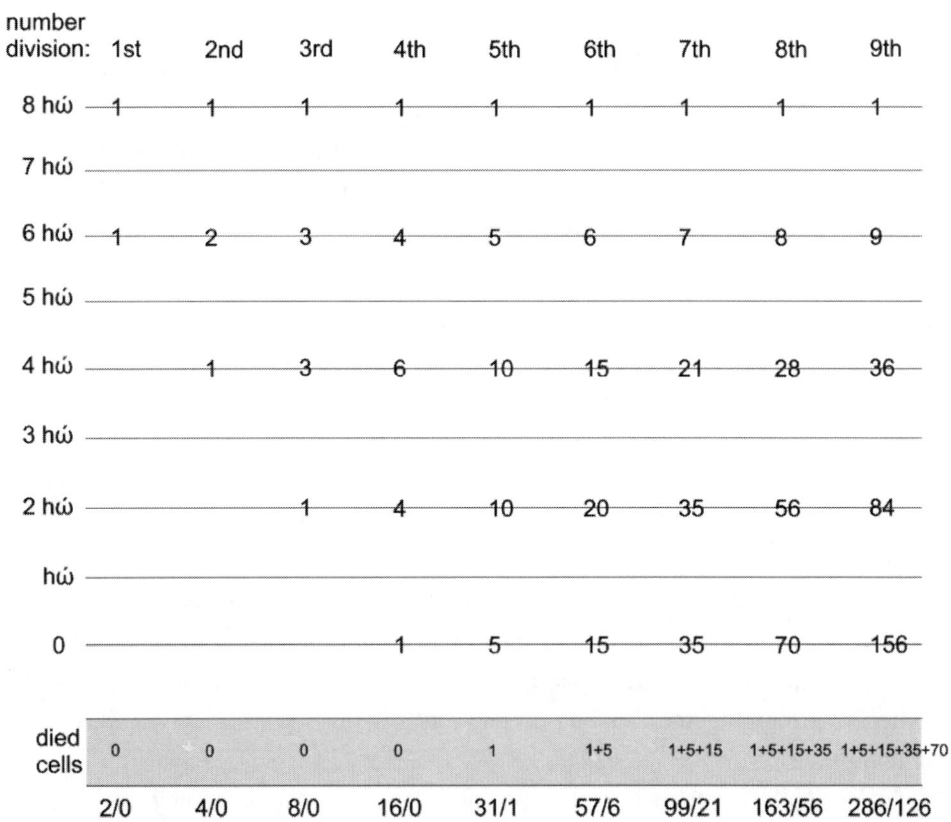

Figure 79. Scheme of 9 divisions of initial cell by the asymmetric mitosis, and the number of living and dead cells after each division.

Usual mitosis is considered non-polar division, when the mitotic spindle can move in different directions with respect to gravity, except in the transverse division. It ends by the formation of two totypotent daughter cells, which differ from the mother cell by the DNA-energy on one energetic level. If we assume that the somatic cell (8th energetic level) will be divided by the usual mitosis, then all its derived cells after 8 divisions reach zero level and after that die.

The excretory function of a tree, with the respect to dead cells, is considered separately for the root, leaf, and stem. A root system is a very dynamic organ of plants. It constantly develops, renews, dies, and moves in the soil substrate. Its growth in length occurs by the divisions of meristematic cells of root apical meristem.

The root apical meristem supports the state of meristematic tissue owed to the asymmetric mitosis. The energetically weak derivatives of this division undergo a certain number of divisions up to Hayflick limit. Then their development follows by a path of deep differentiation, specialization, and death. The most energetically weak cells of these divisions are moving to the apical end of root and form the protective layer around its growing tip. Root cap is found in all plants and has the shape of small conical caver. It consists of loosely thin-walled cells, whose DNA-energy has very low energetic level. Therefore, the cells of root cap do not live long: Only one or several days and then are removed. They are replaced by energetically weak cells of the next layer, and then the next one. The external layers of dying cells excrete mucus because their shell loses integrity. Mucus reduces friction of the root on the soil, promoting its motion and access to new nutrients. This process takes place throughout all plant life. Consequently, the root system solves the problem of energetically weak and dead cells due to their continuous motion towards the root cap; then removes them from the cap in the external environment.

Leaves belong to organs with limited growth. Having reached a species-specific shape and size, they begin to differentiate and cease to grow. The duration of their life is not great in comparison with the life of a tree. As a rule, it only lasts one or few vegetative seasons.

As mentioned above, the leaf primordium originates from the meristematic cells of a shoot apex. It is laid in the peripheral zone of shoot apex, closer to the lower layers of metamer. The level of DNA-energy of meristematic cells in this layer is less than in the upper layers of the metamer. Its development is preceded by the formation of the young non-differentiated leaf primordium which represent a complex of totipotent meristematic cells. It is established that these leaf primordia give rise to plants in the culture *in vitro.* Initial cells during formation of a leaf tubercle are divided only by the usual mitosis and their level of DNA-energy decreases. Then a period of intrarenal development of a leaf begins. In the surface layer of non-differentiated primordium, a group of meristematic cells is localized and the longitudinal axis coincides with gravity direction. These cells initially grow vertically upwards in the form of needle-like outgrowth (apical growth).

They are divided mainly by asymmetric mitosis, elongated and pass to the process of internal differentiation (formation of midrib, leaf plate and petiole) in accordance with their species-specific DNA-information. Cultivation of later differentiated primordium leads to the formation of only leaf blade.

External development of the leaf begins after unfolding of bud. It grows rapidly and the surface of leaf blade increases tens and hundreds of times. The young leaf of many deciduous species, as a rule, has three meristematic zones: An apical meristem on the top and two marginal meristems, which are located at the edge of its blade. In the beginning of the vegetative season, they grow rapidly in the length and width, pass a certain number of usual and asymmetric mitosis in certain sequences, and shortly reach a species-specific shape and size. The life duration of the leaves of one generation is short and approximately the same for all trees of the same species. Their formation begins with the same (sufficiently low) energetic level and has the same length and construction ontogenetic cellular lines.

Cell divisions cease when the leaves reach the maximum sizes and differentiate, but the decrease of the DNA-energy of differentiated cells continues. Signs of deeper differentiation and aging (reduction of photosynthesis, respiration, yellowing or reddening leaves, etc.) are manifested and intensified as their level of DNA-energy decreases. The leaves fall off then the lowest energetic level approaches. Significant parts of energetically weak differentiated and dead cells are removed from the tree crown after leaf fall. Many by-products of metabolism and toxic substances, which are accumulated in the cells of leaves during vegetative season, are removed in this time. Consequently, the tree crown is rejuvenated and purified annually. Energetically strong meristematic cells are located inside of the apical and lateral buds, which are laid during vegetative season.

The main shoot develops from the apical bud. Primary morphogenetic tissue is the apical shoot meristem, from which all trunk tissues originate. Meristematic cells located in the central and peripheral part of an apex periodically undergo asymmetric mitosis. The apical derivatives of these divisions become the meristematic initials. Development of the cells of the basal pole and their derivatives, which lose DNA-energy in the course of asymmetric mitosis, is carried out by the path of differentiation, aging, and death. These energetically weak cells have different destinies: Meristematic cells from the peripheral zone give rise to the leaves; from the central zone - a core of the trunk. The leaves have direct contact with the external environment, and the core, in the form of a thin pivot, is located in the center of trunk and remains on the tree body until its death.

The core tissue has a parenchymal nature. Initially, they are the living cells which have large sizes and thin walls. They occupy the central part of a bud and connect with the core of the stem. The core of a trunk, unlike the leaves, is not differentiated into specialized tissues. Its cells are place for the deposition of reserve nutrients and different by-products of metabolism. With aging the tannic, coloring substances, and resin accumulate in them. At the end of their life, the processes of lignifications of cell envelopes and destruction of cytoplasm occur. The central part of the trunk becomes very firm and impermeable tissue for water and other substances. The core begins to perform a mechanical function, giving the trunk stability. Heartwood of forest trees consists of only dead cells and is found in coniferous and deciduous species, such as pine, oak, cedar, yew, ash, etc.

Divisions of lateral meristem occur by the same scheme as in the cells of apical meristem of the shoot and root. Cambium is a single-layered tissue. It is located under the bark of the trunk and covers the outer surface of its cylinder by a continuous mantle. Cambial cells of each horizontal layer undergo the asymmetric mitosis and give rise to the meristematic cells with different DNA-energy. One daughter cell remains in the cambial zone and the second cell begins to develop into the specialized elements of secondary xylem and secondary phloem. Mitotic divisions of these energetic weak cells happen in a certain order up to the Hayflick limit (Shay, Wright, 2000). Then the proliferating cells pass to a less energy-consuming path of development – a differentiated cell. Further development occurs according to the descending energetic vector (from the differentiated cell to the specialized and dead cell) and inside to the trunk in the form of a radial ring. They are superimposed on the elements of the previously formed wood, giving rise to a new annual layer of xylem. Every year a tree forms one layer, the width of which depends on the species, age, habitat, and climatic conditions of the year. New elements of phloem are laid under the cells of the old tissue of the bast. Their development also proceeds by the descending energetic vector, but outward from the radial ring of trunk.

Xylem and phloem present the complex specialized tissues, consisting of conductive, mechanical, and storage elements. During the vegetative season, the development of energetically weak cells and their derivatives occurs not uniformly. Their differentiation and specialization in the xylem and phloem tissues is carried out strictly in accordance with the species-specific energetic, informational program, the energetic matrix, and informational matrix of every tree. Cells of vessels, tracheids, and mechanical fibers are alive only at the early stages of their development. In the process of cell aging, there is a thickening and lignification of cellular walls and the cytoplasm destruction. During ontogenesis, the dead cells of wood are amassed more and more into the trunk of tree.

It should be noted that the bark, which represents non-conducting dead cells from different tissues, has direct contact with the external environment. Its thickness is less than thickness of the sum of the layers of its constituent tissues, which is due to their radial compression. Only some species of forest trees (eucalyptus, platanus orientalis, etc.) are able to completely dump its bark. The secondary bark of most species thickens and turns into a slightly tensile crust – a retidorm. While the trunk thickens, it cracks and is then discarded. Inner layers of dead wood cells do not contact with the environment, so they remain on the body until the end of its life. In the form of annual rings, they are located around the core and increase the thickness of the trunk every year in the direct proportion to the volume of the tree crown. On the one hand, the dead cells render resistance to the mechanical loading on the trunk, due to the large weight of the crown and its placement in the upper part of tree. On the other hand, the rate of their accumulation increases every year and the proportion of living cells decreases. However, their rate of accumulation increases with age, the proportion of live cells is reduced, and this burden is hampering the normal activity of a tree.

Summary

The ontogenesis occurs in accordance with its individual species-specific corpuscular-wave developmental program and the three-plane matrix of spatial organization of plant. Its soma, with respect to DNA-energy, DNA-particle and DNA-information, is a closed system. Usual and asymmetric mitosis, during which the zygote becomes a multicellular plant, participate in the process of morphogenesis. Energetics is provided by two internal (biological harmonic oscillator and metaphase chromosomes) and two external (photosynthesis and respiration) sources of energy. During ontogenesis two basic states of the genome (energetic and informational) are changing. External morphological manifestations of tree growth and aging are associated with the processes of their directed change. The bipolar type of the matrix supports totipotency and high energetic potential of the meristematic zones (apical shoot meristem, apical root meristem, and lateral meristem), retaining the ability to morphogenesis (differentiation and specialization) during all plant life. If you exclude all dramatic events, tree aging, in accordance with the 2nd law of thermodynamics, will occur because of annual DNA-energy transitions of cambial and both apical meristems on one lower level only. In ontogenesis, the natural selection takes place between phenotypes. In the center of species range, its forms contribute to the survival of the corpuscular-wave norm of species – the phenotypic normal, outbred, and most adapted to regional conditions genotypes.

Chapter 7

PROPERTIES OF SPECIES GENE POOL OF FOREST TREE SPECIES

A gene pool by definition is a set of genes, which are present in individuals of the given population, groups of populations or species. This is a real biogeographical object evolving during the process of species adaptation to the environment (Gene pool and ..., 2000; Hedrick, 2003; Severtsov, 2005; etc.). In the previous chapter, it was shown that a genotypic composition of the population gene pool during development undergoes the uniform changes. This confirms not only the reality of its existence, but also the possession of certain general and specific properties. Therefore, the evaluation criteria of the population gene pool should be the set of common properties – the general or manifested during environmental interactions. However, these data for forest plant species have not yet been systematized and often are contradictory. The complete list of common properties is absent. Therefore, obtaining information about the gene pool of forest tree species in the norm, the description of properties on the population level of their organization are the actual problems of forest genetics. Owing to the specificity the forest plant species are a convenient model to study these questions.

From a theoretical and methodological standpoint, a species and population gene pool, like any biological object, should be analyzed from two positions – the whole (as a system of species integration) and its constituent parts (as a system of internal species organization). Underestimation of the relation between the part and the whole in living systems is still one of the basic problems of modern biology (New Encyclopedia of Philosophy, 2010).Therefore, our purpose is a description of properties of the population gene pool from a position of the whole and the parts; the elucidation of contradictions, which take place after transition of genetic material from the whole to the parts, show how these two approaches are mutually interconnected and mutually complementary. The characterization of properties of the gene pool is represented by the example of forest plant species growing mainly in the Russian plains of European Russia.

Table 9. Pairs of common properties characterized the species and population gene poll and connected with each other as the whole and the part

No.	The whole	The part
1	genetic integrity	discrecity
2	geographical certainty	mosaicity
3	stability	mobility
4	ecological specificity	adaptability
5	continuality	systematicity
6	uniformity	variability

12 common properties of the population gene pool as a result of the analysis and generalization of the genetic, ecological and silvicultural data have been allocated. They are: Genetic integrity, discrecity, geographically certainty, mosaicity, stability, mobility, ecological specificity, adaptability, continuality, systematicity, uniformity and variability. Then, the properties which are connected with each other as a whole and the parts are joined in 6 pairs. In Table 9, the list of connected properties are shown. Analysis and characterization will be carried out in pairs.

GENETIC INTEGRITY – DISCRECITY

Genetic Integrity

Analysis of a relation between the whole and the parts will begin with the first pair of interconnected properties – integrity and discrecity. This is a typical example of a philosophical category of the union of opposites. We can talk about the long-term anti-entropy existence of a species gene pool, only understanding it as the integrity, which is supported at the expense of discretion – the final lives of individual genotypes. The main trend of ontogenesis is the transition of biological systems from an ordered state to a less ordered state until to the thermodynamic equilibrium (death). Despite the fact that forest tree species are characterized by a long period of ontogenesis, their aging and death are inevitable. Simultaneously, with the partial loss of the gene pool a continuous process of its reproduction occurs, which is realized on the basis of discrete genotypes. Therefore, the integrity of the gene pool cannot exist without its opposite – the discrecity of its own parts (genotypes).

Genetic integrity is the fundamental property of any gene pool. Gene pools are characterized by the energetic, corpuscular, informational integrity and species specificity on the species, population, organismic, and molecular levels of organization of their genetic material. It is expressed in the community of different traits (morphological, anatomical, physiological, biochemical, cytogenetic, molecular-genetic, etc.), which are replicated in the generation of plants of one species. The integrity of a species gene pool is provided by the fact that all of its genotypes have a similar corpuscular-wave structure of the genome, the identical construction of a matrix of a reproductive cycle and the same historical path of development. Another distinctive feature of the species gene pool is the huge (but finite) duration of existence on the Earth as compared with the plant life. The main way of overcoming organizational degradation and maintaining of the anti-entropic state is a continuous self-reproduction of genotypes and the ratio between the speed of their reproduction and death.

As a rule, the species separation is caused by the action of different mechanisms of incompatibility (genetic, ecological, morphological, physiological, etc.), which lead to their inability to interspecific crossing. Intra-specific diversity of the gene pool is maintained by mechanisms of meiosis, fertilization, and pollination. As it has been experimentally shown, the population reaction to change in the environmental conditions by means of an orderly restructuring of its own gene pool (as the whole) (Kuznetsova, 2015). In open-pollination, the optimal level of genetic diversity is regulated with the participation of environmental factors, a different system of seed reproduction, and forms of natural selection. All these processes on the different hierarchical levels are well described in the literature (Timofeev-Resovsky et al., 1973; Yablokov, 1987; Ayala & Kiger, 1988; Pukhalsky, 2002; Dynamics of Population Gene Pool ..., 2004; Klug & Cummings, 2007; Inge-Vechtomov, 2010; etc.).

The panmictic population is also a system of genetic integration, but only it is less numerous. As it has been experimentally shown, the population reaction to a change in the environmental condition by means of an orderly restructuring of its own gene pool (as the whole). In open-pollination, the optimal level of genetic diversity is regulated with the participation of environmental factors, a different system of seed reproduction, and forms of natural selection. All these processes on the different hierarchical levels are well described in the literature (Timofeev-Resovsky et al., 1973; Yablokov, 1987; Ayala & Kiger, 1988; Pukhalsky, 2002; Dynamics of Population Gene Pool ..., 2004; Klug & Cummings, 2007; Inge-Vechtomov, 2010; Kuznetsova, 2015, etc.).

Sexual reproduction is the main way of reproduction for forest tree species. This is a complex genetically determined process, which proceeds on the basis of species reproductive cycle. According to our understanding, a reproductive cycle has the dual corpuscular-wave nature, a matrix of which does not change, until a species exists (Kuznetsova, 2013). Integrity and autonomy of the gene pool are determined by the principle of DNA-energy, DNA-particle and DNA-information. Accordingly, the genotypes of certain species are linked to one unit by the energetic (species-specific level of biological energy), the corpuscular (species-specific nucleotide composition) and the informational (the same species informational signal) components of a genome. A DNA-particle is formed on the hereditary line of the reproductive cycle and provides a succession of corpuscular essence of plants in the alternation of generations of one species. At the present time the species specificity by the DNA-particle is confirmed by the large number of genes and markers (Dynamics of Population Gene Pool ..., 2004; Gonzaler-Martines et al., 2006; Krutovsky et al., 2012; etc.). Restoration of a species-specific level of biological energy is carried out on the energetic line of the reproductive line by the transition of a DNA-wave at the maximal species-specific energetic level. The energetic line (number of activation stages, their structure) has a genetically determined structure and provides the energetic succession of plants in the alternation of generations of one species (Kuznetsova, 2002, 2013).A generative cell receives the new informational signal during meiosis. The communication channel of DNA-information includes three basic elements for receiving the space information signal, two types of signal boosters and the intracellular system of its realized species-specific plants of one species.

Discrecity

The gene pool of each species and population consists of the individual genotypes. Discrecity is a fundamental property of their internal organization, which is manifested at all hierarchical levels and reflecting the principle of life existence on Earth. A genotype, as one of the main subsystems (the parts), presents enough of an autonomous genetic system, which has a certain independence with regards to the environment and possesses its own path of individual development (Yugay 1976, Afanasyev, 1986).

Unlike the gene pool, all of its subsystems (populations and plants) are characterized by age. As was mentioned above, during reproductive development in each initial cell of the future plant, three development programs are laid – a corpuscular on the basis of the DNA-particle, an energetic on the basis of a DNA-wave and an informational on the basis of DNA-information. During ontogenesis, these programs are realized independently from each other, giving rise to the many genetically (i.e., the corpuscular, energetic and informational) non-identical plants of one and the same species. In this case, the plant body represents a closed discrete system where these programs are realized in life by the trinity principle of energy, matter and information. Accordingly, all cells of the plant have the same DNA-particles, and different DNA-energy and DNA-information, because they change in a certain way during ontogenesis. The biological energy level decreases from maximal up to the lowest DNA-wave level in accordance with the 2nd law of thermodynamics. The DNA-energy level of a tree depends on the plant age and the tissue type.

In ontogenesis, the corpuscular body of each discrete plant presents homogeneous and continuous space, since all its cells have the same DNA-particle. The energetic and informational field inside its body is uneven. It is organized in accordance with the energetic and informational matrix, which were created during early and late embryogenesis. The principle of DNA-energy distribution and the location of tissues byte DNA information do not change during ontogenesis. The polar distribution of somatic cells by the energetic level is supported by two types of cell divisions – typical mitosis and asymmetric mitosis. During asymmetric mitosis, a DNA-energy of meristematic cells is not changed. Their derivates in other tissues (xylem, periderm, sclerenchyma, etc.) are the differentiated specialized cells, which gradually lose their biological energy in accordance with the 2nd law of thermodynamics, and tend to the lowest ^{h} level or the level of death.

The life duration of any tree is terminal and depends on many factors: The species-specific age limit; genome initial state; growing conditions; and sharp climate fluctuations, catastrophic events, and others. There are many different internal and external reasons for meristematic cells and tissue death too. The external reasons may be various: Mechanical damages, weather and anthropogenic stress, and vermin. The main internal reason is the chromosomal and gene violations of the DNA-particle. It is believed that the transport and synchronizing functions of trees with age are weakened (Gamaley 2015), which may also be the inner cause of death of the apical meristem, top-dry and tree death.

GEOGRAPHICAL CERTAINTY – MOSAICITY

Geographical Certainty

The next pair of properties, connected together as a ratio of the whole and the part, is geographical certainty and mosaicity. All species in genetic integrity systems are characterized by geographical certainty – they possess a definite geographic location, numerosity, occurrence time and developmental history (Zawadzky, 1961; Eriksson & Ekberg, 2001; Jorgensen, 2002). The term geographical certainty has been introduced by Zawadzky (1961) for species evaluation according to these criteria.

According to our three-plane corpuscular-wave conception, the species has an earthly and cosmic origin. A species gene pool can be realized in species-specific permanent plant communities only after the suitable material substance has been created on the planet and there are necessary external environmental and climatic conditions for its existence. Therefore, development of any species occurs in a limited geographical area. The notion of a gene pool is usually associated with the natural habitat or a species range

– the territory on which a given species grows, breeds and develops through the populations and by means of the populations. The range size varies in wide limits and is an important characteristic of each species, which is equal in its own significance with its morphological characteristics. It is established that the spatial distribution of trees and dynamics of their numerosity are closely interconnected (Gilyarov, 1990). Three types of species range depend on the size and shape of the organisms; they are the continuous, the intermittent and the relict species. Accordingly, the range territory depending on the forest growing conditions are divided by the optimal and pessimal zones.

The size of a fundamental and realized niche is the additional information concerning the total volume and state of a species gene pool (Severtsov, 2004, 2012). A fundamental niche represents the geographical space of Earth, limited by the adaptive capabilities of a species with regards to all factors and conditions of its natural habitat. The realized niche is the portion of the fundamental niche, which occupies a concrete species in a certain historical period in real biogeocenosis. The fundamental niches of several species often overlap, which creates the prerequisites for an interspecific competition, an invasion of species, and a shift of realized niches inside the fundamental niche.

The gene pool change (increase, decrease) testifies to the geography of species spread, trends of development and the state of its genetic resources in a certain historical period. Changes of realized niches within the potential ones occur periodically, but the gene pool is always striving to restore the lost volume. Dynamics of linden forests on the territory of the Moscow region confirms this tendency. In the last century, 700 thousand to 1 million of linden trees were cut down (Rysin et al., 2004, 2006). Linden gradually disappear from the forests near Moscow, and mostly oak and pine occupy their niches. Today, there is no demand for linden with regards to other species, and it has begun to recover its position. During 50 years, its number has increased 4 times. Therefore, it is likely that the edificator species will appear once more.

The panmictic population, as the system of intra-specific integration of more low order species, has a common population gene pool, and occupies a certain territory, within which the trees are freely crossed between themselves and carry information about the environmental conditions of its ecological niche. The amount of trees in the populations and their territory can also vary widely. Therefore, information about a total volume of the population gene pool may give such indicators as the number of trees and density of their distribution in the populations, the relationship between fertility and survival of genotypes, etc.

Mosaicity

The compact body in a gene pool (population, species), unlike a plant body, is absent. Continuous species range does not exist. Its territory is not homogeneous by climatic and edaphic conditions, inhabited by many species, and a distribution of populations and trees is complex. Therefore, such a property as the geographical definition, of course, should conclude the opposite from the position of the parts – the certain organization of the space structure of the species gene pool inside a range. In the literature, this phenomenon is described as a mosaicity. The system of a spatial-age mosaic (the parts) arises within any species range as a result of the negative impact of both external and internal factors (Timofeev-Resovsky et al., 1973; Tilman, 1982; Hubbel, 2001, Mirkin & Naumova, 2012; etc.). This property reflects the mutual arrangement of trees, allowing them to use as much as possible the environmental conditions and defines a space volume, which is controlled by these forest ecosystems.

A forest as a mosaic system was first described using the example of a beech tree forest (Walt, 1947). According to current views all natural forests possess the non-uniform, mosaic-tiered formations (Whitteker & Lewin, 1977; Rankle, 1989; Korotkov, 1991; Zaugolnova & Smirnov, 2004; Gilyarov, 2010). Different size fragments are included in each, whereby the plant cover takes a hierarchical structure with lots of populations as elementary changeable units. Their transformations (successions) are treated as the probabilistic transitions of discrete units of a particular system from one state to another. Populations and trees in them are distributed by an unequal manner. The irregularity is manifested in the form of "island" mosaics, separated by space; or in the form of subpopulations of different density (Timofeev-Resovsky et al., 1973). Mosaics are highly variable by a form and a size. This directly relates with the component heterogeneity of their geographical landscape, different spatial-age structure and dynamics of forest ecosystems. Minimum mosaic in the hierarchy is an inter-parcel mosaic, which is formed as a result of falling of one tree (Whitteker & Lewin, 1977; Rankle, 1989). An arising window depends on the zoning, forest type, tree size and its environment.

The latest emergence of zonality and species composition of forests of the Russian plains was formed one to two thousand years ago (Smirnova et al., 2001). Boreal, deciduous and mixed forests grow here, and 11 species occupy more than 80% of all forestland (Romanovsky, 2002). 113 ecotypes, which vary by terrain, humidity, soil composition of soil, etc., are encountered within this area (Romanovsky, 2002; Romanovsky et al., 2009; Bobrovsky, 2010). Micro-heterogeneity forests, formed on the basis of a plurality of mosaics and variability of habitat conditions, have an important role for their stability, renewal processes of genetic resources, preservation of genetic diversity, coexistence of different life forms (animals, plants, and microorganisms), etc. The processes of the sexual and vegetal reproduction, different life expectancy and non-synchronous change of generations enable a plant community to continuously replace the parts of their structure on the local areas (clearings, "gaps") and to preserve a historically established species range, intraspecific genotypic composition and the total volume of a gene pool.

Stability – Mobility

Stability

By the words of Berg (1993), “A living only in the measure of stability is the alive.” The strategic aim of any species, such as the integral formation, is the maintenance of population stable existence within certain territories and a total number (Gilyarov, 1990). In terms of stability, we mean the ability of living systems to preserve their properties and homeostatic parameters under the influence of internal and external stressors, and the ability to reproduce a progeny, corresponding to the species norm (Rosnovsky, 1997). It is considered that a population gene pool can be the stable system, when its genetic material is not fixed, and fluctuates near equilibrium within the norm reaction of a species (Schwartz, 1980). A forest, which is in the equilibrium state and characterizes by a higher stability, is a natural forest undisturbed by a man (Zaugolnova & Khanina, 2004) (Figure 80, 81). Its population gene pool as a whole, with an increasing stress load is transformed into the new state, and will remain in the stable system (Odum, 1975; Berg, 1993; Rosnovsky, 1997; Bekh & Danchenko, 2007; Kuznetsova, 2012, 2015). Up until now, the stability is one of the most prominent problems of biology. We are considering the stability as the main strategy of the development of the population gene pool and main property that characterizes the population and species as the whole.

The general trend of evolutionary development is a movement of species towards the greater complexity (Orr, 2000; Shcherbakov, 2005a). A complex system is more sensitive to degradation. Therefore, more energy is required to maintain their stability. Orr (2000) called this phenomenon the “cost of complexity.” According to its conception, over millions of years only oscillatory systems and processes have been able to survive and persist (Orr, 1998, 2000). In our model, the stability and anti-entropy of biological systems as quantum-mechanical systems is achieved by means of different reasons, including energy of a biological harmonic oscillator, the capacity of which progressively increases during evolution. The analysis of change of plant reproductive cycles showed that the number of activation stages on the basis of which the DNA-wave transition to higher energy levels occurs from one (Ulotrichophyceae, Conjugatophyceae, etc.) to six (Gymnospermae, Angiospermae) stages is increased during speciation.

Figure 80. Natural pine forest in the Tver region of Russia.

Stability is an important property of the population gene pool and its main characteristics, as the system of intra-specific integrity, in the state of dynamic balance. Thanks to the equilibrium state, the viability of forests as the structural units of nature are provided. Various traits and parameters are used for its estimation –from visual estimation of the vital state of trees, populations (Alekseev, 1989) prior to the traits, which indirectly reflect the level of their general and specific resistance (Alekseev, 1989; Polevoy, 1989; Vorobeichik & Hantemirova, 1994; etc.). The gene pool of the dryland forests possess enough stable formation and are able to stand the stress until its intensity is within the limits, to which there are genetically and historically pre-adapted conditions. The best growing conditions are in the center of the range. Here, forest tree species are able to withstand the high external loads (Vasilevich, 1983; Bekh & Danchenko, 2007). As far as the removal from the optimal zone is concerned, a relationship between the genetic system and the environment becomes more and more tense, and, accordingly, their ecological potential and stability of populations (as a whole) are reduced.

Figure 81. Natural oak forest in the Lipetsk region of Russia.

Mobility

In spite of the immobility of trees, their community and populations are constantly moving, occupying and liberating new areas of forest land (Sukachev, 1964; Tilman, 1988; van der Maarel et al., 1995; Palmer & Rusch, 2001; Dynamics and Stability…, 2006, etc.). Mobility is the property gene pool, reflecting incessant random or directional changes of their internal state, the ability of subsystems (populations, trees as the parts) to move in time and space within their species range.

Sukachev (1964) divided all forms of the dynamics of forest ecosystems on the cyclic (diurnal, seasonal, annual, etc.) and succession (autogenous and exogenous successions) parameters. In the first case, the changes proceed on the basis of one and the same, although the time-varying population gene pool is a factor; in the second case, the result of its loss (partial or complete), with the change of species composition (partial or complete) is considered. During the last one to two thousand years, the main species edificators of the Russian plains (pine, oak, spruce, birch, etc.) were characterized by a stable and dynamic enough gene pool (Smirnova et al., 2001, 2004). The processes of annual sexual reproduction, cross-pollination, special devices for the spreading of pollen and seeds provide a receipt of mass quantity of pollen, ovules, embryos and plants (the parts), a rapid rotation of genetic material at all territory of species ranges, and, as a consequence, promote the preservation of integrity and stability of a gene pool (the whole) in time and space.

Stability of forests (as a whole) is ensured in spite of the constant changes in the internal structure (the parts), originating as a death result of the separate or groups of trees and partial or complete breakdown of the population systems. For this reason, their deaths can be different (genetic anomalies, competition, fires, windfalls, etc.). However, if the habitat is not destroyed, the ceaseless sexual and vegetative reproduction, different life expectancy and asynchronous alternation forest generations allow the plant community to continuously replace themselves in local territories the lost parts of their structure (gaps, burned areas), while preserving the historically established region, genotypic and species composition of natural forests. "Gaps" arising after tree death have the favorable microclimatic conditions for the development of young trees. They give rise to forests of a new generation and soften the impact of the environment on the juvenile stage of their formation.

The evolutionary development of a species gene pool is divided, with regards to the properties of stability and mobility, into three stages: 1) The upward vector at the initial stage, when the reproduction level exceeds the mortality rate; 2) stable development, when reproduction and mortality rates are approximately equal; and 3) the downward vector in the end stage, when the mortality rate is higher than the reproduction level. For example, at present *Pinus sylvestris* and *Betula pendula* on the territory of the Russian plains have a comparable size area (Kuznetsova, 2010a). However, their gene pools have different development trends. In the twentieth century, the volume of pine realized as a niche began progressively decreasing (Pismenny, 1970; Rysin, 2006). The most rapid departure of this species takes place in the center of the range. Birch, in contrast, increases its volume in the whole area. In the center of the Russian plains, where it is increasingly replacing the pine.

The comparison of stability and mobility of species ranges shows the unity of opposites, a subordination of the part to the whole. These properties are in inseparable unity and are not opposed to each other but coexist simultaneously in time and space. Mobility has a leading role in the continuous rejuvenation of a population and species gene pool and maintenance of their stability, which is carried out by continuous reproduction and the death of genotypes on the range territory.

ECOLOGICAL SPECIFICITY – ADAPTIVITY

Ecological Specificity

It has long been observed that all species differ by a response to the changes of the external environment (Odum, 1975; Schwartz, 1980; Levitt, 1980; Gilyarov, 1990; Eriksson & Ekberg, 2001; Jorgensen, 2002; Peek & Welch, 2004; etc.). They are the species-specific formations for their capabilities to adaptation. All species are ecologically individual systems in the framework of relationships with the environment and internal possibilities of a gene pool to the adaptation. Their development occurs in the environment, with climatic and edaphic conditions of which the gene pools are interconnected (as the whole) by the principle of the interaction of the "species-environment." Therefore, different species cannot occupy the same ecological niche. It is necessary to take into account the inverse connections – the participation of populations in the creation of their own ecological niche.

A gene pool of any species with regard to the environment and other species acts as the united genetic system, which can maintain its own integrity and activity only in those conditions, which correspond to its inner nature. At the same time, genetic material of populations is able to change within certain limits in accordance with the alterations of environmental conditions. Therefore, on the one hand, a species gene pool as a stable and anti-entropy formation, separates from the environment and is a relatively independent genetic system, and, on the other hand, it, as an ecological species-specific formation, constantly contacts with the environment as "a living with a non-living, "a living with a living" and "a living with a man" scenarios.

A gene pool of any species has the number of common ecological and genetic characteristics, which are often described as the phenomena of pre-adaptation. Ecological specificity or the ability to adapt is the inherent properties of any species. There are different forms (genetic, morphological, and physiological) (Georgievsky, 1974; Iordansky, 2001). Pre-adaptation is not created *de novo*, and is revealed during the interaction of a gene pool with an environment. It promotes the survival of species not only in the optimal conditions and under usual regional stress load, but in the conditions in which it may be in the future. The adaptive possibilities determine by the set of adaptive norm reactions, beyond which the existence of this species is not possible. Between the boundary values of species-specific norm reactions, the gene pool and its ecological niche correspond (more or less) to each other. Ecological specificity depends on such factors as an ecological plasticity, morphological and physiological devices, seasonal and daily rhythms, genetic heterogeneity and natural selection (genotypic and phenotypic), which ensure the preservation and normal development of species in a constantly fluctuating biotic and abiotic environment.

At the present time, the system of eco-cenotic groups on the basis of ecological scales and geo-botanical descriptions work out, which reflects the full spectrum of adaptive capacity of forest tree species on the territory of the European part of Russia (Smirnov et al., 2006).

Adaptability

When a large experimental material is confirmed, all existing species are ecologically specific formations and their genotypic composition is linked with the environment (temperature, humidity, soil type, length of daylight, etc.) by the principle of the interaction "genotype-environment" (Schmalgauzen, 1963; Timofeev-Resovsky, 1973; Schwartz, 1980; Yablokov, 1987; Clausen, 1995; etc.). Due to the property of adaptability, population gene pools are supported, from position of the parts, by the viability of individual trees and forest ecosystems in a complex system of territorial complexes, forming the landscape of a given district under continuous changes and cyclical fluctuations of regional climate.

The forests of temperate latitudes are characterized by increased heterogeneity of microclimatic conditions. Therefore, the strategy of their resumption is associated with the rather wide amplitude of phenotypic variability (Rankle, 1989). In addition to the quality, productivity and stability of forest ecosystems is a function of the degree of compliance with the genetic characteristics of trees and regional conditions of their growing (Wright, 1975).

As it has been shown, genotype-environmental ties are created and corrected during reproductive development and ontogenesis. This process begins from the expansion of genotypic and/or phenotypic variability during meiosis, fertilization, pollination, and seed germination. In consequence, a huge number of genetically and phonetically non-identical pollen, ovules, embryos and plants are formed. Initial genotypic spectrum pollen and ovules is significantly wider in comparison with the gene pool of maternal and final daughter populations. The narrowing of variability takes place in the end of each stage and leads to a closer genotype-environmental relations, as a result of the death of the less adapted energetically weak genotypes (phenotypes) of pollen, ovules, embryos and plants. A changing of stages is also related with the process of selection – with the removal from a gene pool, on the contrary, the most closely linked with the environment energy strong genotypes (Kuznetsova, 2003). Hence, it follows that, firstly, the expansion of genotypic (phenotypic) composition is mainly carried out on the basis of norm-forming genotypes. Secondly, in the normally fluctuating environment from one stage to another, from one generation to another the genotypic norm of a species is preserved.

Ecological conditionality of a phenotype is manifested in the fact that the similar environmental conditions induce the same response (down to details) in plants of different species, forming the set of phenocopies of the same type according to the law of homologous series (Vavilov, 1968). During the integrating of each genotype into its "own" ecological niche, it is revealed that such property of an organism as the ability to the adaptive modifications is present, resulting in a phenotypic diversity of seedlings that is greatly extended (Schmalgauzen, 1963, 1968). In ontogenesis, natural selection is carried out by the phenotypes and narrows their initial phenotypic diversity by 2-4 times (Semerikov et al., 1996; Romanovsky, 2002; Romanovsky et al., 2009). The selective advantage is given to those plants, which are realized in the most optimal way of genotype-environment interactions into phenotypic species norm and the adaptation of populations to the habitat increases with age.

From the position of the whole and the parts, the properties of ecological specificity and adaptability are in indissoluble unity, and between them a causal relationship is traced. We see the unity of the beginning and the result of development during transition of pre-adapted population gene pool (as the whole) that has adapted to their own "micro-niche" genotypic composition of pollen, ovules, embryos and plants (as the parts). Then, the consequence (fertile trees, which are adapted to their habitats, as the parts) again becomes the beginning (pre-adapted population gene pool, as the whole), and so on. As a consequence, a potential pre-adaptation during development of the gene pool each generation turns into its opposite – the realized adaptability of genotypic composition, and it, in turn, again becomes a potential pre-adaptation of the gene pool. This algorithm of movement of a genetic material ensures the preservation of genetic diversity, stability and genotypic norm of species, promotes the mastering of new habitats and adaptation to its conditions.

CONTINUALITY – SYSTEMATICITY

Continuality

Continuity is the property that reflects the equilibrium state of population and species gene pool, the invariance (identity) of their systemic organization, the succession of genetic material in time and its continuity in space. The action of these properties, unlike the 8 characteristics of the gene pool described above, does not extend to the all species ranges, but only to its "core" - the zone of the species optimum. Moreover, all the territories of ecological trouble from this zone must be eliminated, both in terms of natural origin (swamps, highlands) and anthropogenic origin (industrial pollution, technogenic catastrophes, and others).

A comparison of Fst, Gst, Rst, and Nst statistics has allowed us to establish that several thousand years ago, the genetic material of the majority of *Palearctic coniferous* (pine, spruce, fir, larch) has already reached ecological and geographical balance with the environment (Altukhov, 1995; 1989; Dynamics of population gene pools …, 2004). Since then, their intraspecific subdivision is an auto regulated process, and intra-population and inter-population variability are maintained at the optimum level. The homozygote and heterozygote ratio in the populations does not change, because a lack of the heterozygotes and their abundance is equally unfavorable for them.

In phytocenology, one of the most common theories is the concept of continuity of a plant cover (McIntosh, 1967; Whittaker, 1980; Mirkin at al., 2001, 2012, 2015; etc.). In the "core" range, a broad reaction norm, the lack of substantial barriers for gene flow leads to the formation of a continual species gene pools (as the whole), which is expressed as the invariance of different populations by the morphological, cytological, molecular-genetic traits. The Russian plains between northern and southern populations of Scots pine, which has the longest by the latitude range, was detected only by a slight shift of the phenophases of reproductive processes. The difference in the date of flowering and pollen departure is only few days and sometimes the synchronization processes takes place (Elagin, 1976; Sannikov & Petrova, 2003). These dates indicate a lack of phenological isolation between populations of pine and a continuity of its gene pool by the time component (Sannikov & Petrova, 2003, 2007).

The important characteristic of a species is its ability to cross and form viable progeny. Mainly neighboring trees within the population mosaics mutually cross each other, creating a system of overlapping demos (Franklin, 1985). Panmictic demos cover an area of about 1-3 hectares and include 200-1,000 trees (Wright, 1953). Inside them, the intensive exchange of pollen and seeds takes place. In a stabilizing selection, a genetic material of male and female gametes, embryos, and plants in different mosaics get the signs of continuity and invariance. Exchange of genetic information between different populations (particularly remote mosaics) falls sharply, but their ability in terms of the panmixia is not lost. Therefore, the borders of distal (spatial) isolation in the alternation of the previous and subsequent generations of forests can be considered as a factor of a certain time and certain situation, which do not deny such a property of the gene pool as the continuity.

Systematicity

The living world is highly complex and diverse. According to modern notions, all species of plants and animals present the multi-level systems, which are hierarchically ordered and systematically organized (Dynamics of population gene pools, 2004). Therefore, the systemic organization and inner orderliness, or the systematicity property, is, from the parts position, one more characteristic of the gene pool of a species and population.

Species and population gene pool (as the whole) are characterized by the invariant systemic organization, and their parts – genotypes – are strictly individualized genetic systems with an increasing number of internal states of freedom in relation to the external environment (Afanasyev, 1986). Therefore, on the one hand, all plants, such as genetic independent organisms, possess a diverse lifetime and a directional course of individual development – from birth to death in accordance with the 2nd law of thermodynamics. They have different adaptive reaction norms, the range of variability of which overlaps within the range of the adaptive capacities of species. An average value of the reaction norm and the sensitivity to the environment are under genetic control (Connoly & Jinks, 1975; Jinks & Pooni, 1982; Kilchevsky & Khotyleva, 1997).We think that genotypic differences of trees by the adaptive reaction norm (as the parts) promote the preservation of the integrity of populations (as the whole) in a constantly fluctuating environment and the survival of species in a different growing conditions. On the other hand, the invariant systemic organization of a population gene pool presupposes the existence of certain internal relationships between genotypes.

Evidence of systematicity is the orderliness of intra-population structure in optimal conditions and under stressful pressure (Callis, 2000; Kuznetsova, 2010b, 2012; 2015), the subordination of which an organization of the entire system adheres to a certain purpose (Uyomov, 1978; Surmin, 2000). It is shown that a genetic material of different populations at the "core" of the species ranges in the optimal years and is in the equilibrium state (as the whole);it does not change in time and in space by morphological, cytogenetic, and others traits. Genotypic composition of a separate population depending on the adaptive reaction norm of individual trees, conversely, is a mobile structured formation.

Change systemic organization of populations (as the whole) is shown as the result of genotype-environment interactions of individual trees (as the parts) with the environment. In the optimal years, the self-fertility trait in Scots pine varies from 0 to 0.75, and the relationship between self-sterile, partly self-fertile and self-fertile forms is ≈ 8:4:1(Kuznetsova, 2010b). In droughty years, a genetic system comes out from a balanced state and its genotypic composition is rearranged in the direction of adaptation to the new weather conditions. In abnormal years, plants are able to unite in the relatively stable groups of genotypes: Self-sterile and self-fertile forms; sustainable and sensitive to radiation, a drought, frost, etc. Between genotypes of invariant population systems, certain relationships are revealed and a new structured gene pool in the process of selective survival is formed. In relation to the drought, the internal differentiation of a population on the stable, modal and sensitive groups of plants takes place. However, the clear boundaries between these groups of genotypes are absent.

In experiments with model populations, it has been proven that concerning the environmental change they respond by alterations of genome and genotypic composition (Agaev, 1978; Depledge, 1996; Callis, 2000; Kuznetsova, 2015). Monitoring systems of seed reproduction systems during 6 dry years (1991, 1995, 2001, 2007 and 2012) and 12 years of self-pollination experience (1987-1999) on the same 50 trees of Scots pine allow us to trace how the rearrangement of a seed gene pool takes place (Kuznetsova, 2009; 2010, 2012; 2015). It is shown that every year, the genotypic composition of seeds is formed according to the principle of a "genotype-environment" interaction. The number of surviving seeds, the proportion of inbred progenies and progenies from drought-resistant forms directly depends on the strength of weather stress.

From the aforementioned data, it becomes obvious that mutual transition of opposites into each other, as a manifestation of connection of the whole and the parts, also occurs in this pair of traits – continuality and systematicity. We can see the interpenetration of these opposite properties by their characteristics. The population gene pool (as the whole) possesses a certain systemic organization, which is or is not revealed while a genetic system is in the equilibrium state, or has the identical invariant structure in different populations. During the exit of a gene pool from the equilibrium state, its inner organization (as the parts) is detected as a result of individual genotype-environment relationships that develop between a genotype and a group of genotypes with a stressful factor.

UNIFORMITY - VARIABILITY

Uniformity

Uniformity is one more property characterizing a species and population gene pool as a system of species integration. This is the state of minimum genotypic and phenotypic variability, to which all equilibrium populations aspire in the alternation of generations. The uniformity is inherent of the genetic material at all levels of its organization. A distinctive feature of this property is that the inter-population uniformity is achieved with age, as a result of the different forms of natural selection. It should be noted that the action of these properties is limited by the zone of species optimum and delisting from it the territories of ecological trouble (swamps, highlands, technogenic pollution). It should be noted that a forest tree species with long ontogenesis and intensive competition are a convenient model for the study of this property.

Formation and correction of the genotypic and/or phenotypic composition of a gene pool continues throughout all life cycles of population ecosystems. It is established that populations of any species in the central parts of their ranges are characterized by similar parameters of genetic and phenetic traits (Schmalgauzen, 1963, 1968; Timofeev-Resovsky et al., 1973; Yablokov, 1980, 1987; Glotov, 1983; Sannikov & Petrova, 2002, 2003, 2007; Petrova et al., 2014; etc.).Presumably, a reservation of optimal genotypic composition is a prerequisite for the transmission and preservation of genetic information in the alternation of sexual generations (Kolchanov, 2003; Vorobyova, 2006).

A sexual reproduction belongs to the conservative homeostatic processes (Maynard 1981; Eshel, 1991, Shcherbakov, 2005b). Shmalgauzen (1963, 1968) established that in ontogenesis, the selection is always going in the direction of selective advantages of the phenotypic norm. It was calculated that up to 100 thousand seedlings per 1 ha appeared simultaneously on the glades and burnt areas (Semerikov et al., 1996). From them, no more than a thousand of them will survive. During competition they, having occupied the leading position, form the uppermost canopy level of a forest. During this selection, the variability of populations by morphological, cytogenetic, molecular genetic traits is decreased (Zhivotovsky & Duharev, 1985; Romanovsky, 1994, 2002; Nielsen, 2001; Romanovsky et al., 2009, etc.). This leads to a decrease of the variability of vegetative and generative traits, a balancing of functions and processes, giving them characteristic uniformity (Zhivotovsky & Duharev, 1985; Hall, 2001; Gonzaler-Martines et al., 2006).

The study of plain forests of Eastern Europe has shown that differences between upland populations are not significant (Mamaev, 1973; Mamaev & Makhnev, 1988; Sannikov & Petrova, 2003, 2007; Petrova et al., 2014). Analysis of the 24 traits of cones and seeds of Scots pine revealed the small averages (pair-group) of phenotypic Mahalanobis distances between groups of populations, growing in different parts of species ranges. Cytogenetic studies of the central populations of conifers have shown, that they by the criteria of mitotic and nucleolar activity, number and spectrum of mitosis and meiosis pathologies are characterized by approximately the same parameters of intracellular processes (frequency of mitotic pathologies ≈5%) (Abaturova, 1978; Butorina et al., 2001, 2008, etc.). Data about geographical allozyme variability and differentiation of Palearctic conifers (such as *Pinus sylvestris*, *Pinus sibirica, Abies, Larix* and *Picea* species of Palearctic, etc.) was systematized (Altukhov et al., 1989; Goncharenko & Padutov, 1995; Semerikov et al., 1999; Goncharenko & Padutov, 2001; Sannikov et al., 2002, 2003, 2007; Semerikova & Semerikov, 2014; Petrova et al., 2014; etc.). The main conclusion of this series of work is that the weak genetic subdivision of population gene pools by nuclear markers in the central parts of their ranges occurs. Intra-specific differentiation does not exceed a level of 1-6%. At the same time, a direct correlation between the genetic, morphological, and ecological-geographical structure of populations is often detected (Lagerkrantz & Ryman, 1990; Mitton & Duran, 2004).

Variability

As is known, in nature no two identical individuals exist. The population and species gene pool as a whole is enough uniform formation, though from the position of their parts have a variable genetic structure. The variability is the fundamental property of living matter, which reflects the ability of biological systems at all levels of organization (molecular, cellular, organismic, species) under the influence of external effects and internal reasons, which vary within the reaction norm of the species. The classification of variability is accepted to spend by the type of varying traits – morphological, anatomical, physiological, biochemical, cytological, etc. (Mamaev, 1973). At the present time, several types of variability are distinguished: Genotypic and phenotypic; qualitative and quantitative; directed and undirected; etc.

On the example of forest tree species, we consider only two types – genotypic and phenotypic variability. As was mentioned above, during sexual reproduction the expansion of genotypic variability on the basis of DNA-energy, DNA-particle and DNA-information takes place on the basis of species reproductive cycle. On the molecular level, the variability of plants can be estimated by the heterozygosity value. This index in forest trees species is the highest among now living organisms (Dynamics of population gene pools ..., 2004; Brown et al., 2004; Krutovsky, 2006; Gonzales-Martines et al., 2006). Its value is an average of 0.359, which is 2.5 times higher than in annual plants (Mamaev et al., 1988). The highest variability is characterized by the species, occupying a large range and having high fertility (Hamrick et al., 1992). During the study of genetic variability of the dryland populations of *Pinus sylvestris*, species of *Pícea*, *Larix*, etc., general regularity – the main share of gene diversity is the intra-population variability (95-98%) – was confirmed (Goncharenko & Padutov, 1995, 2001; Semerikov et al., 1999; Sannikov et al, 2002, 2003; etc.).

The phenotypic variability is a nonhereditary (modification) change of a genotype, formed as the result of "genotype-environment" interactions. Apparently, the genotypic variability is due to a change of the DNA-particle and DNA-information, and the modification variability is determined only by DNA-information. The study of phenotypic variability and reaction norms formed the basis of the direction of population between genetics and phenetics (Timofeev-Resovsky et al., 1973; Yablokov, 1980, 1987; Mamaev & Makhnev, 1988; Tikhodeev, 2013; et al.). The phene pool of natural populations is a mosaic of inhomogeneous formation, such as the forest cover. Only in this case, as the elementary varying units are not the populations, the individual phenotype of trees and their signs-markers (phens) are considered. The phenotype in comparison with a genotype is more resistant to the environment. When environmental conditions are constant, the levels of a phenotypic trait variability in different species are mostly the same is considered.

Under the stress impact, the equilibrium state of a genetic system is disturbed (Levitt, 1980; Bickhman et al, 2000; Oudalova & Geraskin, 2011; etc.). The instability of a genome is a particular state of the organism, which has a negative effect on the viability, growth and competitiveness of plants. As a rule, the shift of adaptive norm reactions on dependence of the nature of traits and stressors (or/and its intensity) with the extension or narrowing of variability is associated. After the termination of stress action the variability levels are gradually normalized, and the population gene pool returns to the equilibrium state.

The unity of opposites in the transition from the whole to the parts is revealed in the last couple of properties (uniformity and variability). The gene pool and phene pool of the central populations of the Russian plains as a whole are the sufficiently uniform formations; and from the position of internal organization (the parts in all hierarchical levels) they present a very variable genetic structure.

To summarize the above, it should be noted that the evolution of living matter went from simple to complex. The knowledge of biological scientists, in accordance with the technical progress, moved in the opposite direction (from the species level to the molecular). During the second half of the 20^{th}century, molecular biology has more rapidly developed. We think that proceeds from the emergence principle will study the wave properties of the genetic systems of each hierarchical level, their nature and characteristics from the standpoint of the whole and the parts. The biologist should be familiar with different aspects of tree-plane corpuscular-wave plant organization and functioning. Most parts of the properties of species and population gene pool (genetic integrity, discrecity, stability, uniformity, variability, etc.) have a wave nature. For their description, we will also require the knowledge of quantum theory.

CONCLUSION

Quantum biology is the study of biological systems from the position of quantum theory and the fundamental laws of physics. It cannot be understood without the knowledge of classical biology. Therefore, the interpretation of material is constantly turned to the previously obtained experimental data. Interpretations, however, must often be looked at in a new way. A two-plane corpuscular-wave model of a plant cell and plants is described in our previous book. The present book is a continuation of research in this direction. The concept of the three-plane corpuscular-wave nature of the cell and life cycle of higher plants using the example of forest tree species was put forth.

As is known, the number of fundamental laws of physics is small and in corpuscular biology they do not work. In the factual material of quantum biology, we have not found a single fact that would violate the law of conservation of energy, the law of linear momentum conservation, etc. Both books are an application of basic laws to the study of biological phenomena and processes. We tried to show that many of them have a wave nature. The main notions and principles of quantum mechanics, where they were necessary for understanding biological phenomena, were described as simply as possible. The materials that have been described in detail in the first book give a very brief account of this.

Quantum biology is a non-isolated science. In the process of material preparation, it became clear that it could be used to consider other important phenomena and processes (photosynthesis, respiration, geotropism, etc.). The corpuscular-wave approach to the nature of biological systems is reflected at all hierarchical levels of plant organization by the principle of "all in all." The set out materials allow us to confirm the high relevance of this approach to the application of ideas and principles of quantum theory to the most diverse fields of biology. Its importance for the further development of cell theory, a deeper understanding of cell biology, and the nature of prokaryotic to eukaryotic cell systems is absolutely indisputable.

According to modern concepts, the cell is an elementary unit of living organisms, out of which there is no life. Quantum biology fully confirms this statement. It is shown that the prokaryotic and eukaryotic cells are organized by the trinity principle of energy, matter and information in three independent and interconnected planes. The trinity principle is realized on the basis of the DNA-particle-wave at the cellular level, and it is an emergent property of organismic and species levels. Therefore, only a cell can be an elementary and indivisible unit of living matter.

All planes are equal in importance, but differ in nature. DNA-energy and DNA-information possess the wave nature; a DNA-particle is corpuscular. DNA-energy and DNA-particles have an earthly origin. DNA-information is characterized by a cosmic and earthly nature. To divide and join these three essences of living matter is not possible. Therefore, we can give the following definition of a prokaryotic and eukaryotic cell. The prokaryotic cell is a quantum-mechanical system, an open system and an equilibrium system. The eukaryotic cell is a quantum-mechanical system, an open and closed (proliferating cell) or an open (differentiated cell) system and an equilibrium system. The development a cell and its metabolic processes are carried out based on the trinity principle in the energetic, corpuscular and informational planes in accordance with the fundamental laws of physics.

The basic elements of the energetic, corpuscular and informational systems in prokaryotes and eukaryotes are the same, but the level of their structural organization is much higher in the eukaryotic cell. DNA-energy belongs to a central place. The transition of the biological harmonic oscillator to the maximal species-specific energy levels occurs on the basis of the energetic line of the reproductive cycle during sexual reproduction. The reverse process (from the maximal level to natural plant death) is carried out spontaneously in accordance with the 2nd law of thermodynamics.

Knowledge of quantum theory and basic laws is necessary for the further development of evolutionary theory and phylogenetic systematics. In this book, an attempt is made to shed some light on the basic and most general processes of the evolution of the organic world. It is shown that its division into animals and plants took place in accordance with the law of linear momentum conservation.

The genome of each plant can be characterized with 4 of its basic states: DNA-energy, DNA-particle, DNA-information and spin direction. The particular importance in this regard is the decoding of the reproductive cycle of lower and higher plants. Accordingly, the evolutionary process went independently and interconnected by these four components – the energetic, corpuscular, informational and sexual lines. Evolutionary development of plants is a continuous increase of complexity and specialization of their generative and somatic organs, tissues and cells by means of the capacity increase of a biological harmonic oscillator and the high organization of systems for preserving and altering of the DNA-energy, DNA-particle, DNA-information and sexual systems of species. It is necessary to emphasize the important role of the energy protection system, which periodically undergoes essential changes and represents the fifth independent line of plant evolution.

Significant importance of quantum biology is in embryological researches. If we go back, then this work began with a simple question: In the reproductive cycle of higher plants of more than a dozen stages, what is their destiny? Now, it is obvious that sexual reproduction is another unique biotechnology that was created by the living matter. Sexual reproduction is a three-plane corpuscular-wave process by the creation of a multitude of species-specific genotypes on the basis of species reproductive cycles strictly in accordance with the fundamental laws. Embryology is basically a real storehouse of knowledge in the field of quantum biology. Many of its phenomena have a corpuscular-wave nature. Meiosis includes two quantum-mechanical effects (nucleotide exchange between homologous chromosomes – a crossing-over – and nucleotide splitting by the spin direction – a differentiation of gametes based on sex), the wave process of rejuvenating generative cells, and the creation of an active species-specific ring antenna array that ensures the reception and amplification of a species cosmic signal. During embryogenesis, the corpuscular, energy and information matrix of an embryo (or plan concerning plant internal organization) is formed.

Quantum theory is necessary to explain the nature of many phenomena of other areas of biology. It outputs them to a principally new level of knowledge significantly, changing our understanding of the biological world and the principles of its development. Many questions of development and regulation of functional systems of the plants in space and time, the correlations between its vegetative and generative sphere, and the genotype-environmental interactions are still debatable.

It must be recognized that quantum biology is still at the beginning of its establishment as a biological science. There's a lot we don't know yet. The opportunities of the modern instrumental base are very high, which ensures the rapid development of methods of biotechnology and genetic engineering. Therefore, the scientist must be very cautious when implementing scientific achievements.

Theoretical biology lags significantly behind modern features of the experimenter. In fact, the impact on the genetic material of the genome occurs blindly (especially on the generative cells evolutionarily advanced species whose DNA level-highest energy). It is necessary to remember about the existence of risks of such intervention, the result of which can be unpredictable. In the previous and present work, we showed that between the level of DNA-energy and biological defense system there is a genetically deterministic relation which cannot be violated.

REFERENCES

Abaturova M.P., 1978. Research of Elementary Signs of Norway Spruce. In: *Scientific Basis for Selection of Coniferous Woody Species*. Nauka, Moscow, 87-98.

Abaturova, M. P., Duharev V. A., Ryabokon S. M., 2002. Formation of the Fertile Zone of Scots Pine Cones. *Rus. J. For.,* 1, 70-72.

Afanasyev V.G., 1986. *The Living World: Systematics, Evolution and Management*, Politizdat, Moscow.

Agadzhanayn A.M., 1992. To a Question on Evolution of Self-incompatibility in Angiosperms: Gene Mutations. *Rus. Bot. J.,* 77, 19-32.

Agaev M.G., 1978. *Experimental Evolution* (on the Example of Model Populations of Autogamous Plants). Leningrad State University, Leningrad.

Alberts B., Johnson A., Lewis J., Raff M., Roberts K., Walter P., 2002. *Molecular biology of the cell,* 4th edn. Chaffey N. (Ed.), Garland Science, New York.

Alberttsson P.A., 2001. A Quantitative Model of the Domain Structure of the Photosynthetic Membrane. *Trends Plant Sci.,* 6(8), 349-358.

Alekseev V.A., 1989. Diagnosis of the Vital State of Trees and Forest Stands. *Rus. J. Forest Sciences,* 4, 51-57.

Allen G.S., 1946. The Origin of the Microsporangium of *Pseudotsuga Bull. Torrey Bot. Club.,* 73, 247-556.

Allen G.S., Owens J.N., 1972. The Life History of Douglas-fir. Environment Canada, *Can. For. Serv. Headquarters*, Ottawa.

Altukhov Yu.P., 1989. *Genetic Processes in the Populations.* Nauka, Moscow.

Altukhov Yu.P., 1995. Intraspecific Genetic Diversity: Monitoring and Principles of the Conservation. *Rus. J. Genetics,* 31(10), 1333-1357.

Altukhov Yu.P., Krutovsky K.V., Duharev V.A., Larionova A.Ya., Politov D.V., Ryabokon S.M., 1989. Biochemical Genetics of Populations of the Forest Woody Plants. In: *Forest Genetics, Breeding and Physiology of Woody Plants.* Moscow, 16-24.

Alyokhina Yu. V., Balnokin Y. A., Gavrilenko V. F., 2005. *Plant Physiology,* Academia, Moscow.

Andersson E., 1980. Temperature Conditional Irregularities in Pollen Mother Cells of *Picea abies* (L.) *Karst. Heredities,* 92, 27–35.

Ashby W.R., 1956. *An Introduction in Cybernetics.* Chapman & Hall, London.

Ashley T. & Pocock N., 1981. A Proposel Model of Chromosomal Organization in Nuclei at Fertilization. *Genetica,* 55, 161-169.

Asker G.E. & Jerling L., 1992. *Apomixis in Plants.* L.; B.R.: CRC Press.

Bada J.L., 1995. Origin of Homochirality. *Nature,* 374 (6523), 594-595.

Barantseva L.M., 1992. *Cytophysiological Analysis of Archegonium Development in Pinus sibirica Du Tour.* Moscow, Moscow State University.

Barnes B.V., Bingham R.T., Squillace A.E., 1962. Selective fertilization in *Pinus monticola.* II. Results of additional tests. *Silvae Genet.,* 11, 103-111.

Bateman A.J., 1952. Self-incompatibility Systems in Angiosperms. *Ytredity.* 6, 285-310.

Batygina T.B. & Vasileva V.E., 2000.Sexual Reproduction of Flowering Plants: Formation of the Zygote and Types of Karyogamy. *Rus. J. Bot.,* 85(6), 50-66.

Bekh I.A., Danchenko A.M., 2007. *Problem of Stability in Forestry.* Bulletin of the Tomsk State University, 295, 215-219.

Bell G., 1988. Uniformity and Diversity in the Evolution of Sex. In: *Evolution of Sex: an Examination of Current Ideas.* Sunderland, Massachusetts: Sinouer. Pp. 126-138.

Belostotskaya S.H., 1979. Development Peculiarities of the Male and Female Gametophytes of Scots Pine in the Intra- and Interspecific Hybridization. *Rus. J. For.,* 5, 61-72.

Belousov L.V., 2005. *Bases of General Embryology.* Moscow State University, Moscow.

Benken K.A. & Sabaneeva E.V., 2011. Fibrillar actin in the nuclear apparatus of the ciliate *Paramecium caudatum. Tsitologiia,* 53(6), 528-536.

Bennett M.D., 1987. Ordered Disposition of Parental Genomes and Individual Chromosomes in Reconstruction Plant Nuclei, and their Implications. *Somatic Cell Mol. Genet.,* 13, 463-466.

Bennett M.D., 1988. *Parental Genome Separation in F_1 Hybryds between Grass Species.* In. Brandham P.E. (Ed.). Key Chromosome Conference III her Majestry's Stationary Office, London, 195-208.

Berg R.L., 1993. *Genetics and Evolution. Selected Works.* Nauka, Novosibirsk.

Bewley J.D. & Black M., 1982. Physiology and Biochemistry of Seeds in Relation to Germination. Vol. 1. *Development, Germination and Growth.* Springer-Verlag, Berlin.

Bewley J.D. & Black M., 1994. *Seeds: Physiokogy of Development and Germinatiom.* 2nd Ed.Plenum Press. New York, London.

Bhatnagar S.P., Moitra A., 1996. *Gymnosperms. New Age Int.,* New Dehli. Pvt Ltd Publ.

Bickhman J.W., Sandhu S., Hebert P.D., et al., 2000. Effects of Chemical Contaminants on Genetic Diversity in Natural Populations: Implications for Biomonitoring and Ecotoxicology. *Mutat. Res.*, 463, 1, 33-51.

Black M.J., Bewley J.D. & Halmer P., 2006. *The Encyclopedia of Seeds: Science, Technology and Uses.* UK: CABI, Wallingford.

Bobrovsky M.V., 2010. *Forest Soils of European Russia: Biotic and Anthropogenic Factors of Formation.* Associationof scientific publications KMK, Moscow.

Bonch-Osmolovskaya E.A., Atomi H., 2015. Editorial Overview: Extremophiles: from Extreme Environments to Highly Stable Catalysts. *Curr Opp in Microbiol*, 25, 88-96.

Borodin P.M., 1987. Stress and Genetic Variability. *Rus. J. Genet.*, 23(6), 1003–1010.

Bostock K., Sumner E., 1981. *The Chromosome of Eukaryotic Cell.* Mir, Moscow.

Bouta T.J., 2005. Understanding Plant Respiration. In: Plant Respiration: from Cells to Ecosystems. Lambers H., Ribas-Carbo M. Eds. Vol. 18. *Advances in Photosynthesis and Respiration Ser.* Springer, Dordrecht, Netherlands, 177-194.

Brachet J., 1960. *The Biochemistry of Development.* Pergamon Press, London, New York, Paris, Los Angeles.

Brown G.R., Gill G. P., Kuntz R. J., Langley C.H., Neale D.B., 2004. Nucleotide Diversity and Linkage Disequilibrium in Loblolly Pine. *Proc. Natl. Acad. Sci. USA.*, 101, 15255-15260.

Brunkener L., 1973. Beitrage zur kenntrisder fruben Sporangieenentwicklung der Pteridophyten. *Sven. bot. tidskr.*,67, 333-400.

Buchholtz J.P., 1920. Embryo Development and Polyembryony in Relation to the Phylogeny in Conifers. *Amer. J. Bot.*, 7, 125-145.

Buchholtz J.P., 1946. Volumetric studies of seeds, endosperms and embryos in *Pinus ponderosa* during embryonic differentiation. *Bot. Gas.*, 108(2), 232-244.

Butorina A.K., Cherkashena O.N., Chernodubov A.I., Avdeeva I.A., 2005. Variability of the Cytological Indexes of Scots Pine Seeds. *Rus. J. Genet.*, 41(6), 778-788.

Butorina A.K., Ermolaeva O.B, Cherkashina O.N., Mazurova I.E., Belousov M.V., Chernodub A.I., 2008. Perspectives for the Using of Cytogenetic Analysis in Forestry on the Example of Island Coniferous Forests of the Voronezh Region. *Biol. Bulletin Reviews,* 128(4), 400-408.

Butorina A.K., Kalaev V.N., Mironov A.N., 2001. Cytogenetic Variability of the Populations of Scots Pine. *Rus. J. Ecol.,* 3 und der Gymnospermen, 206–210.

Callis H.A., 2000. Environment as a Generator of Adaptive Changes. In: *Modern Problems of Evolutionary Genetics.* Nauka, Novosibirsk, 169-174.

Chadov B.F., Fedorova N.V., Chadova E.V., Khotskina H.A., 2011 Conditional mutations in Drosophila. *J. Life Sciences* (USA), V.5, N3, 2011. P.224-240.

Chamberlain C.J., 1935. *Gymnosperms: Structure and Evolution.* Univ. Press, Chicago.

Chentsov Yu. S., 2004. *Introduction to cell biology.* Akademkniga, Moscow.

Chesnoy L., 1969. Sur la Participapion du Gamete Male a la Constitution du Cytoplasme de L'embryon ches le Biota orientalis [On the Gamete Male's Participation in the Constitution of Cytoplasmic Embryo of Biota orientalis] *Endl. Rev. Cytol. Et Biol. Veg.,* 32(3/4), 273-294.

Child C.M., 1920. Some Consideration concerning the Nature and Origin of Physiological Gradient. *Biol. Bull.,* 39(3), 147-187.

Chin B. & Bernstein I.A., 1968. Adenosine Triphospate and Synchronous Mitosis in *Physarum polycephalum. Bacteriol.,* 96(2), 330-337.

Chizhevsky A.L., 1972. Certain Cosmic Communications Terrestrial Biosphere. *Populated cosmos.* Nauka, Moscow.

Clarkson D.T., 1978. *Ion Transport and Cell Structure in Plants.* Mir, Moscow.

Clausen J., 1995. Population Studies of Alpine and Subalpine Races of Conifers and Willows in the California High Sierra Nevada. *Evolution,* 42(1), 56-68.

Connoly V., Jinks I.I., 1975. The General Architecture of General and Specific Environmental Sensitivity. *Heredity,* 35, 2, 249-259.

Costanzo L.S., 2017. *Physiology.* 6th Ed., Elsevier Saunders, Philadelphia, PA.

Crowe L.K., 1964. The Evolution of Outbreeding in Plants. *1. Angiosperms. Heredity,* 19, 435-457.

Curtis Y.D., Popham R.A., 1972. The Development Anatomy of Long Branch Terminal Buds of *Pinus banksiana. Amer. J. Bot*., 59(2), 194-202.

Cytology and Genetics of Meiosis, 1975. Eds. V.V. Hvostova & Yu.F. Bogdanov. Nauka, Moscow.

D'Amato, 1977. *Nuclear Cytology in Relation to Development.* Cambridge Univ. Press, Cambrige, London, New York, Melbourne.

Darlington C.D., 1965. *Cytology.* J. and A. Churchill LDT, London.

Davis B.D., Dubecco R., Eisen N.N. & Ginsberg H.S., 1990. *Microbiology*. 4th Ed. Lippicott Comp., Philadelphia.

De Nettancourt D., 1977. *Incompatibility in Angiosperms*. Berlin, Springer-Verlag, XIII, 1-230.

De Nettancourt D., 1999. Homomorphic Self-incompatibility in Flowering Plants. In: *Fertilization in Higher Plants*. Springer, Berlin, Heidelberg, Ney York.

De Win A.H.N., Knuiman B., Pierson E.S., Geurts H., Kengen H.P.M., Derksen J., 1996. Development and cellular organization of Pinus sylvestris pollen tubes. *Sex Plant Repr.,* 9, 93-101.

Depledge M.N., 1996. Genetic Ecotoxicology: an Overview. *J. Exper. Marine Biol. and Ecol.,* 200, 57-66.

Dickerson R.E., 1981. Chemical Evolution and the Origin of Life. In: *The Evolution*, Mir, Moscow, 67-107.

Doak C.C., 1935. *Evolution of foliar types, dwarf shoots and cone scales of Pinus.* Univ. III. Bull., 13(3), 7-103.

Dogra P.D., 1967. Seed sterility and disturbances in embryogeny in conifers with particular reference to seed testing and tree breeding in Pinaceae. *Stud. Forest. Suec.,* 5, 1-97.

Dogra P.D., 1978. Morphology, Development and Nomenclature of Conifer Embryo. *Phytomorphology,* 28(3), 301-321.

Drake, J.W., Allen, E.F., Fosberg, S.A., Preparata, R.M. & Greening, E.O., 1969. Genetic Control of Mutation Tastes in Bacteriophase T4. *Nature,* 221, 1128-1132.

Dynamics and Stability of Recreation Forests, 2006. Ed. Ermakov N.B., Association of scientific publications KMK, Moscow.

Dynamics of Population Gene Pool under the Anthropogenic Impacts, 2004. Ed. Altukhov Yu.P., Nauka, Moscow.

Dyuzhikova, 1997. Investigation of Structural and Functional Chromosome Organization during Stress Reactions. *Rus. J. Genet.,*33(8), 1077–1082.

East E.M., 1929. Self-sterility. *Bibliogr. Genet.,* 5, 331-368.

Efimov Yu.P., 2010. *Seed plantation in selection and seed breeding of Scots pine.* Publ. Istoky, Voronezh.

Efimov Yu.P., Chertov N.K., 1976. Influence of Meteorological Factors on Flowering and Fructification of Scotch Pine. In: *Genetics, Breeding, Seed production and Introduction of Forest Species.* Publ. CNIILiS, Voronezh, 3, 64-66.

Eigen M. & Schuster P., 1982. *Hypercycle: Principles of Self-organization of Macromolecules,* Mir, Moscow.

Elagin I.N., 1976. *Seasonal Development of Pine Forests.*Nauka SO AN SSSR, Novosibirsk.

Embryology of Angiosperms, 1984. Ed. B.M. Johri. Springer-Verlag. Berlin. Heidelberg, New York, Tokyo.

Epifanova, O.I., 2003. *Lectures about a Cell Cycle.* KMK, Moscow.

Eriksson G. & Ekberg I., 2001. *An Introduction to Forest Genetics.* SLU Press, Uppsala.

Ermakov I.P., Barantseva L.M. & Matveeva N.P., 1981. Cytochemical Study of DNA in the Process of Oocyte Maturation and Early Embryogenesis of *Pinus sibirica* Du Tour. *Rus. J. Ontogenesis,* 12(4), 339-345.

Ermakov I.P., Barantseva L.M. & Matveeva N.P., 1982. The Structure of Nucleus of Developing Archegonium of Proembryo of *Pinus sibirica* Du Tour. *Rus. J. Ontogenesis,* 13(6), 611-620.

Ermakov I.P., Matveeva N.P. & Barantseva L.M., 1980. *DNA content in the ovicell of Pinus sibirica Du Tour. at different stages of its development.* DAN SSSR, 251(1), 254-256.

Eshel I., 1991. Game Theory and Population Dynamics in Complex Genetical Systems: the Role of Sex in Short Term and in Long Term Evolution. In: *Game Equilibrium Models I.* Springer-Verlag. Pp. 6-28.

Essau K., 1980. *Anatomy of Seed Plants.* Part 1-2, Moscow, Mir.

Evert R.F. & Eichhorn S.E., 2013. *Raven Biology of Plants*. 8th Ed. W.H. Freeman and Company Pabl., New York.

Evert R.F., 2006. *Esau's Plant Anatomy: Meristems, Cells and Tissues of Plant Body: Their Structure, Function and Development*. 3rd Ed. J. Wiley & Sons. Inc., New Jersey.

Eytingen G. R., 1962. *Selected works. Forest Promishlenost*, Moscow.

Fergusson M.C., 1904. Contribution to the Knowledge of the Life History of *Pinus* with Special Reference to Sporogenesis, the Development of the Gametophyte and Fertilization. *Proc. Wash. Acad. Sci.,* 6, 1-202.

Feynman R.P., Leighton R.B. & Sands M., 1963. *The Feynman Lectures of Physics,* vol. 3. Wesley A. (Ed.), Readigg, Massachusetts, Palo, Alto, London.

Filonova L.N., von Arnold S., Daniel G. & Bozhkov P.V., 2002. Programmed Cell Death Eliminates all but one Embryo in a Polyembryonic Plant Seed. *Cell Death Differ,* 9(10), 1057-1062.

Franklin E.C., 1985. Estimation of Gene Flow into two Seed Orchards of Loblolly Pine (*Pinus taeda* L.). *Teor. Appl. Genet.,* 69, 609-615.

Franklin-Tong V.E., 2008. *Self-Incompatibility in Flowering Plants: Evolution, Diversity and Mechanisms*. Springer-Verlag, Berlin, Heidelberg.

Fromm J., 2013. Xylem Development in Trees: from Cambial Divisions to Mature Wood Cells. In: *Cellular Aspects of Wood Formation*. Springer-Verlag Ed., Berlin, Heidelberg, P. 3-39.

Fujita T., 1961. Multinuclear Endosperm Cells in *Chephalotaxus drupacea* Siebolt and Zuccarini. *J. Jap. Bot.,* 36, 29-32.

Gamaley Y., 2015. *Climatic Adaptogenesis/Plants in the Conditions of Global and Local Climatic and Anthropogenic Impacts.* Petrozavodsk: Karelian Research Centre of RAS. 123 pp.

Gebures Th., 1986. Some Results of Inbreeding Depression in Serbian Spruce (*Picea omorica* (Pan.) Park). *Silvae Genet.,* 35(4), 169-172.

Gene pool and Gene Geography of Population, 2000, Ed. Y.G. Rychkov, Nauka, Moskow, V.1.

Georgievsky A.B., 1974. *The problem of Pre-adaptation.* Historical-Critical Research. Nauka, Leningrad.

Georgyev G.P., 1968. The Nucleus. In: *Enzyme Cytology.* Ed. D.B. Roodyn. Academic Press, London, New York.

Gifford R.M., 2003. Plant Respiration Productivity Models: Conceptualisation, Representation and Issues for Global Terrestrial Carbon-cycle Research. *Fundamental Plant Biology,* 30, 171-181.

Gilbert S.F., 2010. *Developmental Biology.* 9th Ed., Sinauer Associates, Inc., Sunderland, Massachusetts.

Gilyarov A.M., 1990. *Population Ecology: Textbook allowance.* MGU, Moskow.

Gilyarov A.M., 2010. In Searchof Universal Regularities of the Communities Organization and Progress Towards the Neutrality. *Biol. Bulletin Rev.,*71(5), 386-401.

Girgidov D.J., 1973.Irregularity of Seed-bearing of the Conifers and Some Aspects of Prediction In: *Sexual Reproduction of Conifers*. Nauka, Novosibirsk, Part 2, 183-186.

Glotov N.V., 1983. Estimation of the Genetic Heterogeneity of Natural Populations: Quantitative Traits. *Rus. J. Ecology,* 1, 3-10.

Goldstein D.B., 1992. Heterozygote Advantage and the Evolution of a Dominant Diploid Phase. *Genetics,* 132(12), 1195-1198.

Goncharenko G.G., Padutov V.E., 1995. *Genetic Structure, Taxonometrical and Phylogenetic Relationships at the Firs of CIS*. Doklady RUN, 342, 1, 122-126.

Goncharenko G.G., Padutov V.E., 2001. *Population and Evolutionary Genetics of Spruces of Palearctic.* IL NANB, Gomel.

Goncharenko G.G., Silin A.E., 1997. *Population and Evolutionary Genetics of Pines of Eastern Europe and Siberia.* Minsk, Tekhnalogiya.

Gonzales-Martines S.C., Krutovsky K.V., Neale D.V., 2006. Forest Tree Population Genomics and Adaptive Evolution. *New Phytologist.,*170(2), 227-238.

Goroshkevich S.N., 2016. The Structure and Development of Siberian Stone Pine (*Pinus sibirica* Du Tour) Annual Shoot. *Vestnic TGU. Biology*, 1(33), 18-38.

Grant V., 1984. *Speciation in Plants*. Mir, Moscow.

Grief C., O'Neill & Shaw P.J., 1987. The Zygote Cell Wall of *Chlamydomonas reinhardii*: a Structural, Chemical and Immunological Approach. *Planta,* 170(4), 433-445.

Grif V.G., Ivanov V.B. & Machs E.M., 2002. Cell Cycle and its Parameters in Flowering Plants. *Tsitology,* 44(10), 936-980.

Griffin A.R. & Lindren D., 1985. Effect of Inbreeding on Production of Filled Seed in *Pinus radiate* Experimental Results and a Model of Gene Action. *Theor. Appl. Genet.,* 71(2), 334-342.

Gurvich A.G., 1932. *Mitogenetic Radiation.* Medgiz, Moscow.

Gurvich A.G., 1944. *Theory of Biological Field.* Sovetskay nauka, Moscow.

Gurvich A.G., 1977. *Selected Works*. Meditsina, Moscow.

Gusev V.A., 2001. About the Energy Source for Preserve Viability and Multiplication of Heterotrophic Microorganism the Condition of Substrate Starvation. I. Formation of Hypothesis. *Rus. J. Biophysics,* 46(3), 862-868.

Hagerup O., 1933. Zur Organogenie und Phylogenie der Koniferenzapfen. [Organogenesis and phylogeny of conifer cones.] *Biologiske Kgl. Dans. Vidensk. Selsk.* 10 (7), 1-82.

Hagman M. & Mikkola L., 1963. Observation on cross-, self- and interspecific pollination in *Pinus peuce* Griseb. *Silvae Genet.*, 12(3), 73-79.

Hall B., 2001. Organic Selection: Proximate Environmental Effects on the Evolution on Morphology and Behavior. *Biology and Philosophy*, 16, 215-237.

Hamrick J.L., Godt J.W., Sherman-Broyles S.L., 1992. Factors Influencing levels of Genetic Diversity in Woody Plant Species. *New Forests,*6, 95-124.

Happer J.L., 1997. *Population biology of plant.* Acad. Press, New York.

Headrick F., 2003. *Genetics of Populations*. Technosphere, Moscow.

Henkel P.A., 1982. *Physiology of Plant Heat-and Drought Resistance*. Nauka, Moscow.

Heslop-Harrison J., 1978. Genetics and Physiology of Angiosperm Incompatibility Systems. *Proc. E. Soc. London. Ser. B.,* 202(1146), 73-92.

Höfler K., 1945. *Physical Chemistry of Cells and Tissues*. P. Blakiston & Co. Philadelphia.

Hubbel S.P., 2001. *The Unified Neutral Theory of Biodiversity and Biogeography*. Princeton Univ. Press, Princeton and Oxford.

Inge-Vechtomov S.G., 2007. Mechanisms of the Modify Variability. *Ecol. Genetic,* V(1), 21–24.

Inge-Vechtomov S.G., 2007. What We Known about a Variability? *Ecol. Genetic,* VIII(4), 4–9.

Inge-Vechtomov S. G., 2010. *Genetics with the Basics of Breeding*. 2-nd Ed. N-L, St. Petersburg.

Iordansky N.N., 2001. *Evolution of Life.* Academy, Moscow.

Isakov Yu.N., 1999. *Ecological and genetic variability and selection of Scots pine*. St. Petersburg, St. Petersburg University Press.

Ivanov V.D. & Dubrovsky I.G., 2013. Longiudinal Zonation Pattern in Plant Roots: Conflicts and Solution. *Trends in Plant Science*, 18, 237-242.

Ivanov V.D., 2011. *Cellular Mechanisms of Plant Growth,* Nauka, Moscow.

Ivanov V.N., 1990. Alteration of Exo- and Endotrophy in the Mitotic Cell Cycle. *Biol. Biology Bulletin Review,* 110(2), 284-289.

Jinks I.I., Pooni H.S., 1982. Determination of the Environmental Sensitivity of Selection Lines of *Nicotiana rustica* by the Selection Environment. *Heredity.* 1982, 49(3), 291-294.

Jockusch B. M., Schoenenberger C. A., Stetefeld J., Aebi U., 2006. Tracking down the different forms of nuclear actin. *Trends Cell Biol.,* 16, 391-396.

John B., 1990. *Meiosis. Developmental and Cell Biology,* Ser. 22, Melburn, Cambridge, Cambridge Univ. Press.

Johri B.B. & Vasil I.K. Physiology of Pollen. *Bot. Rev.,* 27(3), 325-381.

Jonsson A., Ekberg J., & Eriksson G., 1976. Flowering in a Seed Orchard of *Pinus sylvestris* L. *Stud. Fore. Suec.,* 135, 1-38.

Jorgensen S.E. *Integration of Ecosystem Theories: A Pattern.* 3rd edn. Netherlands: Kluwer Academic Publ., Dordrecht, 2002. 428 p.

Joyet-Lavergne Ph., 1931. La Physico-Climie de la Suxualite. *Protoplasma*, Gaston Dion & Cie, Paris.

Juel H.O., 1990. Beiträge zur Kennitnis der Tetradentheilung [Contributions to the cognition of Tetradentheilung]. *J. Wiss. Bot.,* 35(15-16), 626-638.

Kalchenko V.A., Arkhipov N.P., Fedotov I.G., 1993. Mutagenesis of Enzyme Loci, Induced in a Megaspores of *Pinus sylvestris* L. by Ionizing Radiation during the Chernobyl NPP. *Rus. J. Genet*., 29(2), 266-273.

Kalchenko V.A., Fedotov I.G., 2001. Genetic Effects of Acute and Chronic Ionizing Radiation on the *Pinus sylvestris* L., Growing in the Alienation Zone of Chernobyl NPP. *Rus. J. Genet.* 37(4), 437-447.

Kashin A.S. & Kuprijanov P.G., 1993. *Apomixis in Flowering Plants Evolution: Onto- and Phylogenetic Aspects of Problem.* Saratov University Press.

Kato Y. Sacura, 2011. *Archaea: Structure, Habitats and Ecological Significance*. Nova Science Publ., New York.

Kazmin S.D., 1984. *Biochemistry of the Mitotic Cycle of Tumor Cells.* Naukova dumka, Kiev.

Kaznacheyev V.P. & Mikhailova L.P., 1981. *Super-weak Radiation in the Intercellular Interaction*, Nauka, SO AN. Novosibirsk.

Khromova L.V., 1972. Rhythm of Pollen Development of *Pinus sylvestris* L. in the Moscow Region Conditions. *Rus. J. For.,* 1, 73-75.

Khromova L.V., 1985. *Embryological Processes in the Non-pollinated Ovules in the Pine and Anomalies in the Xenogamy*, 2, 47-52.

Kilchevsky A.V., Khotyleva L.V., 1997. *Ecological Plant Breeding*. Techalogiya, Minsk.

Klug W., Cummings M., 2007. *Essentials of Genetics*. 7th Ed. Technosphere, Moscow.

Knoll A.H. & Bauld J., 1989. The Evolution of Ecological Tolerance in Prokaryotes. *Trans. R. Soc. Edinb. Earth Sci.,* 80, 209-223.

Knox R.B., 1990. Pollen Grain. In: *Embryology of Plants: Use in genetics, Breeding, Biotechnology*. V. 1. Agropromizdat, Moscow, 224 - 316.

Knox R.B., 1990. Pollen Grains. In: *Embryology Plants.* Agropromizdat, Moscow, 224 - 305.

Koch G. & Schmitt U., 2013. Topochemical and Electron Microscopic Analyses on the Lignifications of Individual Cell Wall Layers during Wood Formation and Secondary Formation. In: *Cellular Aspects of Wood Formation.* Springer-Verlag Ed., Berlin, Heidelberg.

Kolchanov N. A., Suslov V.V., Shumny K.V., 2003. Molecular Evolution of Genetic Systems. *Paleontologichesky Zhurnal,* 6, 58-71.

Konar R.N. & Moitra A., 1980. Ulrastructure, Cyto- and Histochemistry of Female Gametophyte of Gymnosperm. *Gamete Res.,* 3, 67-97.

Konar R.N. & Oberoi Y.P., 1969. Recent work on reproductive structures of living conifers and taxads – a review. *Bot. Res.,* 35(2), 89-117.

Konstantinov, 1978. *Biology of Individual Development.* BSU Publ., Minsk.

Kormutac A. & Lindgren D., 1996. Mating System and Empty Seeds in Silver Fir (*Abies abata* Mill.). *For. Genet.,* 3(4), 231-235.

Korogodin V.I. & Korogodina V.L., 2000. *Information as the Basis of Life.* Publ. "Fenix," Dubna.

Korogodin V.I., 1983. The Definition of "Information" Notion and the Possibility of its Use in Biology. *Biophysics,* XXVIII(1), 171-177.

Korotkov V. N., 1991. New Paradigm in Forest Ecology. *Rus. J. Biol. Science,* 8, 7-20.

Kosinski G., 1981. Genetic Load and Empty seeds of European Larch (*Larix deciduas* Mill.). *Arboretum Kornickie,* 26, 231-236.

Koski V., 1970. Study of Pollen Dispersal as a Mechanism of Gene Flow in Conifers.*Comm. Inst. Forest. Fenn.,* 70(4), 1-78.

Koski V., 1971. Embryonic Lethal of *Picea abies* and *Pinus sylvesris. Commun. Inst. For. Fenn.,* 75(3), 1-30.

Koski V., 1973. On Self-pollination, Genetic Load and Subsequent Inbreeding in Some Conifers. Metsantutkimuslaitoksen, Juikai Suja, *Comm. Inst. For. Fenn.,* Helsinki, 78(10), 1-42.

Kotelova N.V. & Khromova L.V., 1974. To the Question about the Pollination and Pollen Tube Growth in the Two-needles Pines at the Xenogamy. In: *Physiology, Selection and Gardening of Cities.* Moscow Institute of Forestry, Moscow, 51, 44-50.

Kotelova N.V., 1956. Influence of self-pollination and cross-pollination on the quality of seeds and seedlings of Scots pine. *Scientific and technical information.* Moscow Forestry Engineering Institute, Moscow, 20, 1-74.

Kozubov G.M., 1974. *Fructification Biology of Conifers in North.* Leningrad, Science.

Kozubov G.M., 1986. *Resent Gymnospermae (Morphological and Systematic Review and Karyology),* Nauka, Leningrad.

Kozubov G.M., Trenin V.V., Tikhova M.A., Kondratieva V.P., 1982. *The Reproductive Structures of Gymnosperms (Comparative Description).* Nauka, Lenengrad.

Krutovsky K.V., 2006. From Population Genetics to Population Genomics of Forest Tree Species: An Integrated Population-genomic Approach. *Rus. J. Genet.,* 42(10), 1304-1318.

Krutovsky K.V., Burczyk J., Chybicky I. et al., 2012. Gene Flow, Spatial Structure, Local Adaptation and Assisted Migration in Trees. In: *Genomics of Tree Crops.* Ch. 4. Pp. 71-116. Ed. Schnell R.J., Priyadarshan P.M., Springer Science Inc.

Küppers B. O., 2000. *Information and the Origin of Life.* MA: MIT Press, Cambridge.

Kupriyanov P.G., 1989. *Diagnosis of seed reproduction systems in populations of flowering plants.* Publ. Saratov University, Saratov.

Kuznetsova N.F., 1991. Scotch Pine Male Gametophyte Development in Self- and Open Pollination. *Rus. J. For.,* 3, 27-33.

Kuznetsova N.F., 1994. Electron-microscopic Study of Fertilization and Proembryogenesis in Scots Pine. *Tsitology,* 36 (9-10), 986-992.

Kuznetsova N.F., 1996. Incompatibility and Stages of its Display in Scotch Pine. *Rus. J. For.* 5, 27-33.

Kuznetsova N.F., 2002. The Reproductive Development on Initial Sporophyte of *Pinus sylvestris* L. *Tsytologiya,* 40(10), 841-854.

Kuznetsova N.F., 2003. Trends of Selection in the Life Cycle of *Pinus sylvestris.Rus. J. For.,* 4, 9-16.

Kuznetsova N.F., 2009. Effect of Climatic Conditions the Expression of Self-Fertility in Scots Pine // *Rus.J.Ecol.,* 40 (5), 390-395.

Kuznetsova N.F., 2010a. Population Instability of Coniferous Forests and Ways for Preservation of Forest Plant Biodiversity in Central Federal Region. In: *Principles and Methods of Biodiversity Preservation: Mat.* IV All-Russ. Sci. Conf. Yoshkar-Ola, Mari State University, 2010. P. 367-370.

Kuznetsova N.F., 2010b. Sensitivity of Scots Pine Generative Sphere to Drought. *Rus. J. Forest Sciences,* 2, 44-50.

Kuznetsova N.F., 2012. Phenological Variability of Generative Sphere of the *Pinus sylvesteris* L. in the Voronezh Region. In: *Modern Problems of Introduction and Conservation of Plant Biodiversity.* Publ. Rosa Vetrov, Voronezh, 78-81.

Kuznetsova N.F., 2012. Self-fertility in Scots Pine as System for Regulating Close Relationships and the Species Survival in Adverse Environment. *Advances in Genetics Research.* V. 9. Nova Science Publ., New York, 83–106.

Kuznetsova N.F., 2013. *Corpuscular-wave nature and wave properties of plant cells.* Nova Science Publ., New York.

Kuznetsova N.F., 2015. Development of Specific and Nonspecific Responses to Stress in *Pinus sylvestris* L. at Population Level in a Gradient of Drought Years. *Rus. J. Ecol.,* 46(5), 405-410.

Kuznetsova N.F., 2016. Trinity Genome on the Basis of DNA-energy, DNA-particle and DNA-information: A Procariotic Cell. *Eur. J. Bioph.,* 4(1),1-7.

Kuznetsova N.F., Isakov Yu. N., 1996. Dependence of the Level of Self-Fertility in Scotch Pine on Yearly Climatic Conditions. *Rus. J. Ecol.,* 27, 256-260.

Kuznetsova N.F.,Mashkina O.S., 2011. Stress reaction and it after-effect of Pinus sylvesteris L. in ontogeny and at alteration of generation. *Conifers Boreal Area*, V. XXVIII(1-2), 83-90.

Kvitko K.V., Zakharov I.A. 2012. *Genetics of Microorganisms: The Training Manual.* 2nd Ed. St. Petersburg: St. Petersburg SU Press, 2012.

Lagerkrantz U., Ryman N., 1990. Genetic Structure of Norway spruce (*Picea abies*). *Concordance of Morphological and Allozymic Variation. Evolut.,* 44, 38-53.

Lehninger A.L., 1985. *The Principles of Biochemistry.* V. 1-3. Moskow, Mir.

Lerner I.M. *Genetic Homeostasis.* Edinburg: Oliver and Boyd.

Levis D., 1944. *Incompatibility in Plants* (London). 153(3889), 575-578.

Levis D., 1949. Incompatibility in Flowering Plants. *Biol. Rev.,* 24, 472-496.

Levitt, J. 1980 Responses of Plants to Environmental Stresses. Vol. 1. *Chilling, Freezing and High Temperature Stresses,* 2nd edn. Academic Press, London.

Life of Plants, 1978. V. 4. Grushevitskii, I. V.& Zhilin, S.G., eds. Prosveshenie, Moscow.

Maletsky S.I. & Denisova E.V., 1974. Perspectives of Using of the Inbreeding Method in the Selection of Diploid and Tetraploid Forms of Sugar Beet/*Theoretical and Practical Problems of Polyploidy.* Nauka. Moscow. P. 5-13.

Mamaev C. A., 1973. *Forms of Intraspecific Variability of Woody Plants* (on the example of Pinaceae family). Nauka, Moscow.

Mamaev C. A.,Makhnev A. K., 1988. Using of Phenetics Methods for the Study of Population Structure and Conservation of Gene Pool of the Species of Woody Plants. In: *Phenetics of Natural Populations.* Nauka, Moscow, 92-98.

Mamaev S.A., Semerikov L.F., Makhnev A.K., 1988. About Population Approach in Forestry. *Rus. J. Forest Sciences,* 1, 3-9.

Martens P. & Waterksyn L., 1962. Structure du Pollen "Aile" Chez les Coniferes. *Cellule.,* 62, 172-222.

Masia D., 1961. *Mitosis and Physiology of Cell Division.* Brachet J. & Mirsky A. Eds., Academic Press, New York, London.

Maslakov E.L., Markova I.A., Shestakova T.A., 2001. On the possibility of early diagnosis of tree-leaders. *Rus. J. For.,* 1, 25-31.

Maynard J.S., 1981. *Evolution of Sexual Reproduction.* Mir, Moscow.

McCarthy B. J. & Hoyer v, 1964 Identity of DNA and Diversity of Messenger RNA Molecules in Normal Mouse Tissue Cells. *Proc. Natl. Acad. Sci. US,* 52, 915-922.

McIntosh R., 1967. The Continuum Concept of Vegetation. *Bot. Rev.*, 130, 130-187.

McWilliam J.R., 1958. The Role of Micropile in the Pollination of *Pinus. Silvae Genet.,* 8(2), P. 59-61.

Mergen F., Burley J. & Furnival G.M., 1965. Embryo and Sidling Development in *Pinus glauca(*Moench*)*: Vass after Self-, Cross- and Wind-pollination. *Silvae Genetica,* 14, 188-194.

Merkis A.I., 1990. *The Force of Gravity in the Processes of Plant Growth.* Nauka, Moscow.

Minina E.G. & Larionova N.A., 1979. *Morphogenesis and Sex Expression in Conifers.* Nauka, Moscow.

Minina E. G. & Larionova N.A., 1984. Physiological Characteristics of Pollen Grains. In: *Ecology of Seed Reproduction of Coniferous in Siberia.* Krasnoyarsk, ILID SO AN USSR, 80-87.

Minina E.G., Tretyakova I.N., 1983. *Geotropism and Sex in Conifers.* Nauka SO AN SSSR, Novosibirsk.

Mirkin B. M., Naumova L. G., 2012. *The Present State of the Basic Concepts of Science about Vegetation*. Gilem, Ufa.

Mirkin B. M., Naumova L. G., 2015. Concept of Plant Communities: History and Current State. *Biol. Bulletin Rev.,* 76, 1, 63-76.

Mirkin B.M., Naumova L.G., Solomeshch A.I., 2001. *Modern Science about Vegetation.* Logos, Moscow.

Mirov N.T., 1967. *The Genus Pinus*. Ronald Press, New York.

Mitchison J.M., 1973. *The Biology of the Cell Cycle.* Cambridge, Univ. press.

Mitton J.B., Duran K.L., 2004. Genetic Variation in Pinon pine *Pinus edulis*, Associated with Summer Precipitation. *Molec. Ecology,* 13, 1259-1264.

Modern Microbiology. Prokaryotes., 2005. Ed. Lengeler, J.W., Drews, G., Schlegel, H.G., V. 1, Mir, Moscow.

Monahova M.A., 1990. Cytogenetic Aspects of the Spatial Organization of Interphase Nucleus. *Biology Bulletin Review*, 110(2/5), 163-179.

Muller L., 1977. Untersuchungen uber die Naturliche Selbstbefructung in Bestunden der Fichte (Picea abies (L.) Karst.) und Kiefer (*Pinus silvestris* L.). [Studies on natural self-fertilization in hours of spruce (Picea abies (L.) karst.) And pine (Pinus silvestris L.).] *Silvae Genet.,* 26(5-6),207-217.

Nagl W., 1978. *Endopolyploidy and Polytene in Differentiation and Evolution.* Horth-Holland Publ., Amsterdam, New York, Oxford.

Nekrasova T.P., 1977. Pollen Production of Pine and Cedar. In: *Study of the Forests of Western Siberia.* ILID SB as USSR, Krasnoyarsk, 3-8.

Nekrasova T.P., 1983. *Pollen and Pollen Regime of Coniferous of the Siberia.* Nauka, Novosibirsk.

Nekrasova T.P., Ryabinkov A.P., 1978. *Bearing of Abies sibirica.* Novosibirsk, Nauka SO.

Nevo R., Charuvi D., Tsabari O. & Reich Z., 2012. Composition, Architecture and Dynamics of the Photosynthetic Apparatus in Higher Plants. *Plant J.,* 70(1), 157-176.

New Encyclopedia of Philosophy. 4 part., 2010. Mysl, Moscow.

Nielsen R., 2001. Statistical Tests of Selective Neutrality in the Age of Genomics. *Heredity*, 86, 641-647.

Nikitin M.A., 2016. Origin of Life. From Nebula to Cell. *Alpina non-fiction,* Moscow.

Nobel P., 1993. *Physiology of Plant Cells.* Mir, Moscow.

Noskova N.E. &, Tretyakova I.N., 2006. *Influence of Stress on the Repriductive Ability of Scots pine.* Conifers of the Boreal Area, XXV(3), 54-63.

Noskova N.E., Tretyakova I.N., 2011. *Reproduction of Scotch Pine in the Condition of Climate Global Change and the Strategy of Species Preservation*. Conifers of the Boreal Area, XXVIII(1), 41-46.

Notov A.A. & Zhukova L.A., 2013. About the Rrole of Population-ontogenetic Approach in the Development of Modern Biology and Ecology. *Tver State University Vestnik. Series: Biology and Ecology,* 32 (31), 296-317.

O'Leary B.M. & Plaxton W.C., 2016. *Plant Respiration.* eLS, 1-11.

OdumYu., 1975.*Bases of Ecology.* Mir, Moscow.

Ohniwa R.L., Morikawa K., Ohta T., Takeyasu K., 2010. Nucleoid Architecture and Dynamics Bacteria. In: *Bacterial DNA, DNA Polymarase and DNA Helicases.* Knudsen W.D., BrunsS.S. eds., Nova Science Publ., New York, 91-117.

Olds T.M., 1986. Genetic Load and Heterozygoty in the Pinaceae. *Can. J. Genet. Cytol.,* 28(6), 942-946.

Orr H.A., 1998. Population Genetics of Adaptations: Distribution of Factors Fixed during Adaptive Evolution. *Evolution,* 52, 935-949.

Orr H.A., 2000. Adaptation and the Cost of Complexity. *Evolution*, 54, 13-20.

Orr-Ewing A.Z., 1957. A Cytological Study of the Effects of Self-pollinationon *Pseudotsuga menziesii* (Mirb,) Franko. *Silvae Genetica*, 6, 179-185.

OstapenkoV.I., 1969. Physiological, Biochemical and Genetic Aspects of Sexual Dimorphism and Relative Sexualization in a Higher Plants. *TsGL Works,* 10, 99-104.

Oudalova A.A., Geraskin S.A., 2011. Dynamics and Ecological-Genetic Variability of Cytogenetic Disturbances in Scots Pine Populations Experiencing Technogenic *Impact. Biol. Bulletin Rev.,* 72, 6, 455-471.

Owens J.N., Bruns D., 2000. Western White Pine (*Pinus monticola* Dougl.) Reproduction: 1. Gametophyte Development. *Sex. Plant. Reprod.,* 13, 61-74.

Owens J.N., Fernando D.D., 2007. Pollination and Seed Production in Western White Pine. *Can. J. For. Res.,* 37, 260-275.

Owens J.N., Molder M., 1979. But Development in *Larix occidentalis.* II. Cone Differentiation and Early Development. *Can. J. Bot.,* 57, 1557-1572.

Owens J.N., Simpton S.J. & Moitra M., 1982. Sexual Reproduction of *Pinus contorta.* II. Postdormancy Ovule, Embryo and Seed Development. *Can. J. Bot.,* 60, 2071-2083.

Palmer M.W., Rusch G.M., 2001. How Fact is the Carousel? Direct Indices of Species Mobility with Examples from on Oklahoma Grassland. *J. Veget. Sci.,* 12, 305-315.

Park Y.S. & Fowler, 1984. Inbriding in Black Spruce (*Picea mariana (*Mill*)* B.S.P.) Self-Fertility, Genetic load and Performance. *Can. J. For. Res*, 14, 17-21.

Peek J.R., Welch J.J., 2004. Adaptation and Species Range. *Evolution,* 58, 2011-221.

Petrova E.A., Goroshkevich S.N., Belokon M.M., Belokon M.M., Politov D.V., 2014. Destribution of the Genetic Diversity of the Siberian Stone Pine *Pinus Sibirica* Du Tour., along the Latitudinal and Longitudinal profiles. *Rus. J. Genet.,* 50(5), 467-482.

Pichot C.E. & Maätaoui, M., 1997. Flo Cytogenetic Evidence for Multiple Ploidy Level in the Endosperm of Some Gymnosperm Species. *Theor. Appl. Genet.,* 94, 865-870.

Pierce B.A., 2014. *Genetics: A Conceptual Approach.* 5th Ed. W.H. Freeman and Co Ltd, New York, United States.

Pismenny N.P., 1970. On the Future of the Coniferous Forests of Russia. *Rus. J. Forestry,* 1, 54-58.

Plaxton W.C. & Podesta F.E., 2006. The Functional Organization and Control of Plant Respiration. *Critical Reviews in Plant Science*, 25, 159-198.

Poddubnaya-Arnoldi B.A., 1976. *Cytoembryology Angiosperms* (Basis and Perspectives). Nauka, Moscow.

Polevoy V.V., 1989. *Plant Physiology.* Moscow: Higher School.

Powell G.R., 1977. Patterns of Development in *Abies balsamea* Crowns and Effects of Megastrobilus Production on Shoots and Buds. *Can J. Forest Res.,* 7(3), 498-509.

Pravdin L. F., 1964. *Scots Pine* (Variability, Intraspecific Taxonomy and Selection). Nauka, Moscow.

Protopopov V.V., 1977. Analysis of the Ecological Importance of Dark Coniferous Forest. In: *Problems of Forestry of Siberia.* Science, Moscow, 25-48.

Pukhalsky V.A., 2002. *Introduction to the Genetics.* Kolos, Moskow.

Rankle J.R., 1989. Synchrony of Regeneration, Gaps and Latitudinal Differences in Tree Species Diversity. *Ecology,* 70(3), 546-547.

Raskatov P.B., 1974. *Anatomy of the Vegetative Organs of Woody Plants.* VSU Publ., Voronezh.

Raskatov P.V., 1974. *Anatomy of the Vegetative Organs of Woody Plants.* VSU Publ., Voronezh.

Rathgeber C.B., Rossi S., Bontemps J. D., 2011. Cambial activity related to tree size in a mature silver-fir plantation. *Ann. Bot.*, 108, 429-438.

Raven Chr., 1961. Oogenesis. *The Storage of Developmental Information*. Pergamon Press. New York. Oxford, London, Paris.

Raven P.H., Evert R.F., Eichorn, 2005. *Boilogy Plant*. 7th ed. W.N. Freeman, New York.

Ravin N.V. & Shestakov S.V., 2013. The Genome of Prokaryotes. Vavilov J. *Genetics and Breeding,* 17 (4/2), 972-984.

Redko G.I. & Redko N.G., 2004. *History of the Forest Economy of Russia.* VNIILM, Moscow.

Rees, A.R. & Sternberg M.J.E., 2002. *Introduction to Molecular Biology: From Cells to Atoms,* Moscow, Mir.

Robledo-Amuncio J.J. & GilL., 2005. Patterns of Pollen Dispersal in a Small Population of *Pinus sylvestris* L. Related by Total-exclusion Paternity Analysis. *Heredity*, 94(1), 13-22.

Romanovsky M. G., 1994. *Polymorphism of Woody Plants by the Quantitative Traits.* Nauka, Moscow.

Romanovsky M. G., 1997. *Formation of the Seed Crop of Scots Pine in the Norm and under Mutagenic Pollution*. Nauka, Moscow.

Romanovsky M.G., 2002. *Productivity, Stability and Biodiversity of the Plain Forests of European Russia*, Moscow state forest University, Moscow.

Romanovsky M.G., Khromova L.V., 1992. Seed Formation of Scots Pine in Self-Pollination. *Rus. J. For.,* 5, 3-9.

Romanovsky M.G., Tarkhanov S.N., Schekalev R.V., 2009. The Approaches to the Descriptionof Biodiversity Plain Forests of European Russia. *Rus. J. ForestSciences,*4, 47-56.

Roshchina, V.V. 2014. *Model systems to study the excretory function of higher Plants.* Springer, Dordrecht, Heidelberg, Berlin 200 pp.

Roshchina, V.V. and Roshchina V.D. 1993. *The Excretory Function of Higher Plants.* Springer, Berlin 384 pp.

Rosnovsky I.N., 1997. *Ecosystem Stability: Introduction to the Problem and Research Methods.* Tomsk: Izdatelstvo "Spectr" SO RAN.

Rowley J.R. & Walles B, 1982. Cell Differentiation in Microsporangia of *Pinus sylvestris*. I. The Pre-and Early-Meiotic Periods. *Nord. J. Bot.,* 2, 53-70.

Rowley J.R. & Walles B, 1985. Cell Differentiation in Microsporangia of *Pinus sylvestris*. II. Early Pachytene. *Nord. J. Bot.,* 5, 241-254.

Rowley J.R. & Walles B, 1985. Cell Differentiation in Microsporangia of *Pinus sylvestris*. Diplotene and the Diffuse-Stage. *Ann. Sci. Nat.* (Botanique), 9, 1-28.

Rowley J.R. & Walles B, 1985. Cell Differentiation in Microsporangia of *Pinus sylvestris*. V. Diakinesis to Tetrad Formation. *Nord. J. Bot.,* 13, 67-82.

Rowley J.R. & Walles B, 1985. Origin and structure of Ubisch bodies in *Pinus sylvestris. Acta Soc. Bot. Poloniae,* 56, 215-227.

Roy Chowdhary C., 1962. *The Embryogeny of Conifers: a Review*. Ibid, 12, 313-338.

Rubin A.B., 1999. *Biophysics.* V. 1-2. Nauka, Moscow.

Rutten M.G., 1971. *The Origin of Life by Natural Causes*. Elsevier Bubl. Comp., Amsterdam, London, New York.

Rysin L.P., 2006. Historical Factors in Modern Successional Dynamics Forest in the Center of Russian Plain. *Rus. J. Forest Sciences,* 6, 3-11.

Rysin L.P., Savelyeva L.I., Rysin S.L., 2004. Forest Monitoring at the Urbanized Territories. *Rus. J. Ecol.,* 4, 243-248.

Sacher J.A., 1954. Structure and Seasonal Activity of the Shoot Apices of *Pinis lambertiana* and *Pinus ponderosa. Amer. J. Bot.,* 4(9), 749-759.

Sannikov S.N. & Grishina I.V., 1979. Experimental Study of Pollen Dispersal of the Pine in the Tree Stand. *Rus. J. Ekol.,* 1, 91-93.

Sannikov S.N., Petrova I.V., 2003. *Differentiation of Scots Pine Populations*. Nauka, UrO RAN, Ekaterinburg.

Sannikov S.N., Petrova I.V., 2007. *Phenogeography of Woody Plants Populations: Problems, Methods and the Some Results.* Conifers of Boreal area, 24, 2-3, 288-296.

Sannikov S.N., Petrova I.V., Semerikov V.L., 2002. Genophenogeographycal Analysis of *Pinus sylvestris* L. Populations in the Transect from the Northern to the Southern Boundary of Ranges. *Rus. J. Ecol.,* 2, 97-102.

Sarvas R., 1962. Investigation of the Flowering and Seed Crop of *Pinus sylvestris* // *Commun. Inst. For. Fenn.,* 53, 1-198.

Sarvas R., 1973. Investigations on the Annual Cycle of Development of Forest Trees. I. Active Period. *Commun. Inst. For. Fenn.,* 76(3), 1-110.

Sato T., 1997. Mode of Fertilization and its Individual Variation in *Larix gmelini* var. *Jaronica. Silvae Genet.,* 46, 2-7.

Savidan Y., 2000. Apomixis: Genetics and Breeding. *Plant Breed. Rev*., 18, 13-86.

Scharff M.D. & Robbins E., 1966. Polyribosome Desegregation during Metaphase. *Science,* 151(3713), 992-995.

Schmalgauzen I.I., 1963. *Ways and Regularitiesof Evolutionary Process.* Selected Works.Nauka, Moscow.

Schmalgauzen I.I., 1968. *Factors of Evolution. The Theory of Stabilizing Selection.* Nauka, Moscow.

Schnarf K., 1933. Embryologie de Gymnosperme. [Embryology of Gymnosperm.] *Handbuh der Pflanzenanatomie,* II/2., Gebrueder Borntraeger, Berlin.

Schoen D.E. & Cheliak W.M., 1987. Genetics of the Polycross. 2. Male Fertility Variation in Norway Spruce, *Picea abies* (L.) *Karst. Theor. Appl. Genet.,* 74, 554-559.

Schopf J.W., 1981 The Evolution of the Earliest Cells. In: *The Evolution*, Mir, Moscow, 109-147.

Schwartz S.S., 1980. *Ecological Regularities of Evolution*. Nauka, Moscow.

Sclegel G., 1987. *General Microbiology,* Mir, Moacow.

Semerikov L.F., Isakov Yu.N., Glotov N.V., Tarakanov V.V. & Semerikov V.L., 1996. About a Genetic-Selection Aspects of the Conservation and Improvement of Russia Forests. *Rus. J. Forest Information,* 9, 3-12.

Semerikov V.L., Semerikov L.F., Lascoux M., 1999. Intra- and Interspecific Allozyme Variability in Eurassian *Larix* Mill. *Species. Heredity*, 82, 193-204.

Semerikova S.A., Semerikov V.L., 2014. Molecular and Phylogenetic Analysis of the Genus Abies (Pinaceae) Based on the Nucleotide Sequence of Chloroplast DNA. *Rus. J. Genet.,* 50(1), 7-19.

Severtsov A.S., 2004. Mechanism of Origin and Ecological Significance of Fundamental Niche of Species. *Rus. J. Ecol.,* 6, 403-409.

Severtsov A.S., 2005. *Theory of Evolution,* VLADOS, Moscow.

Severtsov A.S., 2012. Relation between Fundamental and Realized Niche. *Societes,* 5, 323-333.

Shannon C.E., 1949. *A Mathematical Theory of Communication.* University Illinois Press. Urbana, llinois.

Shay J.W. & Wright W.E., 2000. Hayflick, his Limit, and Cellular Ageing. *Nature Rev. Mol. Cell Biol.,* 1, 72-76.

Shcherbakov V.P., 2005a.Evolution as a Resistance to the Entropy.1. Mechanismsof Species Homeostasis.*Biol. Bulletin Rev.*, 66, 195-211.

Shcherbakov V.P., 2005b. Evolution as a Resistance to the Entropy. II. Conservative Role of Sexual Reproduction. *Biol. Bulletin Rev.*, 66, 300-309.

Shimamura T., 1956. Cytochemical Studies on the Fertilization and Proembryo of *Pinus thunbergii. Bot. Mag.* (Tokyo), 69, 524-528.

Singh H., 1978. Embryology of Gymnosperms. *Gebrüder Borntraeger,* Berlin-Stuttgart.

Singhal G.S., Renger G., Sopory S.K. & Irrgang K.D. & Govindjee, 1999. *Conceptions in Photobiology: Photosynthesis and Ohotomorphogenesis.* Narosa Publishing Hause, New Delhi, India.

Smirnov V. E., Khanina L. G., Bobrovsky M. V., 2006. Substantiation of the System of Eco-cinotic Groups of Plant Species at the Forest Zone of European Russia on the Basis of Ecological Scales, Geo-botanical Descriptions and statistical analysis. *Bull. MOIP. Ser. Biology,* 3(21), 36-47.

Smirnova O.V., 2004. Methodological Approaches and Evaluation Methods of the Climax and Successional State of forest ecosystems (on Example of Eastern European Forests). *Rus. J. Forest Sciences,* 3, 15-27.

Smirnova O.V., Turubanova M.V., Bobrovsky, V.M., Korotkov L.G. Khanina L.G., 2001. Reconstruction of Forest Cover History in the Eastern Europe and the Problem of Biological Diversity Maintenance. *Biol. Bulletin Rev.,* 2, 144-159.

Smith J.M., 1980. *Evolution of Sexual Reproduction.* Mir, Moscow.

Snape A., Papachristodoulou D., Elliott W.H. & Elliot D.C., 2014. *Biochemistry and Molecular Biology.* Oxford University Pbl., Oxford, United Kingdom.

Solbrig O. & Solbrig D., 1982. *Population Biology and Evolution.* Mir, Moscow.

Sorensen F.S., 1987, Estimated of Frequency of Natural Selfing in Lodgepole Pine (*Pinus contorta* var. Muriana) from Central Oregen. *Silvae Genet.*, 36(5-6), 215-216.

Sorokina Z.P., Tarlinskaya B.P., Danovitch K.N., 1975. Dynamics of Total and Protein Nitrogen in the Tissues of the Developing Ovule of Siberian cedar. *Inform. SIFiBR Materials*. Irkutsk, Vol. 11, 5-6.

Stegniy V.N., 1993. *Genome Architectonics, Systemic Mutation and Evolution.* Novosibirsk University Publ., Novosibirsk.

Stephenson A.G. & Bertin R.I., 1983. Male Competition, Female Choice and Sexual Selection in Plants. Ch. 6. In: *Pollination Biology.* Acad. Press Inc., Orlando San Diego, San Francisco, New York, London, Toronto, Montreal, Sydney, Tokyo, Sao Paulo.

Stewart K.D. & Gifford E.M., 1967. Ultrastructure of Developing Megaspore Mother Cell of *Ginkgo biloba. Amer. J. Bot.,* 54, 375-383.

Stockwell W.P., 1939. Proebryonal Selection in the Pine. *Forestry,* 37, 6, 541-543.

Stockwell W.P., 1939. Proembryonal Selection in the Pine. *Forestry,* 37(6), 541-543.

Strochkova L.S. & Kazan'ev V.V., 1975. Cytochemical Peculiarities of a Cell Cycle. *Cytology. Mitotic cell division.* V. 2. Pp. 9-66. VINITI AN SSSR, Moscow.

Sukachev V.N., 1964. Dynamics of Forest Ecosystems. In: *Principles of Forest Biogeocenosis.* Nauka, Moscow.

Surikov I.M., 1991. *Incompatibility and Embryonal Sterility of Plants.* Moscow, Nauka.

Surmin Y.P., 2000. Systems Theory and Systems Analysis. *Textbook. MAUP*, Kiev.

Svintsova V. S., 2002. *Variability Characteristics of the Reproductive Sphere in Conditions of Central Russian Forest-Steppe* (Voronezh Region), Voronezh, VSU.

Svintsova V. S., Butorina A. K., 2002. Characteristic of the nucleolar activity of different type cells in the early embryogenesis of *Pinus sylvestris* L. *Tsitology,* 12(44)b 1199-1204.

Svintsova V.S., Kuznetsova N.F., Pardaeva E.J., 2014. Influence of Drought on the Generative Sphere and Pollen Viability of Scots pine. *Rus. J. For.*, 3, 49-57.

Swanson C.P, & Webster P.L., 1977. *The Cell.* 4th Edn. Prentice-Hall Inc., Englewwood, Cliffs, Jersey.

Taiz L., Zeiger E., Moller I.M., Murphy A., 2015. *Plant Physiology and Development.* 6th Edn., Sinauer Associates Inc. Sunderland, MA.

Takao A., 1960. Histochemical Studies on Embryogenesis of *Pinus thunbergii* Parl. *Bot. Mag.* (Tokyo), 73, 379-388.

Teif V.B., Shkrabkou A.V., Egorova V.P., Krot V.I., 2011. Nocleosomes in Gene Regulation: Theoretical Approaches. *Molecular Biology,* 45(6), 1-11.

Thilms L., 2012. Thermodynamics ≠ Information Theory: Science's Greatest Sokal Affair. *Thermodynamics,* 8(1), 1-120.

Thomas M.I., 1973. Etude Cytoloque de L'embryogenese du *Pinus sylvestris* L. 1. Du Proembyon cellulaire en Clivage Embryonaire. [Cytological study of the embryogenesis of Pinus sylvestris L. 1. Cell proembyon in embryonic cleavage.] *Rev. Cytol. et Biol. Veget..*, 36(3-4), 165-223.

Tikhodeev O.N., 2013 Classification of Variability by the Factors Determining the Phenotype: Traditional Views and their Modern Revision. *Ecological Genetics,* 11, 3, 79-92.

Tikhonov A.N., 1996. Mechanisms of Energy Conversion in Chloroplasts – Energy Transducting Organelles of Plant Cells. *Soros Educational Journal,* 4, 24-32.

Tikhonov A.N., 2012. Energetic and Regulatory Role of Proton Potential in Chloroplasts. *Biochemistry* (Moscow), 77(9), 1155-1176.

Tilman D., 1982. *Resource Competition and Community Structure.* Univ. Press Princeton, Princeton, New York.

Tilman D., 1988. *Plant Strategies and the Dynamics and Structure of Plant Communities:* Princeton Univ. Press, Princeton.

Timofeev-Resovsky N.V., Yablokov A.V. Glotov N.V., 1973. *Essay on the Doctrine about a Populations.* Nauka, Moscow.

Tinker P.B. & Nye P.H., 2000. *Solute movement in the rhizosphere.* New York: Oxford University Press.

Tokin B.P., 1987. *General Embryology.* 4rd ed. Moscow. Visshaya shkola.

Tokmanov A.A., Sato K.I. & Funami Y., 2002. Signal Cascade during Fertilization. *Tsitology,* 44(3), 227, 227-234.

Trenin V.V., 1986. *Cytoembryology of Larch.* Nauka, Leningrad.

Trenin V.V., 1988. *Introduction in a Cytoembryology of Conifers.* Petrozavodsk: Karelian branch of AN USSR.

Tretyakova I.N., 1990. *Coniferous Embryology.* Nauka, Novosibirsk.

Tripathy B.S. & Pattaanayak G., 2012. Chlorophyll Biosynthesis in Higher Plants. In: *Advanses in Photosynthesis and Respiration.* V. 34. Springer, Dordrecht, Pp. 62-94.

Troshin, A. S. 1985. *Distribution of Substances between Cell and Environment.* Nauka, Leningrad.

Tsatsenko L.V. & Sinelnikava A.S., 2012. Pollen Analysis in Plant Breeding. *Sci, J.* KubSAU, 77(3), 1-11.

Tupy J., 1959. Callose Formation in Pollen Tube and Incompatibility. *Biol. Plantarum.,* 1, 192-198.

Usmanov I. Yu., Rakhmankulova Z. F., Kulagin A. Yu., 2001. *Ecological Physiology of Plants.* Logos, Moscow.

Uyomov A.I., 1978. *System Approach and General Systems Theory.* Mysl, Moscow.

Van der Maarel E., Noest V. & Palmer M.W., 1995. Variation in Species Richness on Small Grassland Quadrats: Niche Structure or Small-scale Plant Mobility. *J. Veget. Sci.,* 6(5), 741-752.

Van Went J.L. & William M.T.M., 1984. Fertilization. In. *Embryology Angiosperms.* Ed. Johri H.M., Springer, Berlin Heidelberg Ney York Tokyo, pp. 273-317.

Vasilevich V.I., 1983. *Theoretical Essays of Phytocoenology.* Nauka, Leningrad.

Vavilov N.I., 1968. *The Law of Homologous Series in Hereditary Variation,* Nauka, Leningrad, 8-50.

Venalainen M.O., Aronen T.S., Haggman H.M. & Nikkanen T.O., 1999. Differences in Pollen Tube Growth and Morphology among Scots Pine Plus Trees. *For. Genet.,* 6, 3, 139-147.

Veselkin D.V., 2013. *Methodical Features of Evaluation of the Ectomycorrhiza Formation: Variability in Connection with Depth.*Bulletin of Perm University. 3 (7), P. 18-25.

Vidyakin A.I., 2013. *Namber of Cotyledons for Pine Plantatons of Different Age in the Middle Urals.* Urals Agrarian Bulletin, 10, 42-44.

Vishnjakova M.A. Mechanisms of Cross-Pollination in the Family *Fabaceae* / In: *Problems of Plant Reproductive Biology.* Perm: PSU, 1996, pp 13-15.

Vishnjakova M.A., 1989. Structural-Functional Bases of Self-Incompatibility in Flowering Plants. *Rus. J. Bot.,* 74, 137-152.

Vogel G., 1936. *Die Mannliche Haploid-Generation von Pinus sylvestris und Ihre Bezichung zur Hohlkornbildung.* [The male haploid generation of Pinus sylvestris and its relationship to the formation of hollow grains] Thesis, 52.

Volkenshtein M.V., 1983. *Theoretical Biophysisics*. Nauca, Moscow.

Vorobeichik E.L., Hantemirova E.V., 1994. Forest communities' reaction to technogenic pollution: Dependence on the dose-effect. *Rus. J. Ecology,* 3, 31-43.

Vorobyev V. N., Vorobyeva N. A. & Goroshkevich S. N., 1989. *Growth and Sex of Siberian cedar,* Nonosibirsk, Nauka SO.

Vorobyova E.I., 2006. Problem of Organism Integrity and its Prospects. *Biol. Bulletin Rev.,* 5, 530-540.

Walt A.S., 1947. Pattern and Process in the Plant Community. *J. Ecology,* 35, 1-22.

Webster J.D. & Vant'Hof J., 1969. Dependence of Energy and Aerobic Metabolism of initiationof DNA Synthesis and Meiosis in G_1 and G_2 cells. *Exp. Cell Res.,* 55(1), 88-94.

Whittaker R., 1980. Communities and Ecosystems. *Progress,* Moscow.

Whitteker R.H., Lewin A.S., 1977. The Role of Mosaic Phenomena in Natural Communities. *Theor. Population Biol.,* 12(2), 117-139.

Wiener, N., 1983. *Cybernetics or Control and Communication in the Animal and the Machine.* Main editorial office for foreign countries "Science," Moscow.

Wiess P., 1939. *Princeples of Development.* Henry Holt & Co, New York.

Wilkins M.B., 1966. Geotropism. *Ann. Rev. Plant Phisiol.,* 17, 379-408.

Willemse M.T. & Vent J.L., 1990. Female Gametophyte. In: *Plant Embryology.* Agropromizdat, Moscow. V. 1, 184-223.

Williams C.G. & Savolainen O., 1996. Inbreeding Depression in Conifers: Implication for Breeding Strategy. *For. Sci.,* 42, 143-147.

Williams C.G., 2008. Selfed Embryo Death in *Pinus taeda:* a phenotypic Profile. *New Phytologist*, 178, 210-222.

Williams C.G., 2009.*Conifer Reproductive Biology.*Sprinhger, Berlin.

Wright J.W., 1975. *Introduction to Forest Genetics*, Acad. Press, New York. San Francisco, London: Acad. Press.

Wright S., 1953. Pollen Dispersion Studies: Some Practical Application. *Forestry*, 51, 114–118.

Xiao Shi, Xiao Han & Tie-gang Lu, 2016. Callose Synthesis during Reproductive Development in Monocotyledonous and Dicotyledonous Plants. *Plant Signal Behav.,* 11(2): e 10621196.

Yablokov A.V., 1980. *Phenetics*. Nauka, Moscow.

Yablokov A.V., 1987. *Population Biology.* Nauka, Moscow.

Yagodinsky V.N., 2015. Alexander Chizhevsky. Institute Russian Civilization. Moscow.

Yockey H.P., 2005. *Information Theory, Evolution and the Origin of Life.* Cambridge Univ. Press, New York.

Yugay G.A., 1976. *Philosophical Problems of Theoretical Biology*, Mysl, Moscow.

Zacharov V.M. & Klark D.M., 1995. *Bio-test: Integral Estimation of the Health of Ecosystems and Separate Species*. Moscow Department of Intern. Fond “BIOTES,” Moscow.

Zaugolnova L.B., Khanina L.G., 2004. Parameters of Forest Biodiversity Monitoring of Russia at the Federal and Regional Levels. *Rus. J. Forest Sciences*, 3, 3-14.

Zawadzky K. M., 1961. *Doctrine about the Species.* Leningrad University Publishing, Leningrad.

Zhao J.P., Davidson D., 2014. Chromosomes into Discrete Groups in Colhicine-Induuced C-Metaphases of Barley. *Caryologia,* 37(4), 331-342.

Zharskaya O.O. & Zatsepina O.V., 2007. Dynamics and Mechanisms of the nucleolus Reorganization during Mitosis. *Tsitologiia,* 49(5), 355-369.

Zhimulev I.F., 2003. *General and Molecular Genetics.* Novosibirsk, Sibirskoe universitetskoe izdatelstvo.

Zhivotovsky L.A., Duharev V.A., 1985. "Compression" of Genotypic Variability under Stabilizing Selection, and its Manifestation at early stages of ontogenesis. *Biol. Bulletin Rev.,* 6(1), 32-40.

Zholkevich, V.N., Gurvich, A.L., Borisova, T.L., Livanova, T.N. & Shlyakhtina L.S., 1973. *On energy discharge prior to cell division*. DAN SSSR, 209(4), 961-965.

Zhukova L.A., Dorogova Y.A., Turmuhametova N.V., 2010. *Ecological Scales and Methods of Analysis of the Plant Environmental Diversity.* MarSU, Yoshkar-Ola.

ABOUT THE AUTHOR

Nina Fedorovna Kuznetsova was born April 9, 1950 in Novograd-Volynsk, the Zhitomir Region of USSR (now Ukraine). In 1967, she graduated high school and got into the Biophysics Department of the Biology and Soil Faculty of Voronezh State University. In 1972, Fedorovna finished the University and then went to the genetics laboratory of Central Research Institute of Forest Genetics and Selection in Voronezh. At present, she works in the same Institute (All-Russian Research Institute of Forest Genetics, Breeding and Biotechnology) as Leading Researcher and Head of Laboratory of Ecological Genetics. Fedorovna's specialty is a cell biology, embryology, biophysics, population and ecological genetics of forest tree species (pine, oak, poplar, etc.). The understanding of the double nature of reproductive cycle occurred in the process of the work on the dissertation "Embryogenesis Peculiarities of Scots Pine in the Self-Pollination and Open-Pollination (Light and Electron Microscopy)", which she defended in 1988. The result of further researches in this direction is the book "Corpuscular-Wave Nature and Wave Properties of Plant Cell" (2013). Now, Fedorovna is the author of 71 scientific works.

INDEX

B

C

D

F

G

H

I

K

L

M

N

O

P

Q

R

S

T

U

V

W

X

Y

Z

Tree Growth: Influences, Layers and Types

Editor: Wesley P. Karam

Series: Environmental Science, Engineering and Technology

Book Description: The amount of height growth in any one season depends on hereditary factors, immediate past environmental conditions and present environmental conditions. Diameter growth also proceeds basipetally and is much more related to current foliage and present environmental conditions.This new book presents the most up-to-date research in the field.

Hardcover ISBN: 978-1-60741-784-2
Retail Price: $195

Corpuscular-Wave Nature and Wave Properties of Plant Cells

Author: Nina F. Kuznetsova (Laboratory of Genetics, Research Institute of Forest Genetics and Breeding (NIILGiS), Voronezh, Russia)

Series: Genetics - Research and Issues

Book Description: Quantum biology is a wide area of research closely connected with almost all parts of biology. It is based on experimental data of biological sciences and the fundamental laws of physics (de Broglie law of corpuscular-wave dualism of the matter, the conservation laws, including the laws of thermodynamics).

Hardcover ISBN: 978-1-62618-785-6
Retail Price: $179